Acclaim for GEORGE JOHNSON's

FIRE IN THE MIND

"Undeniably fascinating. . . . Johnson is masterful at explaining complicated ideas and fitting them into the framework of modern science."
—Jill Sapinsley Mooney, *San Francisco Examiner & Chronicle*

"Subversive. . . . Johnson has veered away from the pack with a brilliant new book, one that raises unsettling questions about the claims of science to truth. . . . Readers are unlikely to finish the book without undergoing a crisis of faith."
—John Horgan, *The Sciences*

"Clear and thought-provoking. . . . An intellectual and cultural journey through the landscape of northern New Mexico. . . . An excellent book."
—David K. Nartonis, *Christian Science Monitor*

"Remarkable and eloquent. . . . Original and revealing . . . Johnson's desire not only to explain but to understand the urge to explain infuses *Fire in the Mind* with its own fire."
—Seth Lloyd, *Scientific American*

"Here is a book in the spirit of *Zen and the Art of Motorcycle Maintenance.* Compression is the essence of science [and] Johnson proceeds to compress with utter clarity, almost casually tap-dancing his way through particle physics, quantum theory, cosmology and evolutionary biology. . . . *Fire in the Mind* is a connoisseur's gazetteer. . . . Vibrant and exhilarating and even inspirational."
—Ian Watson, *New Scientist*

"One of the most stimulating books of popular science to have been written for some time."
—Ray Monk, *The London Observer*

"A spectacular tour of the most compelling theories of current science."
—Jon Turney, *The Financial Times of London*

GEORGE JOHNSON

FIRE IN THE MIND

George Johnson writes about science for *The New York Times*, and has written regularly for *The New York Times Book Review* and *The New York Times Magazine*. He is the author of three previous books: *Architects of Fear, Machinery of the Mind,* and *In the Palaces of Memory.* A former Alicia Patterson Fellow and the recipient of a Special Achievement in Nonfiction award from the Los Angeles chapter of PEN, Mr. Johnson grew up in New Mexico and now lives in Santa Fe.

Books by GEORGE JOHNSON

Fire in the Mind:
Science, Faith and the Search for Order

In the Palaces of Memory:
How We Build the Worlds Inside Our Heads

Machinery of the Mind:
Inside the New Science of Artificial Intelligence

Architects of Fear:
Conspiracy Theories and Paranoia in American Politics

FIRE IN THE MIND

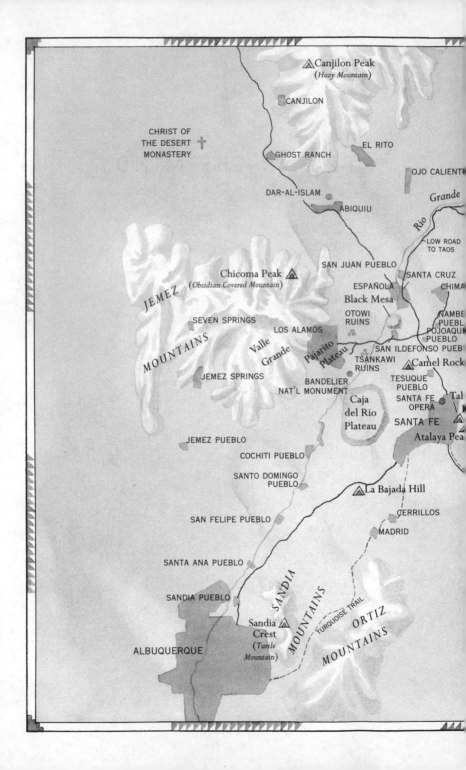

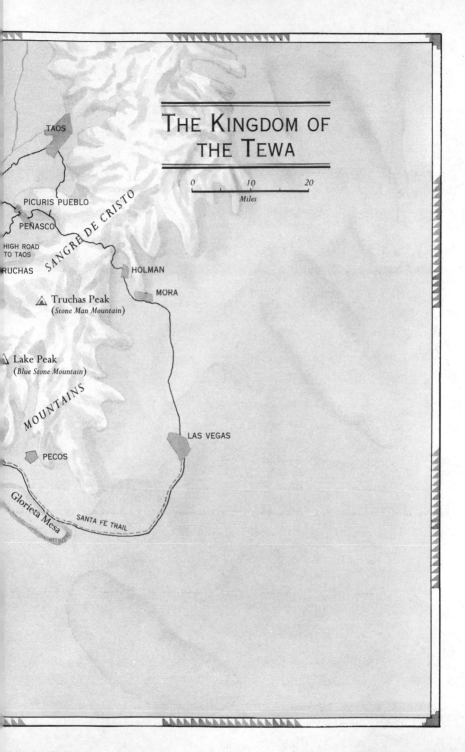

THE KINGDOM OF
THE TEWA

0   10   20

Miles

TAOS

PICURIS PUEBLO

PEÑASCO

SANGRE DE CRISTO

HIGH ROAD
TO TAOS

RUCHAS

HOLMAN

MORA

△ Truchas Peak
(Stone Man Mountain)

△ Lake Peak
(Blue Stone Mountain)

MOUNTAINS

LAS VEGAS

PECOS

Glorieta Mesa

SANTA FE TRAIL

FIRE IN THE MIND

SCIENCE, FAITH,

AND THE SEARCH FOR ORDER

GEORGE JOHNSON

VINTAGE BOOKS

A DIVISION OF RANDOM HOUSE, INC.

NEW YORK

FIRST VINTAGE BOOKS EDITION, SEPTEMBER 1996

The Library of Congress has cataloged the Knopf edition as follows:
Johnson, George.
Fire in the mind: science, faith, and the search for order /
George Johnson.
p. cm.
Includes bibliographical references and index.
ISBN 0-679-41192-5
1. Religion and science.   2. Hermanos Penitentes.   3. Tewa philosophy.
4. Science—Philosophy.   5. Johnson, George.
6. New Mexico—Description and travel.   I. Title.
BL240.2.J547   1995
215—dc20   94-38382
CIP
Vintage ISBN: 0-679-74021-X

Random House Web address: http://www.randomhouse.com/

Printed in the United States of America
10 9 8 7 6 5

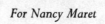

For Nancy Maret

When all the stars were ready to be placed in the sky First Woman said, "I will use these to write the laws that are to govern mankind for all time. These laws cannot be written on the water as that is always changing its form, nor can they be written in the sand as the wind would soon erase them, but if they are written in the stars they can be read and remembered forever."

—FROM A NAVAJO CREATION STORY

CONTENTS

FIRE IN THE MIND

KIVAS, MORADAS, AND THE SECRETS OF THE NUCLEAR AGE

Several years ago, on a visit home to New Mexico from my self-imposed exile in New York City, I was driving through the predominantly Catholic village of Truchas, on the high road from Santa Fe to Taos, when I rounded a corner and was startled to see a tiny adobe church with a makeshift steeple of corrugated green and yellow plastic (the kind used to cover carports and swimming pools) and a sign that read "Templo Sion, Asambleados de Dios"—Zion Temple, Assembly of God.

I have always felt a little uneasy driving through Truchas. Most of the small towns on the high road to Taos—Chimayo, Cordova, El Valle, Ojo Sarco, Trampas, Peñasco—are nestled comfortably in valleys, sheltered from the elements. Truchas is more like a Tuscan village, sitting high and exposed in an austere mountain meadow in the Sangre de Cristo (Blood of Christ) Mountains, with an uninterrupted free-fall view down to the Rio Grande. The town has long had a certain reputation for unfriendliness to outsiders, whether Anglos from Santa Fe or New York, or Hispanics from the next village over the rise. One occasionally hears stories of visitors stopping for a drink at the local bar only to be lured into a fight they are destined to lose, or of hikers parking in the nearby National Forest for a walk to the Trampas Lakes or an assault on the Truchas Peaks, and returning to find their tires slashed or maybe parts of their engine gone. But the legend is probably exaggerated in the retelling. Most often, the people of Truchas are simply trying to protect their quiet mountain life. Like people all over the world, they are wary of strangers and sometimes prefer to be left alone.

Especially jealous of their privacy are the Hermanos Penitentes, the "Penitential Brotherhood," a Catholic lay society known not only for its acts of charity and kindness but for privately practicing flagellation and other self-inflicted punishments so as to better appreciate the suffering of Christ. Near the edge of the village, across town from Templo Sion, is the adobe morada, the meeting place where the Hermanos perform their secret rites. Not even the wives of members of the order are allowed to know what goes on inside the morada during the long nights of Holy Week. Truchas is one of the remaining outposts of this fierce distillation of Catholicism, and so it was especially surprising to find the village invaded by an upstart fundamentalist church called Zion Temple, run by the sort of Protestants who, in their less generous moments, are known to declare that the Catholic Church is the Whore of Babylon in the Book of Revelation.

Colonized by art galleries from Santa Fe and even a bed-and-breakfast, Truchas seems a shade friendlier these days. When I returned a few years later, Templo Sion was boarded up and there was a For Sale sign on the door, but it lingered in my memory as an emblem of New Mexico's stark contrasts and strange juxtapositions, which make this such a weird and fascinating place to live.

Thirty miles, as the eye flies, across the Rio Grande Valley from Truchas is another city of secrets, perched on a mesa top in a different mountain range, the Jemez. So sterile and modern for such a spectacular setting, Los Alamos, named for its cottonwood trees, is known for giving the world the atomic bomb. The days are long gone when this laboratory city officially existed only as a post office box in Santa Fe. Weapons work is now slowly being eclipsed by theoretical physics, cosmology, nonlinear mathematics, biology, immunology, and the monumental task of cleaning up the defense industry's toxic nuclear mess. But in many quarters of the city, the sense of secrecy endures. Drive through the canyons and mesas around Los Alamos and your eye is constantly assaulted by distinctive blue-and-white signs—Tech Area 39, Tech Area 33, Tech Area 49—marking makeshift buildings, the white elephants of the cold war, still surrounded by guardhouses and chain-link fences. Some areas are marked with signs that warn:

DANGER
EXPLOSIVES
KEEP OUT

A few sites are marked with the three converging triangles meant to warn people of all languages of radioactivity. What went on in Los Alamos' technological moradas? No one will say for sure.

Between the Tech Areas of Los Alamos and the moradas of Truchas lie still more temples with their own secrets: the adobe kivas of the Tewa Indians which dot the valley of the Rio Grande, the dry expanse of piñon and juniper trees that stretches between the Jemez and the Sangre de Cristo Mountains. The pueblos of San Ildefonso, Santa Clara, San Juan, Nambe, Tesuque, and Pojoaque hold occasional public dances as a concession to the curious, but their most sacred rituals are still carried out behind closed doors.

Arcing diagonally across the Rio Grande Valley are the sparks of yet another polarity, generated by the city of science and the city of arts: Los Alamos, hard-edged, made of concrete and steel, and Santa Fe, with its soft skyline of adobe houses and galleries, million-dollar parodies of the Tewa's traditional homes. Today the Royal City of the Holy Faith of St. Francis of Assisi is more New Age than Catholic, a bazaar offering every heresy under the sun. Santa Fe is also becoming a city of science—what some of its practitioners like to think of as a gentler, more open kind. In recent years, the Santa Fe Institute, which sits amid the foothills that roll from the mountains to the town, has become the center of a search for laws of complexity, which seek to explain how our unfeeling universe gives rise to life, mind, and society. Some of this work is closely allied with the School of American Research, a Santa Fe institution that has long puzzled over the rise and sudden fall of the Anasazi civilization at Chaco Canyon and Mesa Verde, whose remnants seem to have washed ashore to help form the pueblos that now line the Rio Grande.

New Mexico has long billed itself as the land of three cultures—Indian, Hispanic, and Anglo—but when one includes the various subcultures of science and religion, the diversity is overwhelming. It was soon after my unexpected encounter with Templo Sion that the idea for this book began germinating. This sea of immiscible bubbles, where beliefs new and ancient bump up against one another, seemed just the place to think through some of the questions about science, religion, and philosophy that long have troubled me—the very kinds of mysteries that are being explored by some of the thinkers at Santa Fe and Los Alamos:

How could the universe arise from pure nothing?

How does the hard-edged material world we experience arise from the indeterminacy of the quantum haze?

How does life arise from the random jostling of dead molecules?

How does the mind arise from the brain?

And, the single mystery arching over the rest: Are there really laws governing the universe? Or is the order we see imposed by the prisms of our nervous systems, a mere artifact of the way evolution wired the brain? Do the patterns found by the scientific subcultures of Santa Fe and Los Alamos hold some claim to universal truth, or would a visitor from a distant galaxy consider them as culturally determined as those divined by the Tewa and the Penitentes?

With its jigsaw puzzle of world views and its long tradition of attracting those on the intellectual and spiritual fringes and frontiers, northern New Mexico seemed the perfect perch for exploring the penumbra where science's shining light fades into darkness, for plumbing the depths of what we know—or think we know—about this world in which we find ourselves. For a variety of reasons, historical and geographical, northern New Mexico has become a node in a network of people the world over who are beginning to question some of science's most deeply entrenched faiths. I found that rather than hop a plane to the West Coast, the East Coast, or another continent, I could sit like a spider in the middle of this web and wait as Santa Fe attracted some of the most interesting thinkers in the world. While some offer arresting new ways to think about physics and biology, others are turning their sights inward and contemplating the built-in limits of their enterprise.

There seems to be something about the altitude here and the stark relief between mountain and desert that pushes speculation to the edge and makes even the most sober of scientists more reflective, more willing to turn science back on itself, to theorize about what it means to theorize— about how we make these maps of the world. A theory can be thought of as the fitting of a curve to a spray of data. One can always simply go from point to point, connecting the dots like those in a child's coloring book.

But all that is left is a meandering line with little explanatory power; there is no way to predict how future points are likely to fall. Science is the search for neat, predictable curves, compact ways of summarizing the data. But there is always the danger that the curves we see are illusory, like pictures of animals in the clouds. As we draw our self-propelling arcs, some points will inevitably lie outside the line—those that must be dismissed as random error or noise. So we are left with a gnawing dissatisfaction: Are we missing something? If we looked at the points a little harder, graphed them a different way, would a more elegant order emerge?

There are two opposing ways to view the scientific enterprise. Almost all science books, popular and unpopular, are written on the assumption that there actually are laws of the universe out there, like veins of gold, and that scientists are miners extracting the ore. We are presented with an image of adventurous explorers uncovering Truth with a capital T. But science can also be seen as a construction, a man-made edifice that is historical, not timeless—one of many alternative ways of carving up the world.

In our society, we make a distinction between the history of science and the history of everything else. In the history of a country or an individual, there is no necessary pattern that things have to follow. We play games in which we imagine what the world would be like if John F. Kennedy had not been assassinated or if Ronald Reagan had; or what our own life would be like if we had taken a different plane or a different class in college and never met our husband or wife.

The history of science is supposed to be different. The only contingencies most physicists would admit are things like who made a discovery or when it occurred. The names of the particles are historical contingencies—"electron" is from the Greek word for amber, "quark" alludes to a line by Joyce—but certainly not the particles themselves. It is assumed that there is a gold standard backing up the value of our scientific currency: the way the universe really is. Venture too far from the straight and narrow and you will be snapped back by reality. For most scientists this vision of an objective world—governed by platonic laws of nature existing somehow in a realm beyond everyday space and time— is a deep though seldom stated hypothesis. In a way, it is the basis of their religion.

But what if science is as historical a process as anything else, a labyrinth of branching possibilities? Perhaps in putting together our picture of the world, there are many paths we could have taken. How, though, could we ever tell? We can think of each experiment and its interpretation as a fork in the road. Decision by decision, we are pushed into new regions in the space of possibilities. Before long, we have ventured so far in one direction that it is all but impossible to go back. Our search for truth has carried us along a single branch of the tree of knowledge until we are so far out on a single twig at the end of a certain limb that we are powerless to imagine how it could be otherwise. What if, at the end of many other twigs, there are equally valid—maybe better—ways of explaining the world? We would never know. We can't jump from our leaf to the next, leaping across the terrifying vacuum of empty conceptual space. To get to another leaf, we would have to retrace our steps, go back down the twig, the branch, the limb, perhaps all the way to the trunk, and start the climb all over again.

Just as there are many ways to write a book, and one is channeled in certain directions by decisions made early on, perhaps there are many ways to construct a science. With an unfinished book it is possible to go back and tear up the whole thing, to start over again. But with thousands of scientists all working together on the same manuscript, it is all but impossible to go against the flow.

This book is unusual, I think, in that it takes an agnostic stance—between the extremes of science as discovery and science as construction. In the end, there is no way to know whether science is converging on a single truth, the way the universe really is, or simply building artificial structures, tools that allow us to predict, to some extent, and to explain and control. This dilemma hovered in the back of my mind as I explored New Mexico's patchwork quilt of cultures, talking with people up against the edge of knowledge, and of what it is possible to know.

The tension between history and science, contingency and timeless natural law, runs throughout these pages. Traditionally, biology has been seen as a historical science, while physics is regarded as a search for absolutes. Physicists seek that which is constant throughout the universe. Biologists are supposed to be content to pick their way through the accretion of mechanisms and mechanisms built on top of mechanisms that evolution happened to lay down on earth, to describe natural artifices—organisms—that, with a different roll of the Darwinian dice, would be unrecognizable to us. One of the themes of this book is that this demarcation between physics and biology is becoming blurry. We will see biologists looking for timeless truths, principles of complexity—laws of the organism that might be reflected in all creatures, domestic or extraterrestrial, and even in metaorganisms like societies and economies. Conversely, we will see physicists seeking signs of contingency in the way the universe happened to crystallize from the big bang. Perhaps the particles and forces we observe and the laws they obey are "frozen accidents," just like biological structures. If so, it would be no more required that we have neutrinos than that we have hemoglobin, no more necessary that we have four fundamental forces than twelve ribs and thirty-three vertebrae.

What I propose to provide between these covers is a tour of some of the edges of twentieth-century science that are being explored in the laboratories of northern New Mexico. After a panoramic sweep of the physical and intellectual terrain, Part One will present a bird's-eye view of particle physics and astronomy, the science of the very small and the science of the very large. By retracing the history of these disciplines in a different way—viewing them more as artful constructions than as excavations of preexisting truth—these chapters will set the stage for Parts Two and Three, which describe what struck me as some of the most entrancing projects at Los Alamos and the Santa Fe Institute. Part Two will describe an attempt to recast physics and cosmology by climbing back to the trunk of the tree of knowledge (or at least to the base of one of its limbs) and taking a somewhat different branch, in which the seemingly ethereal concept of information is admitted as a fundamental quantity as palpable and real as matter and energy. One of the goals of this alternate way of carving up the world is to better understand how the certainty of our material world arises from the randomness of quantum theory, and how an unfeeling universe gave rise to creatures like us, who feed so voraciously on information. In Part Three, ideas about information will be marshaled to illuminate another mystery: how something as complex and self-sustaining as life could have emerged from the random turmoil of the primal seas. Once this earthly infection began (a "fever of matter," Thomas Mann called it), how did it increase in complexity to the point where it could ponder its own beginnings? Is the random variation and selection of Darwinian evolution enough to explain this phenomenon? Or could there be a deeper source of order?

Sifting order from randomness—from the very beginning, this has been the driving force of life, organizing haphazard collections of molecules and cells into these creatures with their sciences and their faiths. For science is only half the story. In keeping with the strange juxtapositions and stark contrasts of this haunting land, the tour will include an occasional side trip to other New Mexican subcultures, which have developed very different ways of finding and imposing order in a sometimes dishearteningly capricious world. In the course of all of this, we will try to see science as part of a larger story: the drive to find a place for ourselves in a universe into which we never asked to be born.

I came back to New Mexico to see if it was possible for someone like myself, a nonscientist who is passionately interested in science, to develop a feel for the contours of our current knowledge, a map of the terrain, a picture that would fit comfortably inside my head. But like the Spanish conquistadores who wandered into this mysterious northern hinterland from their empire to the south, I soon found myself in uncharted territory, the wilderness the mapmakers call terra incognita.

PART ONE

FOUR MAGIC MOUNTAINS

The only laws of matter are those which
our minds must fabricate, and the only laws of mind
are fabricated for it by matter.

—*James Clerk Maxwell*

1

PHAEDRUS'S GHOSTS

In the evening, just as their planet is about to complete another revolution, small bands of earthlings gather in the foothills of the Sangre de Cristo Mountains and engage in a ritual that is probably as old as humankind. Like their fellow creatures who assemble at the pier in Key West, Florida, at the pyramids of Chichén Itzá in the Yucatán, or on the observation deck of the World Trade Center in New York City, they are seeking a vantage point, a place where they can watch the sun go down. Breathing the thin, cool air that sustains life in the high altitudes of northern New Mexico, they stand on a crumbling red foothill overlooking the stylized adobe architecture of downtown Santa Fe and gaze west across the mesas and buttes of the Rio Grande Valley, struck perhaps by the way an unfamiliar angle of sunlight can illuminate a new geography, making the familiar suddenly appear so strange.

By day the Jemez Mountains, the million-year-old volcanic eruption whose congealed lava and compacted ash form the western horizon, sit unobtrusively in the background, like an idea taken for granted, or an undemanding friend—no rival for the imposing majesty of the Sangre de

Cristos, those vastly older peaks of Precambrian granite that long sheltered Santa Fe from all things east. But in the day's finale of twilight, the
Jemez get their fifteen minutes of fame. As the sun begins to move behind them, background suddenly becomes foreground, two dimensions
are projected into three. Overhead the clouds seem to glow like cinders,
scattering low-frequency reds. And the hidden geometry of the mountains unfolds like an origami blossom, revealing peaks and mesas and
canyons you didn't know were there.

But the vision of this newly glimpsed terrain is as fleeting as the mountains of cumulus clouds piled temporarily in the sky. Just as the eye begins its explorations, the sun moves lower, the detail disappears; three
dimensions collapse back into two, until all that is left is a silhouette.
Backlit by the rays of the setting sun, the jagged peaks look flat and black,
like a hole chipped in the bottom of the sky.

On some nights the Sangre de Cristos glow in these final moments
with the blood-red glory that gives them their name. And slowly, one by
one, the stars come out. In winter, Orion and his dog Sirius sprawl across
the heavens, following the sun behind the mountains; in summer, Scorpio appears on the southern horizon, sent by Apollo, the Greeks used to
say, to chase Orion through the sky.

Down below, in the arroyos, the lights of Santa Fe begin to link into
their own constellations, shimmering geometries that extend farther
each year as the developers stamp their blueprints onto the land. Far to
the south, an artery of car lights—red corpuscles alternating with
white—stretches toward the megawatt glow of Albuquerque, where the
Sandia Mountains rise cold and silent, marked by the blinking lights of the
television towers, radiating their invisible signals to the creatures who
live in this sea of incandescence, and up to the uncomprehending stars.

In the creation myth of the Tewa Indians, descendants of the lost tribes
of Anasazi whose pueblos line the northern Rio Grande, this land is the
center of the universe, the place where life began. Ascending through a
lake from the underworld, the first people walked north, south, east, and
west, returning to declare that only a small square of this hostile surface
was fit for habitation. Four sacred mountains—Sandia Crest, Chicoma
Peak in the Jemez, Truchas Peak in the Sangre de Cristos, and Canjilon
Peak, a hazy presence in the north—mark the edges of this world, a tiny
enclave where the gods said people could live in harmony.

Whether or not one believes that northern New Mexico is the center
of creation, it is easy to sympathize with the desire for a more orderly,

circumscribed world. There is something comforting about knowing the names of the mountains, living under a familiar sky. Just as nature abhors a vacuum, the mind abhors randomness. Automatically we see pictures in the stars above us; we hear voices in the white noise of a river, music in the wind. As naturally as beavers build dams and spiders spin webs, people draw maps, in the sky and in the sand.

Standing in the hills above Santa Fe, looking out on the uneasy mix of civilizations—Indian, Spanish, and Anglo—that has grown up within the space marked off by the Tewa's four mountains, one naturally begins to speculate about this most basic of human drives: the obsession to find and impose order. Whether the orders we invent are geographic, religious, or scientific, inevitably, it seems, we come to identify the map with the territory, to insist that the lines we draw are real.

It was from beyond the southernmost magic mountain that the Spanish came four hundred years ago, bringing horses, guns, and Catholicism. Backed by the power of Spanish soldiers, the Franciscan priests forced their rituals on the Indians, supplanting their spirits with the Church's own. From then on, the Corn Mother would be addressed as the Blessed Virgin Mary; the holy beings known as kachinas would be replaced by saints. On the surface, anyway. Secretly the Indians continued to draw strength from their own pantheon. On a hill above Santa Fe, a tall white cross commemorates the Franciscan fathers who died in the Pueblo Revolt of 1680, when the Indians—convinced after years of disharmony and drought that the friars' magic was no better than their own—rose up against the Spaniards and drove them back down the Camino Real, as far as El Paso. But the rains still refused to fall. Twelve years later, Don Diego de Vargas's reconquest met only muted resistance, and Catholicism was ascendant again. Just east of the plaza that forms the focus of Santa Fe stands St. Francis Cathedral, chiming out the hours, dividing the day with its sound.

Transported to a different part of the galaxy, we would be startled to see our constellations stretched and squeezed, distorted by a new vantage point. But how hard it is to appreciate that one person's distortion can be another person's reality, that we look at the world through different eyeglasses, that there are different ways of carving up the sky.

Instead of the Big Dipper and Cassiopeia, the Navajos, whose kingdom lies just beyond the Jemez Mountains, see First Man and First Woman. These constellations are also called Whirling Male and Whirling Female because of the way they dance around Polaris, the North Star. The tail of

Scorpio combined with stars in Canis Major becomes Rabbit Tracks. There are also the Porcupine, Red Bear, Thunder, Big Snake, Horned Rattler, Monster Slayer, Born for Water, Corn Beetle, Turkey Tracks, the Wolf, the Eagle, the Lizard, the Lark Who Sang His Song to the Sun Every Morning. And there is Black God. In one version of Navajo cosmology, it was Black God who carefully arranged the constellations in the heavens and set them on fire. But before he came close to completing his task, Coyote stole the pouch of star crystals, scattering them randomly through the sky.

One needn't travel to Alpha Centauri to see the universe from a different angle. Part of the magic of the land around Santa Fe is the astonishing number of cultures, both ancient and new, that have been drawn by the New Mexican light. After the Spanish, the Americans began arriving with their own peculiar ideas. At first they came from the east in a trickle: traders, settlers, and adventurers wearing the first grooves in the long strip of erosion that would become the Santa Fe Trail. Then, in 1846, they came in full force, as soldiers, charged with taking the mountains and mesas for the United States. The people of these northern hinterlands had barely noticed when a revolution down south had led to independence from Spain in 1821. The occupiers might call themselves Mexicans instead of Spaniards, but everything else had remained pretty much the same. The United States' victory over Mexico left a deeper impression. Colonel Stephen Kearny and his troops marched into Santa Fe, planting the third flag to flap in the northern New Mexican wind (the Tewa had never felt a need for one). Sitting at the crossroads of the Santa Fe Trail and the Camino Real, the Spanish, like the Indians before them, now strained to see the world through alien eyes.

With the harnessing of a powerful new science called thermodynamics, the wagons of the Santa Fe Trail were replaced by steam engines pulling trains. A new conduit was open, and all kinds of strange notions came pouring in. Follow the zigzag peaks of the Jemez northward and you reach the Pajarito Plateau, where in the early 1940s the secret city of Los Alamos appeared like an outpost from another planet. Picked for its beauty as much as for its isolation, the location of the nuclear laboratory may have been secret to most of the world, but the people of San Ildefonso pueblo, a Tewa settlement on the Rio Grande, knew that something funny was going on in the mesas above them. For as long as they could remember, the dirt road that led past their village and up the side of the Pajarito Plateau, to a remote boys' school and the scattered ruins of their Anasazi ancestors, had carried little traffic. But suddenly there came a steady stream: trucks carrying building supplies and laboratory equipment from Santa Fe, buses coming down the hill to ferry San Ildefonso men to help with the construction. But the most important commodity traveling up the road from Santa Fe was impossible to detect: some of the deepest ideas of Western physics, encoded in the neural webs of scientists from all over Europe and the United States—the full-time researchers like Robert Oppenheimer, Hans Bethe, Edward Teller, and the young Richard Feynman; the prestigious visitors like Enrico Fermi, I. I. Rabi, John von Neumann, and Niels Bohr.

While these luminaries of another land walked the trails of the Pajarito, discussing fission cross sections, the hydrodynamics of spherical imploding shock waves, and other esoterica, the Indians of San Ildefonso remained immersed in a world animated by spirits. Throughout the year, as the days grew longer and then shorter again, they gathered near their ceremonial kivas to dance the world back into balance, to ensure the return of the seasons and the sun. Now as then, the pueblo's dusty plaza is at the center of an ancient cosmology. It is here that the *sipapu*, or spirit hole, leads like a wormhole to another universe—the world beneath the lake where the Cloud Beings, or kachinas, are said to live and where the souls of the departed go.

In the minds of believers, this ceremonial nexus is connected not only to the earth below but to the sky above, acting like a lens focusing energies emanating from shrines atop the four magic mountains, one for each point of the compass. It is at the *sipapu* that the four horizontal directions come together, each associated with a color—blue for north, yellow for west, red for south, white for east. And each color is associated with an animal, a god, and a sacred lake. The sun rises from the eastern lake and sets in the western lake; like the people themselves, it begins and ends in the underworld. Everything is connected in a great celestial tissue: people, animals, plants, spirits, stars.

If one part of the net is disturbed, it is said, the ripples radiate throughout the whole. Opposing spirits must be kept in balance: the female earth and the male sky, the forces of hot and cold, ripe and unripe, magic and witchcraft. Beneath the calm stoicism so many Anglos see, the pueblo people live in a world of order and control. Dances are scheduled according to a complex calendar based on the positions of the sun and moon. Each member of the tribe occupies a precise position in an intricate hierarchy. Dozens of spirits must be propitiated, with the right cer emonies performed at the right time and with the right frame of mind. Dressed in costumes like those they have used for centuries, the people of the tribe dance the turtle dance, the buffalo dance, the eagle dance, the rainbow dance, the corn dance—moving round and round, circles within circles, imposing their own geometries as they sing to the rhythm of the drums. There are ceremonies to melt the snow, calm the wind, ensure the fertility of crops, animals, sons and daughters. By dancing to reenact the seasons, moving in resonance with the earth's own circles, they hope to maintain order, cast out the chance occurrences—a child born dead, a spring without rain—that come when the world is allowed to drift out of kilter.

This elaborate order runs according to a variety of time that is more cyclical than linear. Look beyond the distracting details and the rhythm of life is very much the same, season after season. Knowledge is passed, largely unchanged, from generation to generation. Though people die and rejoin their ancestors, they will return in some future revolution. But the Tewa do not think of themselves as the passive pawns of gods. They are participating in the control of the cosmos.

It is a terrible responsibility, keeping the universe running. But unlike so many of those enmeshed in their self-constructed systems, the pueblo people have not lost their ability to laugh—not just at their enemies but at themselves. Among the players in the seasonal rites are the Kossa clowns, painted head to toe with black and white stripes, who bumble their way through the carefully choreographed ceremonies, shattering the solemnity. Sometimes they make fun of the pueblos' old foes, the Navajos and the Spanish. They perform parodies of the Catholic mass, of other tribes' dances, and even of their own rituals.

It was into this precisely delineated universe that the physicists came, bringing their own kind of order. They too saw a universe of dualities carefully balanced: the positive and negative charges within the atom, Einstein's equation of mass and energy (with that astonishing constant, the speed of light squared, which showed that a nucleus toppled would release ungodly power). While the Indians had their quiver of colored directional arrows radiating from the *sipapu,* the physicists were laying the groundwork for what would one day be known as the standard model, with its up quarks, down quarks, red quarks, blue quarks—tiny shards of mathematics that, the theorists tell us, make up the cores of atoms.

To the physicists, time was linear, not cyclical. They believed with all their might that knowledge is not fixed, an inheritance from the past, but something that grows, lumbering forward year by year as they increased their control over nature. While the world of San Ildefonso was centripetal, a whirlpool revolving around the village kiva, to the scientists there was no center. With his special theory of relativity, Einstein had upset the notion that there is a privileged position from which to view the stars. Any inertial frame would do. And while the Indians danced to keep the world synchronized and struggled to keep their thoughts pure, the scientists were coming to Los Alamos to learn how to unleash the very forces that helped hold the world together.

To the Tewa, terrorized by Catholicism, it may have seemed appropriate that it was at a place called Trinity Site, in southern New Mexico, that the words of the physicists' equations took on substance, melting sand into a green, glassy crater on the Jornado de Muerto, the most dangerous stretch of the Spaniards' old Camino Real, the royal road to occupation. If the Tewas' magic had been so powerful, they might have picked the same location to detonate a bomb.

In the half century since this mathematical transubstantiation, Los Alamos has continued its paradoxical quest of unlocking nature's secrets for use in making ever more destructive weapons. But it has also turned its sights to more peaceful pursuits, elaborating the conceptual filters we use to sift order from randomness and make sense of the world.

In searching for the most economical way of mapping the universe, scientists have slowly eliminated earth, air, fire, and water as fundamentals, converging on the twentieth-century vision in which all is made of mass-energy interacting in an arena of space and time. The pinnacle of this quest is often said to be quantum mechanics, which provides such precise forecasts of the way subatomic particles behave, but which seems to suggest that observers are necessary to conjure our rock-solid world of classical Newtonian physics out of the uncertainty of the quantum realm. In quantum theory, a particle exists in a juxtaposition of possible states; only when it is measured does it take on definite qualities, like position or momentum. Repelled by the potentially mystical overtones of this anti-Copernican twist, some physicists, like Wojciech Zurek of Los Alamos' Theoretical Division, have gone looking for a less anthropocentric approach. The problem, they believe, is that in carving up the world scientists have omitted an important ingredient: information. Once this new piece is added to the puzzle, along with mass-energy, some of the spookiness may be expelled from quantum theory. Trading ideas with Murray Gell-Mann, the inventor of the quark, who lives in Santa Fe, Zurek and some of his colleagues are trying to recast quantum theory in a way that doesn't require the existence of observers.

Los Alamos is also at the forefront of research in nonlinear dynamics—chaos, as it is loosely called. In the days of the Manhattan Project, some of the equations for designing nuclear bombs were so difficult to solve that mathematicians like Von Neumann, Stanislaw Ulam, and Nicholas Metropolis used what were called Monte Carlo methods, feeding equations with strings of random numbers and observing how they behaved. To do so they had to devise ways to coax computers, the most deterministic of beasts, to generate random strings of numbers. The result was some of the first concentrated work on what are now called chaotic equations, which look simple on the surface but generate patterns so complex that they are difficult to distinguish from randomness. More recently, the tables have been turned. Looking through the opposite ends of their telescopes, mathematicians at the Los Alamos Center for Nonlinear Studies are among those trying to use the tools of chaos theory to find hints of order hiding behind phenomena once dismissed as random.

Going beyond chaos, Los Alamos has joined with an interdisciplinary think tank called the Santa Fe Institute to develop what is often described as a new science of complexity, which seeks to explain why, against all odds, order seems to arise in the universe. Of course, an important question is to what degree the orders we observe are out in the world and to what degree they are imposed by our nervous systems, the invisible spectacles that refract everything we see. To some extent, complexity may be in the eye of the beholder. But many of the scientists at the Santa Fe Institute believe they can divine fundamental rules that apply to complex systems of all kinds—cells, organisms, brains, societies, galaxies. All, they believe in their boldest moments, might obey the same universal laws. Every other year, at the Artificial Life 4-H show, computer scientists and biologists gather to demonstrate their prowess as creators of simulated ecologies in which beings made of pure information evolve. A few—taking to heart the heretical notion that information is as fundamental as matter and energy—go so far as to say that the simulated beings in their simulated worlds really are alive.

In this land of strange juxtapositions, ancient ideas continue to coexist with the new. Like pieces of a skeleton, the shards of an old world view can become fossilized and carried from century to century. Across from

Los Alamos, in the Tewa's easternmost holy mountains, members of the secretive Penitential Brotherhood, the Hermanos, believe that only pain like Jesus felt can wash their sins away. In recent years the prying eyes of curiosity-seekers have driven even more of their rituals underground. But a few photographs survive. Marching in their annual Good Friday processions, shouldering heavy wooden crosses and lashing one another with whips, they look like the bands of medieval flagellants that roamed Europe during the Black Plague.

In their own fight against chaos, the Hispanics who cling to life in the Sangre de Cristos, one of the poorest regions in the United States, have always taken their Catholicism very seriously. Not all Spaniards were conquerors. Placed at the northeastern fringe of New Spain as a barrier against Comanches sweeping in from the Great Plains, the villagers sought solace in their religion, but they received little supervision from the Catholic Church. All but forgotten in the empire's hinterlands, they developed a folk religion whose obsession with death and suffering is still evident in the bloody icons that populate the old adobe churches: Christ writhing in agony on the cross, hemorrhaging from the gashes in his chest, the nail holes in his hands and feet; his brow slashed by the crown of thorns. It is no wonder these are called the "Blood of Christ" Mountains. Death itself is represented by a grinning skeleton, La Muerta, riding in a cart and shooting a bow and arrow.

But for the people who live here, this is also a land of miracles. In the village of Chimayo, just down the road from Truchas, pilgrims arrive daily at the Santuario, making their way to a tiny room at the side of the chapel where they fill envelopes and jars with the miraculous soil they believe has the power to heal all wounds. On the plastered walls of the old adobe building hang crutches, said to have been cast off by those who were cured. In these modern times, the Catholic Church sometimes finds this kind of devotion an embarrassment. In the 1970s the priests were asked by the Archbishop to stop sending samples of the soil to cancer victims who wrote to the church seeking a cure; he was afraid they might forgo more effective treatments. At about the same time, word began to spread that in Holman, a village on the other side of the mountains, an image of Christ made a nightly appearance on an old adobe wall, just after the town's sodium vapor streetlight switched on. Two bullet holes formed his eyes. Also reported among the cracks and swirls of the stucco were sightings of the Virgin Mary and John F. Kennedy. For months people gathered in crowds to gaze at the wall, seeking patterns in the coincidences of shadow and light. Finally the Church sent a team of investigators to Holman, who concluded that nothing supernatural had occurred, that the only miracle was the faith of the people.

There are few places on earth that so many people have claimed as holy and where so many people see the world in different ways. In recent years the Catholic churches and Penitente moradas of the Sangre de Cristos have been joined by makeshift churches erected by fundamentalists from the Assembly of God and other denominations, some of whom believe that the Pope is the Antichrist, and that the wars and rumors of war on the nightly news are part of a Manichean struggle, light versus darkness, that is neatly laid out in the pages of the Bible. In the same area is an outpost of Sikhs, followers of the order of holy warriors that guards the Golden Temple in India. Like the pueblo Indians, they believe that northern New Mexico is sacred land. From a Penitente morada in the hills above Abiquiu, Georgia O'Keeffe's old haunt, one can gaze across the Chama River at an Islamic mosque rising from a semiarid moonscape of wind-carved volcanic ash that might pass for northern Africa. Several mesas beyond the Dar-al-Islam community lies the Presbyterian church's starkly majestic Ghost Ranch retreat, and beyond that, at the end of a long dirt road, is the Christ of the Desert monastery. From here one can jump across the mountains to the village of Jemez Springs, known throughout Christendom as a hideaway for Catholic priests with emotional and spiritual troubles, including pedophilia. Here the Servants of the Paraclete and the Handmaids of the Precious Blood share the village with the Bodhi Mandala Zen Center. Zen Buddhists, Tibetan Buddhists, Trappist monks, Benedictines—each has found a niche in the cultural ecology that has taken root in the deserts and mountains of this jarring terrain. All, in their different ways, are trying to see patterns in the swirl around them, to find a home in a universe that sometimes seems oblivious to our existence.

The result of all this is not a melting pot but a turbulent, chaotic boil. One can spend a day in September watching the corn dance at San Ildefonso, then drive to Santa Fe to catch the candlelight procession from St. Francis Cathedral to the Cross of the Martyrs honoring the victims of the Pueblo Revolt. The procession marks the end of Fiesta, a week-long celebration of the day de Vargas and his army recaptured Santa Fe from the Indians. In a spirit of brotherly revisionism, the festival has been edited in recent years into a sometimes halfhearted celebration of multiculturalism. In the annual Fiesta pageant, the Hispanics portraying the conquistadores now arrive on foot instead of on horseback and no longer wear menacing suits of armor. In a recent Fiesta, their arrival at Santa Fe's plaza was accompanied by a war dance performed by Picuris pueblo, one of the most avid supporters of the 1680 revolt. Some pueblos have declared the date when the revolt broke out an official holiday—their own Fourth of July. For most, the highlight of Fiesta is the burning of Zozobra, a towering effigy that, torched by the seductive fire dancer, writhes and moans as it goes up in flames. Though sometimes described as an ancient Aztec custom, Zozobra was actually invented in 1926 by Anglos who felt they should have a ritual of their own.

Whether or not we believe that the future can be influenced by the circular rhythms of the dance or foretold from an analysis of Bible verses, or that from a few underlying physical laws we can generate a cosmos, we all share a faith that lurking beneath the world's complexity is simplicity. Psychologists have found that if you put people in a room with a contraption of lightbulbs wired to blink on and off at random, they will quickly discern what they believe are patterns, theories for predicting which bulb will be next to blink. Once a person becomes enmeshed in an ideology or a scientist in a hypothesis, it is difficult not to see confirmation everywhere. Our brains are wired to see order, but we are prisoners of our nervous systems, cursed with never knowing when we are seeing truths out there in the universe and when we are merely inventing elaborate architectures.

For all the checks and balances of the scientific method, this inevitable confusion extends to the laboratories as well as to the kivas, moradas, churches, and cathedrals. Years after the initial excitement over cold fusion turned to ridicule in most quarters, a retired scientist at Los Alamos insists that he is among those few who have replicated the experiments of the chemists Stanley Pons and Martin Fleishman, conjuring nuclear energy from a jar of water. Most experimenters who have tried to replicate cold fusion have failed, and most theorists continue to declare that fusion, the energy powering the sun and stars, cannot ignite at room temperature. But a few believers still insist that there is something funny going on in their flasks.

In another corner of the laboratory, scientists are struggling with a far more respectable mental edifice called string theory, in which the confusion of particles that scientists momentarily bring to life in their accelerators can be thought of as different notes played by infinitesimal strings, vibrating in ten dimensions. Is this physics, pure mathematics, or, as some detractors insist, theology?

Some of the world's best physicists, including Murray Gell-Mann, hold out hope that string theory will someday be used to tie particle physics and cosmology into a single package, forming the ultimate creation myth. But they know it is necessary to proceed with caution. Santa Fe is filled with reminders of what can happen when the search for order becomes an end in itself. In the city's New Age shops, one can buy occult books filled with esoteric knowledge recycled generation by generation since medieval times. In these baroque intricacies, planets resonate with colors which resonate with crystals which resonate with numbers, all trading energies undetectable by even the most sensitive instruments at Los Alamos. Signs advertising herbalists, psychics, astrologers, and homeopathic physicians line the streets radiating from the town plaza. The newspapers include advertisements for a psychic surgeon, who says she can painlessly cure your ills by applying her scalpel to your body's spiritual double, in an operating room on the etheric plane.

On a spring evening not too long ago, Stuart Kauffman, a biologist at the Santa Fe Institute, waited in line to hear the Dalai Lama address a standing-room-only crowd of four thousand people in the gymnasium of Santa Fe High School. Looking at the audience, drawn from any number of conflicting spiritual persuasions, he recalled spending a recent afternoon discussing science with two Nobel laureates (the economist Kenneth Arrow and the physicist Philip Anderson, who fly in frequently from Stanford and Princeton to visit the Santa Fe Institute) and the evening having dinner with two neighbors: a crystal healer and a channeler, who believes she can communicate with the dead. Though firmly anchored in the rationalist tradition, Kauffman himself was working at the fringes of biology. For the last few years he and a few colleagues at the Santa Fe Institute have been trying to devise a less dispiriting version of Darwin's theory of evolution, in which principles of self-organization join with random variation and selection in generating the order we see in the biological world. Some of Kauffman's colleagues, like Brian Goodwin, a regular visitor from England, go so far as to deny that natural selection is largely responsible for biological structure; they insist that the patterns in the living world are not imposed from outside by Darwinian pressures but are generated from within, as the organism obeys internal laws of its own. The lesson of all this is that life as we know it may not be a fluke, but something expected—that we are, as Kauffman likes to put it, "at home in the universe."

The idea that evolution is significantly shaped by hidden harmonies is anathema to most biologists. But since the days when Thomas Edison leased land in the Ortiz Mountains, just south of Santa Fe, hoping to use electricity to extract gold from ore, northern New Mexico has been a haven for iconoclasts. At the Prediction Company, located across the street from an herbalist in downtown Santa Fe, a former Los Alamos physicist, Doyne Farmer, and his colleague Norman Packard are committing their own heresy. Most economists hold that the day-to-day fluctuations of the financial markets are essentially unpredictable, a random walk. But the physicists at the Prediction Company believe they can use computers and mathematics to divine hidden orders in the zigzag lines of currency fluctuations and, they sometimes hope, the Dow Jones Industrial Average. Their dream is to beat the market and achieve the kind of financial independence that would let them explore questions that mainstream science has long ignored. The day after the Dalai Lama's visit, Farmer sat in a restaurant on Santa Fe's Canyon Road explaining the need to find a new law of physics—a kind of inverse of the second law of thermodynamics—that would explain why complexity arises. Occasionally Farmer even entertains what for most physicists is the ultimate heresy: the belief that quantum events may not be truly random, that if you look deeply enough you will find a hidden order.

In the time since people began gazing at the sky, an insignificant amount of starlight has fallen into our eyes and our instruments. But from that meager signal we extrapolate a cosmos, in which great spiraling galaxies join to form galactic clusters and clusters of clusters so huge that we give them names like the Great Attractor and the Great Wall. We tell of quasars at the edge of creation pouring out the energy of a trillion stars, of black holes so unfathomably deep that they gobble up light. And when asked how it all began, we tell the story of an explosion that occurred some ten to twenty billion years ago, creating all matter and energy and even the dimensions—time and space—in which it expands. An explosion so powerful, we're told, that we can still detect its afterglow in the hissing of our radiotelescopes and amid the snow on our TV screens.

But we are finite creatures contemplating the infinite, and there is always the danger of confusing our maps with reality itself, of seeing more order than is really there. The standard model, our theory of matter and energy, is built from particles—the quarks of Murray Gell-Mann—that, according to the very tenets of physics, can never be isolated and directly observed. Intertwined with the standard model is our theory of cosmology, the big bang. But the big bang cannot begin to explain the structure we see around us—the galaxies and clusters of galaxies—unless we make auxiliary assumptions, the most popular of which holds that much—perhaps nearly all—of the universe is made of something called dark matter, which can be inferred but not seen.

Scientists are quick to argue that there is a preponderance of indirect evidence for quarks and dark matter. Why should the universe be made only of components that our eyes, aided by our instruments, are attuned to see? But why, for that matter, should the universe be comprehensible to us at all? Compelled by our faith in the brain's ability to see to the core of creation, are we simply filling in the fractures of our imperfect theories? Are quarks and dark matter discoveries or are they inventions, artifacts of the brain's hunger for symmetry?

In any of our grand creations, there may be cracks in the foundations, niches in which heresies can grow. And it is by overturning monuments that new pictures of the universe emerge, new orthodoxies whose cracks will form different patterns. Whether there is hope of converging asymptotically on something called truth depends on whether you believe, like Plato, that mathematics and natural laws exist in and of themselves on some ethereal plane, or whether you believe they are human inventions—at best, an intersection between the way the world is and the way our nervous systems happened to evolve.

In *Zen and the Art of Motorcycle Maintenance,* Phaedrus, the author Robert Pirsig's alter ego, is sitting outside a motel room in the West, drinking whiskey with his traveling companions and listening to his son, Chris, tell ghost stories. "Do you believe in ghosts?" Chris asks his father. "No," Phaedrus says. "They contain no matter and have no energy and therefore, according to the laws of science, do not exist except in people's minds." Then he pauses and reflects: "Of course, the laws of science contain no matter and have no energy either and therefore do not exist except in people's minds."

Pushed up against this edge, science often retreats into platonism. Here on earth there may be no such thing as a perfect circle, but we recognize the rough approximations because we somehow have access to the perfect Circle, a pure idea existing in a separate ectoplasmic realm. And so we are left with a duality between mind and matter, ideas and things.

Some followers of the information physics being pursued in Los Alamos, Santa Fe, and elsewhere suggest a way of bridging the divide: the laws of the universe are not ethereal, they say, but physical—made from this stuff called information, the 1s and 0s of binary code. And so they seek to turn science back on itself and use information theory to understand where the laws of physics lie (in both senses of the word—where they reside and what their limits are). If information is as physical as matter and energy, and if ideas and mathematics are made of information, then perhaps they are rooted in the material world. But the price for banishing platonic mysticism may be a dizzying self-referential swirl: the laws of physics are made of information; information behaves according to the laws of physics. Everything begins to seem like ghosts.

For a while, modern science was content with explaining the how of existence, leaving the why to religion. At least, that is what we were taught in school. But sometimes it seems that in the final act of the millennium, science the world over is reaching beyond its old self-imposed limits. With its grand unification theories and cosmological schemes, it is seeking answers so fundamental that they border on theology. Why is there something instead of nothing? Why does the universe seem to operate according to mathematical laws? What is consciousness—a biological artifact, an accident of evolution, or something deeply woven into the warp and woof of the universe? But as our scientific cathedrals become ever grander, the ideas are that much more difficult to test; it becomes harder and harder to tell whether the orders we see are real. Is information truly fundamental, or just an artifact of brains parsing the universe in their own peculiar way? Does complexity arise inexorably, or is the concept just a human invention? One reads time and time again about how the human brain is the most complex device in the known universe. But it is the brain making this immodest judgment, elevating itself and complexity—whatever it is—to the pinnacle of creation. We may be like fish up against the edge of the aquarium; the shapes and colors that dazzle us could simply be our own reflections distorted by the glass.

With its scientists, both orthodox and heterodox, and its subcultures both ancient and new, the world around Santa Fe is a particularly interesting part of the aquarium. The tensions concentrated between New Mexico's four magic mountains run throughout Western civilization. This passion for an all-encompassing order, this raging cerebral fire, is nothing less than humanity's driving force, a unifying spirit that extends across continents and throughout time. The Gnostics believed that the mind was like a flame, a little glint of starlight trapped inside earthbound flesh. And so they looked upward, longing for the day they could ascend from this prison as pure spirit, pure understanding, streaming toward the sky.

The Pythagoreans gazing into the confusion around them and insisting that all is made of number; the Aztecs chipping their cosmology onto a disk of stone; the Australian aborigines guided on their walkabouts by the invisible maps they call songlines; Paracelsus with his elaborate system linking the body's illnesses to the movements of the stars; the Scholastic philosophers contemplating their universe of concentric crystalline spheres—everywhere we look we see the flame that drives the search for a science that would not only explain why the universe is the way it is, but why we are in it; why, through what seems an incredible chain of coincidences, we are sitting here as conscious beings striving to contemplate the whole.

In New Mexico, living side by side with the scientists, who see themselves on the expanding circumference of the bubble of knowledge, are cultures whose ancient beliefs still preserve the first sparks of the Promethean flame. The descendants of the Anasazi dancing in resonance with the seasons, the fundamentalists with their attempts to predict the future through biblical interpretation, and the physicists and biologists with their search for hidden harmonies are battling over the same spiritual ground. All are trying to make sense of life's overwhelming complexity, to come to terms with the fact that, for all our well-laid plans, we are buffeted about by contingency and chance. Each of these subcultures, in very different ways, is trying to replace randomness with order, to spin webs of ritual and reason, to try to convince itself that if we don't actually live at the center of creation, at least we can comprehend it— that there is reason to believe that the human mind can pierce the universal panoply. Each is trying to answer the question of why we are here, as a species, a society, and as individuals. In both science and religion, we seek creation myths, stories that give our lives meaning.

For those who stand before science's airy cathedrals with a mixture of wonder and skepticism, but who have found little satisfaction in the other matrices of belief that crisscross the land, northern New Mexico seems the perfect place for a search. The frightening clarity of the stars invites one to look upward and wonder at the pictures science has drawn of the heavens. And the wind-torn earth, so inhospitable to life that the bare bones of geology show through, invites one to gaze downward at the ground and into the rocks themselves, wondering at the concentric circles of matter—quarks within nuclei within atoms within molecules—that we have drawn to represent this hidden world. Standing between earth and sky, Los Alamos and Santa Fe, where people are rethinking some of the most basic beliefs of science, invite one to gaze inward and wonder if the maps could be drawn differently, if there is one or many ways of slicing up the sky.

In the end, there is no guarantee that any of our system-building will protect us from the fall. On the mesa tops of the Pajarito Plateau, near the one on which Los Alamos stands, lie the ruins of Tsankawi, one of the abandoned stone cities that the San Ildefonso consider their ancestral homes. Sometime after the mysterious collapse of the civilizations at Chaco Canyon and then Mesa Verde by 1300 A.D., the Anasazi built these smaller settlements, on the Pajarito and elsewhere, only to leave them two hundred years later for a simpler life on the banks of the Rio Grande. Sometimes it seems that complexity can rise only so high before it collapses, and it may never reach such lofty heights again.

It's a short hike to the top of Tsankawi Mesa, along a trail so traveled that in some places footsteps, ancient and modern, have worn a groove a foot deep into the stone. The mesa, a wedding cake of beige and white layers of fallen ash from the Jemez volcano, is so soft that the Anasazi scooped out rooms in the cliffsides or cut the rock into blocks to build villages, some with hundreds of chambers several stories high. Today all one hears standing on the mesa is wind and the sound of crows. All that is left of Tsankawi is mounds of buried rock where buildings once stood, and shallow depressions where kivas long ago fell in. The geometry has all but melted back into the earth, leaving only the ants to build new structures.

On the south, sun-drenched face of the mesa, the cave dwellings still have soot on their ceilings, and carved all over the cliffs are petroglyphs: roadrunners, turkeys, little birds (in Spanish, *pajaritos*), an occasional spiral. Some of the figures look like a man, hunched over, playing a flute—the being that the Hopi, the Tewa's cousins to the west, call Kokopelli. A chiseling of a man with lines sprouting from the top of his head looks like the clowns of Santo Domingo, a pueblo south of Santa Fe, who wear their hair done up with cornhusks. Other figures look like the Tewa clowns with their black-and-white-striped antennae. And some look so much like the popular image of extraterrestrials that one can hardly keep from entertaining fantasies like those of Erich von Däniken, who argued in best-selling books like *Chariots of the Gods?* that the Great Pyramids and other magnificent edifices were built with the help of aliens. The more you look, the more creatures seem to crawl out of the cliffs, until, like the faithful at Holman seeing Christ on the adobe wall, you're left to wonder: How much of the patterns are in the rock, how much are in your head?

Ultimately, all of us are faced with the same dilemma: the pictures we draw, the systems we build, can never fully embrace the richness and the unruliness of creation. Yet it is endemic to our species that we keep trying, huddled on our tiny planet, shining our flashlights into the darkness.

2

THE DEPTH OF THE ATOM

In 1540, after Coronado and his troops marched into New Mexico from the kingdom to the south, searching for the seven golden cities of Cíbola and finding the pueblos of the Zuni Indians instead, an expedition of his soldiers rode northwest as far as they could go, until they were stopped dead in their tracks by that mile-deep gash in the earth we now call the Grand Canyon. The land had opened up beneath their feet, revealing the layers of history they had been riding upon. The journals from the expedition don't record what went through their minds as they gazed upon those stripes of fossilized time. Without a theory to explain what lay before them, it seems likely that they saw the canyon mostly as an obstacle, an inverted mountain range that, like it or not, defined the northwestern limit of their newfound world.

It would be some three centuries before people were equipped with the conceptual lenses that we now use to make sense of so overwhelming a sight. Actually there were (and are) two sets of geological spectacles, their lenses ground according to which one holds most sacred: the accuracy of Genesis or the timelessness of natural law.

To fundamentalist Christians, the accumulated age of the generations in the Bible—all those Old Testament begettings—yields a world no more than six thousand years old. For those compelled by Holy Writ to believe in a planet so young, only great, sudden catastrophes could leave so deep a scar; the Grand Canyon is taken as evidence of the Great Flood, and the fossils embedded within its tiers are seen as the remains of victims that didn't make it onto Noah's Ark.

Of course, catastrophes could happen because of natural causes—the impact of a meteor, for example—and some catastrophists came to believe that the geology we find beneath our feet was made by a series of such cataclysms. But by the early nineteenth century, the assumption that the earth we see was formed by catastrophe after catastrophe began to strike many scientists as awfully arbitrary—one ad hoc assumption piled on top of another. They preferred to believe that the geological stripes were laid not by a great flood, or even a series of disasters, but by the same gradual forces that continue to operate to this day: erosion, deposition, compression. This view required a leap of faith in a different direction. In place of random (or ordained) catastrophes, one was forced to posit vast geological time. Through this set of spectacles, we now see a Grand Canyon whose layers were laid down over hundreds of millions of years and exposed a mere ten million years ago, when the Colorado River began cutting its way through.

Floating down the Colorado by raft from the vicinity of Page, Arizona, where the canyon is so shallow that you can begin the adventure by driving to the water's edge, an explorer can follow the earth's geological progression, compressing what we have come to believe about early Paleozoic time into a journey of several days. In the early part of the expedition you glide past the uppermost, Permian layers, deposited some 250 million years ago (newer layers have been eroded away): Kaibab limestone, consisting of the remains of shells and skeletons from animals that swam in a long-gone sea; Coconino sandstone, made from desert solidified into rock; petrified mudflats called Hermit shale. By the second day of the trip, Permian time looms high above as you descend another hundred million years, through the older Pennsylvanian and Mississippian layers: the Supai stripes of sandstone alternating with siltstone sitting atop the Redwall limestone—more fossilized sea. Then come the half-billion-year-old Cambrian layers: Bright Angel shale (more ancient mudflats) and Tapeats sandstone (desert again). By the time five days have passed, the raft is a mile below the surface of the earth, traveling the lay-

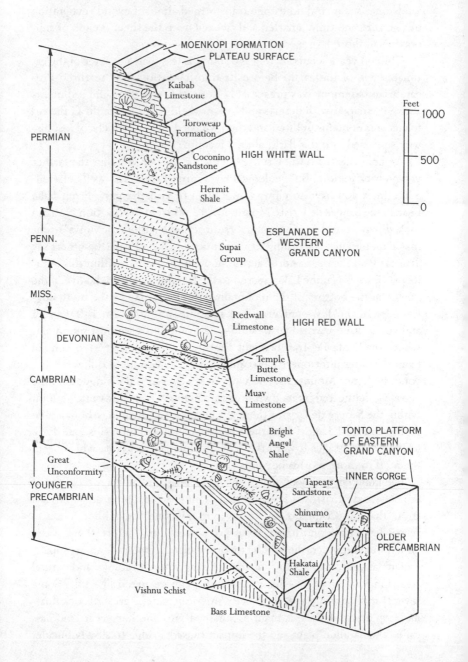

erless black rock of the billion-year-old inner gorge: the Vishnu schist—sandstone, shale, and limestone transformed almost beyond recognition by pressure and time, gnarled and twisted from the sheer weight of supporting all those layers.

A billion years seems so long ago, but it is less than a quarter of the estimated age of the earth. Before its transformation, this tortured Precambrian basement was presumably as neatly arrayed as the sediments above it, striped with deserts and oceans, fields of volcanic flow, the remains of mountains uplifted and then whittled away—epochs of lost history squeezed into this dark, almost formless rock.

We hold this truth to be self-evident: the deeper you dig, the farther you go back in time. But geological layers aren't always so well behaved. Sometimes you have to climb a mountain to visit the past. From Lake Peak in the Sangre de Cristo Mountains, 12,400 feet in elevation, you can look down upon the whole universe inscribed within the Tewa's four magic mountains. According to a few pueblos, Lake Peak (the others say Truchas Peak) is the eastern sacred mountain. Cradled within this granite precipice is Nambe Lake, sacred to the Nambe Indians, who live in the pueblo at the bottom of the mountain. Lake Peak may be the mythological edge of the Tewa universe, but stand upon it and turn 180 degrees and you can see a world far beyond—east of the easternmost mountain, where the Santa Fe Trail brought the Americans who wrested control from the older interlopers who came up the Rio Grande. Look west, beyond the Jemez Mountains, and you can see over into the kingdom of the Navajos, delineated by its own four mountains. From this vantage point within the Sangre de Cristos there are universes all around. About a mile to the south of Lake Peak, the microwave towers atop Tesuque Peak beam their messages toward the reflectors on Sandia Crest, which beam them off to more worlds beyond.

The knife edge of rock that forms Lake Peak and all the precipices of the Sangre de Cristos is Precambrian granite, from the same geological strata that the Colorado had to dig one mile and ten million years to reach. Lake Peak and its neighbors were here long before there was a Grand Canyon. According to the theory of plate tectonics—an organizing scheme as central to geology as the big bang is to cosmology and natural selection is to biology—this strange inversion occurred like this: Some seventy million years ago, the North American plate, a crust of rock floating atop a sea of viscous mantle, smashed into another great landmass called the Farallon plate and the impact caused a ridge to slowly buckle, eventually giving rise to the Rocky Mountains. The Sangre de Cristos form their southern tip. Precambrian rock was thrust from the layers below into the twelve-to-thirteen-thousand-foot mountains of northern New Mexico: Lake Peak, Tesuque Peak, Santa Fe Baldy, Pecos Baldy, Truchas Peak—all made from granite (which is nothing more than hardened magma), topped with just a frosting of sediments deposited later on by Mississippian and Pennsylvanian seas. Ice-age glaciers later sharpened the peaks, leaving behind lakes reflecting rock and sky.

At the foot of this mountain range lies the red, eroded waste of the Española badlands, dotted with herds of piñon and juniper that crowd toward the arroyos like deer in search of water. This rough erosional landscape was formed by sands deposited by the Rio Grande just tens of millions of years ago, a geological yesterday. Climb down from Lake Peak to these newer formations and look back up at where you were standing: the past towers overhead—the world turned upside down.

Creatures with less imagination than *Homo sapiens* might have been content to believe that the earth is as it must always have been. There is rarely evidence of geological change within a human life span, or even within the life span of a civilization. An occasional earthquake, a volcano erupting here or there—surface blemishes on what is otherwise terra firma. But as we see patterns in the landscapes around us, we are driven to explain how they came to be. And soon what we take for granted, the hard surface on which we wander, takes on a new strangeness. We can stand on a mountaintop and marvel at what the geologists tell us was once the bottom of an ancient sea.

In trying to understand how science finds order in the surrounding confusion, what better place to start than with geology? Exploring the way science has carved up the physical realm—mountains into minerals and minerals into particles made of more particles still—will quickly lead into the most austere abstractions, a world of patterns so subtle that some can be resolved only through the finest mathematical lenses. We start our search with our feet on the ground, but we will find that even something as seemingly solid as geology is a tissue of artful assumptions woven from threads that lead deep into the tangles of twentieth-century physics.

How can we, captives of a world view no more than a few hundred years old, speak so confidently about the ages of the earth? Determining *relative* age seems fairly straightforward: it is common sense to assume that, left undisturbed, newer layers of rock will lie above older layers; that when we find one formation thrusting its way through another, the

intruder must be the younger of the two. But as soon as we try to calibrate the clock—determining absolute ages—we leave the realm of the
commonsensical and the obvious. Suddenly we must call upon a dense
network of educated guesses and beliefs. According to our theories of
nuclear decay, certain atoms—the radioactive isotopes—are so top-
heavy with neutrons that they spontaneously disintegrate into simpler,
more comfortable forms. While quantum theory tells us that it is impossible to predict when a single nucleus will break down—it happens at
random—we can say with some certainty what a large population of nuclei will do. With the uncertainties averaged over millions of atoms, we
can calculate the likelihood that a mass of mother elements will break
down into simpler daughter elements, stepping their way down the periodic table at a steady clip: uranium to lead, potassium to argon, rubidium to strontium. On its surface, a rock, unlike an organism, gives few
hints of its age. But we can measure the ratio of mother to daughter elements. Then, taking into account the rate at which we believe one was
transformed into the other, we can guess how old the rock is, or at least
how much time has passed since it cooled off and hardened enough to
trap the telltale elements inside.

But how do we know that all of the daughter element we measure was
actually produced by nuclear decay of a mother element? If there was already some daughter element in the rock when it was formed, it would
throw off the ratio we are using to determine its age. The rock would
seem older than it really is. Drawing upon the strata of theory on which
they have built a field, geologists try to imagine the initial state of the
rocks they are dating and any changes that might have occurred during
their existence. After a rock is formed, cosmic rays can bombard stable
nuclei, turning them into radioactive ones, adding more mother element
and making the rock seem younger; through leaching, mother or daughter elements can leak away, twisting the hands of the geological clock.

In reckoning geological time, no single observation can stand on its
own; it is dependent on all the others in the web that has been strung together over the years. Loosen a strand here, tighten another there, and
we have a kink in the fabric of knowledge; the rest of the weave must be
readjusted to make it lie flat again. If we take a rock from a layer we already have reason to believe is a certain age, and radioactive dating tells
us it is older, we can toss the sample onto the discard pile and dismiss that
particular datum as experimental error, random noise. But if we find a
lot of these rocks, we might be compelled to revise our assumptions

about how much mother and daughter elements were in the samples to begin with.

Thread by observational thread, geologists have reached the tentative conclusion that the age of the earth is about 4.5 billion years. At the same time, physicists, chemists, and evolutionary biologists are working on their own pieces of fabric. Sometimes the squares mesh neatly, sometimes they are stitched together like a patchwork quilt. In one experimental clash, physicists, using their own methods, calculated that the half-life of rubidium (the time it would take half a sample to decay) was sixty billion years; the geologists said it must be fifty billion years to agree with established dating techniques. We construct the best models we can. When we inevitably find gaps in the fit between theory and theory and between map and territory, we make auxiliary assumptions, readjustments of the net. In the Grand Canyon there seems to be a 250-million-year gap between the Tapeats sandstone and the Precambrian gorge—the Great Unconformity, it is called, a formation that can also be found in the Pecos Wilderness above Santa Fe. Rather than abandon the geological timescale we have so carefully constructed, we suppose that, for some reason, whole epochs left nothing behind them, or that the remains were somehow eroded away.

In building a geology, certain fundamentals are rarely questioned: the theory of radioactivity, for example, or the assumption of uniformity— that the laws of physics have not changed over the earth's lifespan. These can be thought of as the foundational threads of the fabric—or even as the loom on which it is woven. Some things must be taken on faith. To biblical creationists, who believe the earth is just a few thousand years old, the whole geological edifice is suspect. The journals of creation science are filled with papers entertaining the possibility that, in ancient times, radioactive decay occurred at a faster rate; tweak the variables just so and it is vast geological time, not Genesis, that turns out to be an illusion.

The deeper we dig, the farther we go back in time. If we could drill into the silicon and oxygen atoms that form the granite of the Sangre de Cristo Mountains, past the recently deposited layers of electrons, to the older nuclei, then into the even more ancient protons and neutrons, we would finally arrive at the Precambrian bedrock of matter—the quarks that once buzzed free in the seething energy of the big bang but are now frozen together so tightly that they can't ever be pried apart. As we dig into the nucleus with our accelerators, smashing particles with energies closer and closer to those said to have existed in the first moments of creation, we search for the fossils of that legendary time, particles that no longer survive in this frigid universe.

Trying to understand the pieces our world is made of has required daring leaps of imagination and a belief that beneath the rough surface we see lie hidden geometries. Plato, in the *Timaeus,* proposed that the world is composed of perfect polyhedra, the five platonic solids—tetrahedra, cubes, octahedra, dodecahedra, icosahedra. Today we speak instead of atoms as the geometric building blocks, and in the chemistry books the structure of silica and the other molecules that make up the earth look as symmetrical and carefully constructed as Plato's perfect forms.

But in the world as we find it, the geometry is often obscured. Pick up a piece of pink granite from a mountain trail, and you see hints of an inner measure: planar surfaces of mica and feldspar, the corners and edges of what almost look like buried cubes and pyramids, as though little geometric solids were trying to push their way through. Like the Prisoners, those half-finished Michelangelo statues in the Accademia in Florence, perfect forms seem trapped in imperfect rock, struggling to break free.

The chemists tell us that molecules may indeed be thought of as polyhedra—symmetrical arrangements of atoms held in place by resonating chemical bonds. But once the tiny molecular crystals join into larger shapes, random disturbances distort the pattern. Captured in the wild, a garnet will indeed have six-sided faces, reflections of its molecular structure. But the hexagons are lopsided, the harmony disturbed. Under a microscope, tiny cubes of salt—sodium chloride—are pitted with flaws. Broken symmetries, the physicists call them, as though lurking within the messiness of the world is geometry as pure as Euclid's. The world would be mathematical if only reality didn't mess it up. Occasionally, a piece of quartz will crystallize with nearly perfect hexagonal faces, or pyrite as a seemingly flawless cube, hinting at the atomic scaffolding underneath. Water crystallizes from the sky in the six-lobed symmetries of snowflakes. But nature is rarely so obliging in expressing what we take to be its laws. The more symmetrical the geological shapes we find, the more likely it is that a mind has been at work, making earth into arrowheads, beads, bricks, and bowls, marking it off in checkerboards, smoothing out roughness, turning the world into circles and lines.

How did we come to believe in a world built on a scaffolding of buried symmetry? An early hint came in the eighteenth and early nineteenth centuries, when chemists found that elements combine in ratios according to their weights. Break down water by electrolysis and you get twice as much hydrogen as oxygen. H_2O. Reviving ancient speculations, the English chemist John Dalton proposed that this symmetry could be explained if the world was composed of atoms. A drop of water may look smooth and seamless, but it is made from invisible "atoms" of water—molecules, we now would say—which are, in turn, each made from precisely two hydrogen atoms and one oxygen atom. Further support for this hidden architecture came when Dmitri Mendeleev, the Russian chemist, found in 1869 that when the known elements were arranged in rows by increasing atomic weight, they lined up, as though by magic, into columns with similar properties—the periodic table of the elements. It was troubling that there were gaps in the table, but soon the missing elements were discovered, the holes filled in. A man-made system seemed to predict things in the real world. The feat was especially impressive considering that the table was constructed long before the discovery of electrons, protons, and neutrons, which we now consider the generators of the order.

In the next decades, scientists sought clues for what was going on behind the orderly facade of Mendeleev's table. Take a gas-filled tube with metal plates at either end, connect each to the opposite pole of a battery, and the apparatus will generate rays, one traveling from the negatively charged plate to the positively charged plate, and one traveling in the other direction. As scientists studied these emanations, they found that it didn't seem to matter what kind of gas filled the tubes—electricity broke it apart into whatever was in those countervailing beams. The rays seemed more like streams of particles than waves: an obstacle placed in their path cast sharp shadows; if the obstacle was a little paddle wheel, the beam would make it spin around.

By placing magnets and electrically charged plates around the tube, one could make the rays bend. In 1897, using such an apparatus, J. J. Thomson, director of the Cavendish Laboratory in Cambridge, England, calculated the ratio of charge to mass of what he assumed were beams of invisible particles. Putting the results together, he was able to make a strong claim for the existence of lightweight negatively charged particles called electrons. A decade and a half later, the American scientist Robert Millikan devised an ingenious way to measure the charge of Thomson's particles and found (within the bounds of experimental error) that they were all about the same. By scrutinizing oil drops quivering in electrostatic fields, he found that there was a minimal amount of negative charge that a drop could carry, and that stronger charges were integral multiples of this value. Thomson had calculated the ratio of charge to mass; Millikan had measured charge. All it took was simple arithmetic to calculate the mass of the electron.

Today, beginning in the earliest years of elementary school, students are instilled with the atomic theory of matter. Even people with little knowledge of science have a picture of an atom burned into their heads—that old Atomic Energy Commission symbol with the electrons whirling around the nucleus. Looking at the early experiments in retrospect, we see Thomson and Millikan converging on their prey, uncovering a hidden, preexisting order. But imagine not knowing how the story would come out. Try to see the experiments through their eyes, open-ended, with nothing but mystery ahead, as they tried to manipulate the invisible, teasing new structures out of strings of ambiguous data, developing a feel for what probably could be accepted as a legitimate observation and what could be dismissed as experimental error.

In similar experiments, the complementary beams that showed up in the vacuum tubes were found to be composed of heavy particles with charges opposite to that of the electron. And it was discovered that the two rays existed not only in these artificial situations—gas-filled tubes hooked to wires. They were also emitted by radioactive isotopes—nuclei breaking apart and shooting out positive and negative shards, what we now call alpha and beta rays. The world seemed to be confirming that it was built somehow from these unseen particles, these constructs put together piece by piece from a growing body of circumstantial evidence.

Since opposite charges attract, scientists first imagined that the positive and negative particles, the protons and electrons, were clamped together in an invisible little ball. But in an experiment in which he shot helium nuclei at a piece of gold foil at the Cavendish Laboratory, Ernest Rutherford showed that while a few bounced back, most sailed right through. He concluded that atoms were mostly empty space: a hard, positively charged nucleus surrounded by electrons. For every proton in the nucleus there was an electron orbiting around it, offsetting its charge. And it was the arrangement of the electrons that determined the atom's chemical behavior—the machinery behind Mendeleev's table laid bare.

But soon discrepancies began to fuzz up this pleasing picture. For one thing, there seemed to be atoms with the same charge (the same number of orbiting electrons) but with different weights. If the chemists' classification scheme was right, neon, for example, came in two varieties: one seemed to have a mass twenty times that of a hydrogen atom, the other seemed to have a mass of twenty-two. But both carried identical negative charges. The rule of one electron for one proton seemed to be violated. One way out of this mess would be to assume that electrons could exist within the nucleus as well as around it. One kind of neon would have twenty protons; the other would have twenty-two, along with two nuclear electrons to cancel out the extra positive charge. Rutherford suggested a better solution: perhaps the extra weight was caused not by additional protons but by particles as heavy as protons but with no charge. And so the neutron was conjured into existence. Both varieties, or isotopes, of neon would have ten protons, offset by ten orbiting electrons; but one nucleus would have in addition ten of these uncharged particles, the other would have twelve.

Neutrons didn't remain hypothetical for very long. In the history of physics, we find case after case in which a theory requires a particle and nature forthwith seems to oblige. In 1932, a colleague and former student of Rutherford's, James Chadwick, found particles that couldn't be deflected by magnets or charged plates (they had no charge), but when fired at nuclei they dislodged protons—they seemed to have a similar mass. So neutrons were admitted into the order. From just three particles one could generate the complexity of all the earth's ninety-two naturally occurring atoms. And just a few of these elements—carbon, hydrogen, nitrogen, oxygen, phosphorus, and several others—are combined into the molecules of life.

For a while, anyway, the world looked geometrical. In the picture that Rutherford drew, atoms were like little solar systems with electronic planets and nuclear suns. Inspired by this vision, one science fiction writer told the tale of a man who takes a shrinking potion, becoming smaller and smaller until he descends into the very atoms that make up a metal block on a scientist's laboratory bench. Falling through this inner atomic space, he descends into an atom and lands on an electron—a planet where a tiny civilization has taken root. All the while, of course, he keeps shrinking—into the atoms that make up the electron planet. He lands on another electron, and there a new adventure unfolds. Layer by layer, he descends through this hierarchy of embedded solar systems until he finds himself on an electron with big blue seas, green continents—this visitor from a metaworld beyond has landed on Planet Earth, and is ready to descend into still deeper realms.

No one seriously believed that our planet was some giant atom's electron, or that an electron in a particular hydrogen atom might support a tiny kind of life. With protons, neutrons, and electrons, scientists in the early twentieth century were pretty sure they had reached bedrock. Matter was made from three particles and shoved around by two forces —gravity, which held the planets in orbit around the sun, and electromagnetism, which held the electrons in orbit around the nucleus. But as it turned out, these earthly images could be stretched only so far. We try to explain the strange in terms of the familiar, but sometimes it just won't stop being strange.

As Rutherford, Thomson, and their cohorts were piecing together a theory of matter, others were trying to understand the invisible phenomenon called energy. In the early 1800s, Hans Christian Oersted in Copenhagen had demonstrated that a coil of wire hooked to an electrical current acted like a magnet—it reached across space and made a compass needle move. In England, Michael Faraday later showed that nature was capable of producing a reciprocal effect: moving a magnet back and forth inside a coil generated an electrical current. It was left for James Clerk Maxwell to reveal the architecture behind this intimate link: electricity and magnetism were mutually perpendicular shadows cast by a single more fundamental force, electromagnetism. A moving electrical charge generates a moving magnetic field; the moving magnetic field causes the electrical charge to move . . . which generates a moving magnetic field— and so on. Cast into the form of a diagram, Maxwell's equations revealed a beautifully symmetrical image: magnetic fields vibrating at right angles to electrical fields, like the fringes of an arrow—an electromagnetic ray shooting through space.

The spark that flies off your finger when you cross a carpeted room and touch a doorknob and the lightning that flashes over mountains are intertwined with the pull of a compass needle. And they generate waves that, the equations predicted, would move at a speed very close to that which had been measured for light. So Maxwell made another connection: Light was a form of electromagnetism. From the warp and woof of electricity and magnetism are woven the light beams that stream from the sun.

But Maxwell's equations also spoke of a world beyond the narrow window of our direct experience. There were other kinds of electromagnetism than what human retinas register as light. One of the marvelous things about nineteenth-century physics was the skill and aplomb with which people began to interact with the invisible. In 1800, William Herschel in England used a thermometer and a prism to measure the temperatures in a rainbow of sunlight. When he placed the thermometer beyond the red edge of the spectrum, he was surprised to find that the temperature did not fall off—he was measuring the energy of invisible infrared light. A year later another scientist, Johann Ritter, used silver chloride to make a crude kind of photographic paper and showed that it turned black even when he laid it beyond the violet edge of the spectrum, registering what we now call ultraviolet light. In the search for patterns within nature, light—something rooted in our everyday experience—was becoming generalized, extrapolated into realms we barely had access to. Toward the end of the century, in the late 1880s, the German Heinrich Hertz used coils and currents to generate radio waves and showed that they behaved much like light beams—they moved at the same velocity, they could be reflected and refracted. It was not long before it was shown that the x-rays that shoot from disintegrating nuclei and cast bones as silhouettes are yet another note on the electromagnetic octaves and can be described using Maxwell's laws. His equations, it seemed, had taken on a life of their own, describing phenomena he had not imagined.

In the history of science there hadn't been a grander synthesis since Newton's linkages between mass, force, and acceleration. And so it was disturbing that when this theory of energy was laid next to science's other great creation, the atomic theory of matter, the two didn't mesh. Whether electrons were moving in a coil of wire or in the outer reaches of an atom, they were supposed to radiate electromagnetic waves. If an electron was indeed orbiting a nucleus, it should be constantly losing energy. Why didn't it come crashing down into its nuclear sun? If the invisible machinery humming beneath the surface of the world indeed operated according to a master plan, then something about our picture of the atom or our picture of electromagnetic waves had to be readjusted.

Reconciling this disturbing contradiction ultimately required nothing less than abandoning the deeply grooved distinction between particles and waves. Hints that there might be something wrong with this way of carving up the world came from studying how matter at different temperatures emanates different colors of light. Cooked in a furnace, an iron bar begins radiating invisible infrared waves; as it grows hotter it becomes red and then orange as it climbs up the electromagnetic scale. What is the relationship between temperature and frequency? To understand this phenomenon, scientists studied the ideal case, a so-called black body radiator, for which the frequency and intensity of the radiation depend only on the temperature of the object and not on the material from which it is made. But no matter how they looked at the problem, the equations of classical physics seemed to predict that regardless of their temperature, black body radiators should emit an infinite amount of ultraviolet light. Obviously this is not what happens in real life. Playing around with the mathematics, the German physicist Max Planck found in 1900 that the absurd discrepancy between theory and reality arose because of the assumption that energy was continuous, infinitely divisible like the segments of a line. His way out of what had come to be called "the ultraviolet catastrophe" was to postulate that the energy of a radiating object changed discontinuously in packets—quanta—whose size could be gauged with a number now known as Planck's constant. Then Einstein found that he could use quanta to explain the photoelectric effect. When light is shone onto a metal surface, it causes it to emit electrons; but the energy of these particles depends not on the intensity of the light, as we might suspect, but on its frequency. As Einstein showed in 1905, this phenomenon makes sense if we are willing to postulate that light comes in packets, what we now call photons. Planck showed that energy was emitted in packets; Einstein showed that energy was absorbed the same way.

Captivated by this heterodoxy—particles of electromagnetism—Niels Bohr in Copenhagen suggested an overhaul of the Rutherford atom: suppose, he said, that the electrons around a nucleus are allowed to exist only at certain levels. In these "orbits" (if one could still use that word), the electrons would emit no energy. But when they jumped to a new level, farther from or closer to the nucleus, they would absorb or eject a photon of light. According to quantum theory, these packets of energy were the real "atoms"—that which was indivisible. There could be no such thing as an electron hovering in between atomic shells, or passing smoothly from one to the other.

Since ancient Greece, philosophers have debated whether the substance of reality is discrete or continuous. On a hike outside Göttingen, Bohr explained to his German colleague Werner Heisenberg a compelling reason for carving up the world in a different way—abandoning the assumption of continuity, so that electrons could only be at certain energies, and could essentially jump from one level to another without traversing the space in between. In the classical laws, Bohr reasoned, nothing could explain the one truth most central to our existence: the stability of matter. It is true that we don't see much regularity as we gaze out over the Española badlands. But no matter how rough the erosional terrain, it seems to be made from a finite variety of elements. And we find that these substances invariably have the same properties, which seem to depend on how their electrons are arrayed. How could this be if electrons were allowed to assume any position around a nucleus, depending on how they had been buffeted around by nature—if an atom were allowed to have a history and an individual identity, as an organism does? Imagine that instead of ninety-two elements from hydrogen to uranium there was an infinity of gradations, a continuum. In place of neatly arranged cells, Mendeleev's table would become a continuous band. But if the energy levels of electrons are not continuous, if only certain values are allowed, we can explain why the same forms keep turning up in nature.

It gradually became clear that the image of electrons jumping from orbit to orbit was simply an imperfect metaphor, and scientists began to think of the levels more abstractly, as energy states. Bohr's model also made a quantum leap in another way: he proposed that in trying to explain the world inside atoms we might have to resort to concepts different from those of the everyday world. To preserve the idea of stability, he had to sacrifice that which made common sense, an atom whose behavior we could picture in our heads. The price was the notion that the strange could be explained in terms of the familiar. It was a short step to the quantum weirdness that has become second nature to physicists and a puzzle to readers of popular-science books: waves that behave as particles, particles that act like waves—and not waves made of some underlying stuff but waves of probability, a halo of pure possibility that goes unrealized until it is "collapsed" by an observer, who almost seems to conjure the particle into existence out of a mathematical haze. And even then, there is the complication called Heisenberg's uncertainty principle: once a particle is observed, one cannot determine its position and momentum simultaneously; the more precisely you measure one, the fuzzier the other becomes—a residual, inescapable uncertainty that is measured by Planck's constant, now considered one of the fundamental parameters of the universe. Before we knew it, we were explaining the familiar in terms of the strange, what Heisenberg called "this peculiar mixture of incomprehensible mumbo jumbo and empirical success." Difficult as it was to interpret, quantum theory continued to surprise its own inventors by how accurately it seemed to predict the behavior of the invisible subatomic world. The symmetries apparently went deeper than anyone had supposed.

In building a theory of the world, it helps if one's vision is a little blurry. It was finer and finer observations that forced Ptolemy to add filigrees to his geocentric universe—the notorious epicycles that preserved the illusion that everything moves in perfect circles (and circles within circles) around the earth. On the other hand, it was a kind of astronomical myopia that led Kepler to devise his simpler elliptical geometry. Planets don't really move in either circles or ellipses. If the astronomer Tycho Brahe's numbers had been more precise, Kepler might not have been able to see what appeared to be a signal amid the noise: the ellipses that planets would *ideally* follow if they weren't perturbed by the gravitational pull of their neighbors, which are perturbed by their neighbors, which are perturbed by more neighbors still. The symmetries are always a little lopsided.

Once the vision of an atomic solar system had been overhauled to mesh with quantum theory, it wasn't long before the result, the Bohr atom, had to be bedecked with filigrees of its own. One of the attractive things about Bohr's model was how neatly it explained the emission and the absorption of light by matter. In the mid-nineteenth century, chemists found that if they burned an element and examined its flame with a prism, it would yield a rainbow of its own, a spectrum overlaid with a characteristic pattern of light or dark lines. In Bohr's theory, the lines were caused by photons that were either emitted or absorbed as the atoms' electrons jumped from state to state. A bright line signified the tiny packet of energy emitted when an electron jumped from a higher to a lower level; a dark line signified a photon that was absorbed to propel an electron from a lower to a higher state.

But there were still some discrepancies to explain. When one looked closely at the spectral lines, it was clear that they were not single spikes, that they had what came to be called a "fine structure," as though electrons in the same orbit were not perfectly quantized but could have slightly different energies after all. What could this mean?

To resolve the problem, physicists stretched their imaginations further, supposing that within a single orbit electrons could move in circles, in ellipses of various elongations, or even in more elaborate patterns. In Bohr's original model, electrons were given quantum numbers, 1, 2, 3,

4, to describe what shell they were in. To account for these new "sub-orbitals," more quantum numbers were added. But even these weren't enough. When the flames of some elements were allowed to flicker within a magnetic field, what was supposed to be a single electron in the outer shell generated multiple spectral lines. Even hydrogen, which was supposed to have only one electron, exhibited what for want of a better term was called "two valuedness." To explain these strange gyrations, the Dutch physicist George Uhlenbeck proposed that in its orbit around the atom, an electron could also rotate on its axis, clockwise or counter-clockwise. In each case it would have a slightly different energy in the imposed magnetic field and project a slightly different spectral line.

The hard-core realists, who liked to think that electrons and orbitals were more than just convenient mental constructs—that atomic theory really described actual objects in an invisible world—found problems with Uhlenbeck's image. Electrons were generally thought of as dimensionless points. What exactly, then, was spinning? Even if electrons were considered tiny spheres, to account for "two valuedness" they would have to be whirling around at many times the speed of light, a seeming violation of Einstein's special theory of relativity.

But the idea of spin was so elegant that the objections were overcome, and physicists began talking of particles with clockwise or "down" spins (the axis of rotation pointed downward), and counterclockwise "up" spins, as though they were little tops. But this was not the ordinary spinning we see in our world. To remain consistent with quantum theory, spin itself had to be quantized: it could only assume certain values—½, 1, 1½, 2. The concept of spin became even more confusing as the theorists went on to establish that an electron can be thought of as either a particle or a wave. What does it mean for a wave to spin?

What had once seemed so simple—a lightweight particle with a single unit of negative charge—was becoming horrendously complex as all these quantum epicycles were added. But the model the physicists were building seemed to do such a good job of predicting the behavior of atoms that there was little incentive to climb back down the tree of knowledge and see if they, like Ptolemy, had begun with a fatal foundational error. Instead, science groped for a language to describe an abstract quality we cannot directly experience, but one that helped make sense of the sub-atomic world. In his book *A Brief History of Time,* Stephen Hawking explained how to understand spin in a way that is far more general and abstract than we are accustomed to. The key concept is symmetry—how an object can be viewed from different perspectives and still appear the same. Hawking used the example of playing cards. A face card like a queen or king can be turned upside down and appear unchanged. Particles with this property—they can be rotated a half turn and retain their identity—are said to be of spin 2. (Gravitons, the hypothetical carriers of the gravitational force, are believed to be an example.) However, if you rotate the ace of spades a half-turn, it will be upside down; it requires a full turn to retain its appearance. The ace of spades, like the photon, is said to be of spin 1. So far, so good. But what about particles like electrons and quarks, said to be of spin ½? There is no playing card to correspond to them, for they must be rotated *twice* to return to their original position.

Physicists began by carving up the world into categories we could intuitively understand: force, mass, velocity, acceleration, voltage, charge, frequency. The introduction of spin marked a turning point in which scientists would start with the familiar, the image of a spinning top, and abstract it into barely imaginable realms. It is a mystery that we can build things mathematically that seem far removed from the world in which we dwell. It is even more mysterious that they sometimes turn out to be so useful.

If we take a transparent cube and project it onto a two-dimensional surface, its shadow is a square suspended inside a square with diagonal lines. So imagine a cube inside a cube, suspended with planes—a three-dimensional shadow of a four-dimensional "cube," a tesseract. So would a tesseract suspended inside a tesseract by cubes represent a four-dimensional projection of a five-dimensional object? We have now passed the point of mental visualization, but we can follow the logic into these purely abstract spaces.

And so it is with particle physics. To explain how protons and neutrons are held together to form nuclei, scientists took the already elusive notion of electron spin and abstracted it even further. Neither of the two known forces, gravity and electromagnetism, could explain how nuclei were held together—gravity wasn't strong enough, and by all rights it seemed that electrostatic repulsion should push the protons apart, not glue them together. And what would make a neutral particle, a neutron, stick to a proton? There was nothing in physics to explain why the nucleus exists. Since nuclei are invisible anyway, lesser creatures might simply have abandoned the attempt to construct a theory of matter. Instead, the physicists made yet another conceptual leap. A third force of nature was invented—the strong nuclear force, a mathematical fluid in which protons and neutrons became different manifestations of a hypothetical particle called a nucleon. Spin it one way and it becomes a proton, spin it the other way and it is a neutron. But with this so-called isospin there was barely a hint of a physical interpretation. The "spinning," or whatever it was, took place not in the space we move in, but in an imaginary mathematical realm.

We invent spaces all the time, so compulsively that the talent seems to be wired into our nervous systems, an outgrowth of the evolutionarily advantageous ability to picture mentally what is not immediately before our eyes. Whenever we think of ourselves as moving up in an organization or moving closer to a goal, we are abstracting the notion of physical space. Mathematicians do this in a more precise manner. One can think of two-dimensional space, the surface of a sheet of paper, as represented by the perpendicular axes of a graph. Take any point on the plane and measure its distance from the horizontal and vertical axes, and you have two numbers, or coordinates, precisely describing the position. Add a third axis, perpendicular to the other two, and you can plot any point in three-dimensional space. If we allow ourselves to keep adding axes, we can imagine spaces with any number of dimensions. The labels "height," "width," and "depth" have already been exhausted, so we might call the fourth dimension "time." Suppose we are using a graph to represent the characteristics of a room full of helium balloons. Each balloon is represented by a point with four coordinates describing its position at a certain moment; as it moves through the room, it traces a four-dimensional "line." Add another axis and we can represent a balloon's color (imagine, for the sake of argument, that it is constantly changing); add another for the frequency with which it vibrates in the wind. In fact, there is no reason why the first two or three axes of a graph have to represent spatial dimensions. Suppose we are studying the characteristics of an electrical circuit, a black box filled with components and wires. Applying a range of voltages to the input, we measure the current flowing out the other end. Then we plot voltage against current and study the shape that emerges as the circuit behaves in this imaginary space whose dimensions are current and voltage.

Plotting parameters against parameters, the early atomic theorists created new spaces so compelling that it was hard not to think of them as real. Within these universes beautiful patterns like isospin emerged, each described by a new quantum number. Abstracting from the familiar to the purely mathematical, from cubes to tesseracts, science was taking the geometrical symmetries we are familiar with here on earth as special cases of something more general. We can hold a ball in our hands and appreciate the symmetry, how it appears identical no matter which way we rotate it. We can marvel at the hexagonal symmetry of snowflakes, or that in every triangle drawn on a plane, the angles total 180 degrees. From geometry that pleases the eye, physics was moving to geometry that pleases the mind—but only minds honed to appreciate the subtlest of orders.

Descending into the artificial world of isospin space, imagine reaching inside a nucleus and randomly flipping over the spinning subatomic tops. Clockwise becomes counterclockwise and counterclockwise becomes clockwise as we change protons to neutrons and vice versa. We would find, according to our physics, that despite all the flipping, the nucleus remains stable. It doesn't matter to the strong force whether a nucleon is a proton or a neutron; it treats them just the same. They are said to be symmetrical to the strong force just as a ball is symmetrical to rotation.

In fact, in the language of the new physics, the symmetries are said to give rise to the forces. In this new way of carving up reality, it is these mathematical harmonies that are now considered the physical bedrock, with the forces regarded as secondary epiphenomena. So deep is our passion for geometry that we have come to believe that symmetries we cannot directly experience are more fundamental than the forces we feel whenever we lift a rock or touch a bare electrical wire. Of course, where the symmetries themselves came from was another question. Were they woven into the universe as platonic essences or generated by the human mind as it strives to find order?

The feats of imagination needed to explain how just three particles, the electron, proton, and neutron, manifested themselves to our scientific instruments were heroic enough. But then, as the century progressed, more and more particles began to appear. Some came from the sky as cosmic rays bombarded the atmosphere, showering the earth with muons, pions, taus. But most of the new particles, by far, were generated by machine—the accelerators that smashed particles together, the detectors that examined the shards. Where there had been three particles, soon there were hundreds—rhos, sigmas, thetas, xis. Finding a system to accommodate them required the invention of more strange mathematical qualities. In fact, it was not long before mathematics seemed to be "discovering" particles before the scientists did.

In 1931 the English theoretical physicist Paul Dirac tried to explain away the annoying fact that his equation describing the behavior of the electron had two solutions, one negative and one positive. The implication was that there was a particle the same size as an electron, except with positive charge. Several months later, Carl Anderson at Caltech, studying cosmic rays with a device called a cloud chamber, observed a particle that left a track like an electron's but was curving in the opposite direction, the way a positively charged particle would go. "The equation was smarter than I was," Dirac later said. The positron was born, eventually blossoming into the view that every particle has a counterpart in a mirror world of antimatter. If a matter particle and an antimatter particle chance to come together, they annihilate each other in a flash of photons.

At about the same time the positron came along, Enrico Fermi was trying to make the puzzling phenomenon called radioactivity fit into the subatomic scheme. In the process, scientists had to invent an even more elusive particle called the neutrino. Ever since 1896, when the French physicist Henri Becquerel discovered radioactivity, finding to his astonishment that salts of uranium would cast images of themselves on photographic plates, even when they were wrapped with sheets of black opaque paper, scientists had puzzled over how matter could shoot out invisible rays. One could envision how a disintegrating nucleus might eject protons and neutrons and high-energy photons (in the form of x-rays and gamma rays). More mysterious was the case of beta decay, in which a nucleus shot out electrons. These particles were supposed to hover around a nucleus, not be embedded within it. Even worse, a number of experiments seemed to imply that beta decay violated the principle known as conservation of energy, as well as conservation of angular momentum, or spin. Ever since the invention of the steam engine, scientists had convinced themselves that energy could neither be created nor destroyed. The heat emanating from the coals that vaporized the water in the boiler would show up as the energy turning the turbine; by attributing any shortfall to friction or heat radiating into the atmosphere, the symmetry could always be maintained. It is quite a leap to assume that what is true for steam engines is true for nuclei. We hope for a world in which law reigns at every level. But the energy going into the nuclear reactions didn't seem to match the energy coming out. In an attempt to put the world back in order, Wolfgang Pauli suggested, in 1930, that in addition to the electron, beta decay must produce an all-but-invisible particle (Fermi named it the neutrino), which would conveniently carry away just enough energy and spin to balance the books. Three years later, Fermi proposed that beta decay occurred when a neutron (which is slightly heavier than a proton) somehow turned into a proton and an electron. In the process, he conjectured, one of these things called neutrinos was emitted. Never mind that nothing like neutrinos had been detected. Suppose that these new particles were chargeless and massless, flying through the most sensitive detectors as though they were not there. (It was later decided that it made more sense to call the missing particle in beta decay an antineutrino rather than a neutrino.)

But pulling new particles out of thin air was not enough to explain what in the world would cause a neutron to turn into a proton in the first place. Answering this question required the invention of nothing less than a fourth force of nature: the weak nuclear force. Many orders of magnitude weaker than the electromagnetic force (which helped explain why it had gone unnoticed), the weak force converts neutrons into protons by tampering with their insides—or, as later scientists would come to believe, by changing their quarks.

It would be two and a half decades before there was evidence that neutrinos existed. Sometimes their only role seemed to be to make the energy and angular momentum equations balance. In their defense, experimenters would plead that the particles were, after all, chargeless and perhaps massless (though there is now evidence that they have a tiny mass). It would be almost impossible to make a neutrino interact with another particle, which is the only way we would know of its existence. A neutrino could shoot through the earth, it was said, as though it were not there.

To build a neutrino detector, one first had to do more theorizing about the newly devised fourth force of nature. Perhaps the weak force could work backward: if a neutron could decay into a proton, giving off an electron and an antineutrino, then why, for reasons of aesthetics and symmetry, couldn't a proton and a neutrino come together to produce a neutron and a positron? It would be as though the original reaction were reflected in a mirror. Given the ghostly nature of the neutrino, the chances of one actually colliding with a proton were vanishingly small, making the reaction almost impossible to detect. But in 1951 a Los Alamos scientist named Frederick Reines proposed that in a nuclear reactor enough neutrinos would be produced by the decay of radioactive fuel for there to be a sporting chance of detecting the rare conversion. Two years later he and a colleague, Clyde Cowan, rigged up what they hoped was the right detecting apparatus and placed it next to a reactor at Hanford, Washington. Occasionally, they ventured, a neutrino from the core of the reactor would strike a proton in their carefully designed detector and generate a neutron and a positron. The positron would then collide with a nearby electron. Collisions of antimatter and matter, theory tells us, spark off photons. At the same time, the neutron produced in the reaction would be absorbed by another nucleus, a reaction that also generates photons. According to the suppositions on which the detector was built, this pattern of twin bursts of light, one occurring five microseconds after the other, would stand for neutrinos. It was not until 1956, at Savannah River, Georgia, that the two scientists finally detected what they felt sure was the double photon signature. A quarter century after their invention, neutrinos were declared to be real, and they are now generated and measured with nearly as much confidence as electrons are.

Science had come a long way from such simple devices as Thomson's gas discharge tubes, or even the cloud chamber in which the positron left its curving trail. To "see" a neutrino, one had to submerge the mind into an increasingly elaborate atmosphere of theory and assumption. Once accepted as real, neutrinos could be used to make sense of other phenomena. And so they became woven tighter and tighter into the mesh, the gauze of theory that lay between us and the nuclear world.

As both the mathematics and the laboratory machinery grew in power, the symbiosis between theory and experiment became that much harder to untangle. Theoreticians would describe the particles they needed to make their models work and experimenters would design elaborate devices to pick them from the debris of colliding accelerator beams. Or experimenters would find particles in the showers of cosmic rays that bombard the earth and the theoreticians would ponder what they could possibly be. Soon there were hundreds of candidates for particlehood. Those that were admitted into reality had to be classified, so more quantum numbers were invented. In addition to charge, spin, and isospin, particles were said to exhibit qualities called parity and—perhaps the most appropriate name of all—strangeness. But even with all these new parameters, these symmetries, one could find little order. What was needed was another Mendeleev, who finally appeared under the name Murray Gell-Mann.

In 1961, Gell-Mann, then at Caltech, found that by using strangeness (which he had invented and named) and isotopic spin as axes, he could create a space in which most of the plethora of particles fell into the slots of a neat geometric chart. The oldest rock at the bottom of the Grand Canyon is named for a Hindu god—Vishnu schist. Gell-Mann jokingly called his mathematical bedrock the Eightfold Way, after the Buddhist prescription for an enlightened life. Like Mendeleev, he found his order created gaps—particles, like the omega-minus, that later showed up in the accelerators. But what was the generator of this inner harmony? While chemists had to await the discovery of the electron and proton before they could begin to fathom the periodic table, Gell-Mann was daring enough to invent the particles he needed. He proposed that his scheme would make sense if there were particles within particles— "quarks," a name he found even more appealing when he learned that it was contained in a line from that most obscure of writings, James Joyce's *Finnegans Wake*: "Three quarks for Muster Mark." ("Quark," he insists, is properly pronounced with a Dublin accent: "quorrhk.")

In their own use of language, Gell-Mann and his fellow theorists were as inventive as Joyce. In addition to charge, something we experience in our world, quarks were said to have abstract qualities called "flavor" and "color." Of course, these terms have nothing to do with flavor or color; they might just as well be called "piety" and "talent," or "envy" and "glut-tony." Their beauty lay in the new, ever more rarefied symmetries they seemed to reflect.

For the quarks that make up ordinary matter, the flavors were called "up" and "down." Two up quarks and a down quark make a proton with its upward isotopic spin; two downs and an up make a down-spinning neutron. (By flipping up quarks and down quarks, the weak force can convert protons into neutrons and back again.) Devising a structure for the strange particles that show up in cosmic rays or particle accelerators required, in addition, the so-called strange quark. And, as theory demanded, there had to be a whole shadow world of antiquarks. A quark and an antiquark would come together to make a particle called a meson. (Electrons and neutrinos, which belong to the family called leptons, are not made from quarks.)

But that was just the beginning of the embellishments to this ingenious scheme. According to a dictum of quantum mechanics called the Pauli exclusion principle, no two particles could be in the same state at the same time. Or, to put it another way, they could not have the same quantum numbers. Otherwise, it was said, all the electrons in an atom would take the path of least resistance and bunch together in the lowest energy shell. All atoms would be identical; there could be no chemistry. In contemplating the world through the prism of the Eightfold Way, physicists found that in many cases particles apparently contained two quarks with the same quantum numbers. Rather than adding a footnote to the exclusion principle ("All particles, except quarks . . ."), scientists postulated yet another subatomic quality. What if quarks also came in three "colors"? Then two seemingly identical quarks would not really be the same after all. And so it was proposed that there were "red," "green," and "blue" quarks, the labels more fanciful than calling blue north, or yellow west, as the Tewa Indians do. When all three of these "colors" come together to make a particle, it is colorless, like a beam of white light. And so, by definition, particles never exhibit their color.

Scientists were now classifying the universe in terms of qualities that were in principle unobservable. This maddeningly abstract "color charge" was no less than the generator of the strong force—the very glue that held the quarks together to form nucleons and nuclei. Just as electromagnetism was carried by photons, color was said to be carried by particles called gluons, which came in eight varieties, with names like green-antiblue and red-antigreen. Holding together the quarks in combinations of threes and twos, the gluons helped form the particles that make up the nuclei that make up the molecules that make up the world.

Even three flavors and three colors weren't enough to bring order to the unruly subnuclear universe. As the accelerators manufactured more particles, more quarks were required: not just up, down, and strange quarks, but quarks called "charm," "bottom," and "top." The latter two were also sometimes called "beauty" and "truth"—the names a joke on how far we had come from spin and isospin to qualities that only a mathematician could see. By stretching further and further beyond the parochial viewpoint of the senses, science seemed to be unearthing a dazzling array of symmetries.

At first there was some concern that one never detected quarks in the accelerator blasts. But theorists later showed that the strong force, carried by the gluons, was unlike any force imagined before. While the forces we can feel, gravity and electromagnetism, decrease with distance—the notion almost seems to be built into what we mean by force and distance—the force between quarks grew stronger as they were pulled apart. We might picture this as a spring, but a spring that could never break. Accelerator experiments in which electrons are scattered off protons are taken as showing, indirectly, that there is structure to the proton—that there are indeed little point masses inside. But quarks and gluons exist, in principle, in a world from which we are forever excluded. Pulling a quark from a proton is said to be something like trying to cut the north pole from a magnet: chop as many times as you like, you will always be left with a whole magnet.

Electrons, as we've seen, got their name from the classical Greek word for amber, in recognition of the discovery by ancient philosophers that the substance took on a mysterious attraction when rubbed with a piece of wool. Quarks are echoes from a line in Joyce. Both are theoretical constructs, and their existence depends on not being contradicted by experiment. In the early 1970s, in experiments far more subtle, difficult, and expensive than the one that first detected neutrinos as twin flashes of light, two teams of physicists observed fleeting events that could be explained as the decay of a "charmed meson," made, as theory required, of a charm and an anti-charm quark.

There was something almost stunning about how well this expanding set of ever more elaborate filters made sense of a tiny, distant world, predicting and explaining phenomena that seemed to be confirmed in the spray of patterns left in the detectors by the colliding accelerator beams. With every passing year, it was becoming all but impossible to see through the subatomic confusion without this finely woven mesh. And it was becoming all the harder to tell how much of the order was truly woven into the world and how much was imposed by the brain's hunger for pattern.

Philosophers of science tell us that there is no such thing as a naked observation. How we design an experiment and how we interpret the results are embedded in a constantly expanding web of beliefs and assumptions. As Einstein said, it is the theory that allows us to see the facts. Still, when we see two foil leaves in a jar on a laboratory bench push each other apart as though repelled by like charges, or a dot on a plate in a vacuum tube glowing as though struck by a particle beam, it seems perfectly reasonable to believe in electrons. This phenomenon called electricity arises in so many contexts and can be so simply manipulated—all you need is batteries and wires—that it seems very much a part of our world.

But as we go from electrons to positrons to neutrinos to quarks, and from gas-filled discharge tubes to cloud chambers to accelerators shooting beams at detectors so elaborate that they are among the most complex, delicate devices ever made—it becomes harder to be sure that experimenters are simply observing what the theorists predict. There is a wide gulf between the beholder and the beheld, consisting not only of millions of dollars' worth of detecting equipment but of the complex of theories with which the equipment is designed, its results interpreted.

In even the simplest experiment, there are the random disturbances we call noise, and one must always make a judgment of when enough pains have been taken to reduce it. To register the outcome of rare, vanishingly tiny events that last fractions of fractions of seconds, a detector must be as sensitive as possible. But the more sensitive the equipment, the more subject it is to noise. There is a constant trade-off between capturing the feeble signal that you seek and drowning it out. When the results are finally in, a judgment must be made. Which is data, which is noise? Experimenters trust in their ability to distinguish signal from noise, but there is always the danger of seeing pictures in the clouds.

Even if we can separate foreground from background, we are still left to wonder: Did nature cause the reading, or was it somehow hidden in the design of this very complicated machine? A theory requires a particle and there is a race to find it. The detector is built and then tuned and re-tuned until, lo and behold, the predicted effect is observed—the effect, not the particle itself, which might not live long enough to leave a track. The best we can say is that the hypothetical particle, acting according to theory, interacted with other hypothetical particles, whose existence is also built from a long chain of inferences, and at the end of this series of hypothesized reactions, photons or electrons were produced—the two particles we understand the best. Photons cloud photographic plates or collide with photoelectric cells, producing electrons. And it is electrons that drive our gauges, whose readings we take by bouncing photons from the dial into our retinas, where they generate electrons again, sending signals to the brain. Everything, it seems, comes down to a dance between these particles of electricity and these particles of light.

Looking back, we assume that the newfound phenomenon was there all along and the experimenters cleverly ferreted it out. Or, if the quarry is never found, the theory that predicted it still might be saved, for a while at least. Perhaps the particle was just too massive to produce in existing accelerators—or in any machine that could be conceivably built. In this domain the connection between map and territory is very subtle. Some philosophers worry how easy it is for so delicate a science to become infected by what they call retrospective realism. The experimental design that produced the right result is retrospectively taken to be correct; the ones that failed to find the phenomenon are judged to be mistaken. The particle exists because it was verified by experiment; the experiment is deemed to have been designed correctly because it found the particle.

In an accelerator experiment, more data are produced than we could ever hope to interpret; this information is sifted with computers programmed to look for the patterns the theorists have decided are important. Theory restricts the search space. But maybe more important truths lie in what we thought was noise. To take the most skeptical stance, there is always the possibility that we are simply building big machines, more complex than we can completely understand. We are studying their behavior under different conditions. When an experiment is replicated at another accelerator, it is simply a matter of building or tuning another machine until it behaves in a similar manner. Perhaps we are interpreting machines, not nature.

And perhaps, in our inevitably imperfect ways, we are inching steadily toward a better and better picture of the subatomic world, a realm so alien that it is amazing our minds can enter it at all. Physicists are quite aware that their instruments are not transparent windows on nature. As we move further and further from everyday phenomena, which are mysterious enough, to studying events that last for fractions of a microsecond—and then only under the most assiduously controlled conditions—it takes faith as well as ingenuity to unearth these hidden orders. The faith is driven by the universal passion to find symmetry in the world.

Having developed separate theories of electromagnetism, the weak force, and the strong force, theorists felt compelled to combine the pieces into a single whole, a grand unified theory. Maxwell showed that electricity and magnetism were different faces of a new force called electromagnetism. In 1979, the physicists Steven Weinberg, Abdus Salam, and Sheldon Glashow received a Nobel Prize for showing that, viewed from the proper perspective, electromagnetism and the weak force can be unified into something called the electroweak force. And they went on to explain how this symmetry became broken, so that we experience two forces with different characteristics. But so far, the strong force has resisted being squeezed comfortably into the mold. Physicists have had to content themselves with the so-called standard model, which consists of electroweak theory and the theory of the strong force sitting side by side but only partially connected. And way off in the distance is the theory of gravity: Einstein's general theory of relativity, which bears no resemblance whatsoever to the standard model.

According to this picture of the subatomic world, matter is made from two kinds of particles: (1) the quarks, which combine to form protons, neutrons, and hundreds of other so-called hadrons; and (2) the leptons, which include electrons and neutrinos—particles that are not made from quarks. One thing leptons and quarks, together called fermions, are said to share is a spin of ½ and the fact that they obey Pauli's exclusion principle. These particles are further divided into three families: the world we experience every day is made up of electrons, neutrinos, and the up and down quarks. But with great effort it is also possible to generate, for the briefest instant, two much heavier versions of the electron, the muon and the tauon, which have their own neutrinos and are associated, respectively, with the strange and charmed and the top and bottom quarks.

In addition to the fermions there are the bosons, which carry forces. Bosons have integral spins (1 or 2) and obey rules laid down by the Indian physicist Satyendra Nath Bose; they are exempt from the Pauli exclusion principle. Photons carry electromagnetism, gluons carry the strong force. The weak force is carried by particles called vector bosons, named W^+, W^-, and Z^0. It is also supposed that, for reasons of symmetry, there must be particles called gravitons that carry gravity.

Like the crystals found in nature that never turn out to be as symmetrical as the ones pictured in the encyclopedia, the standard model seems flawed. The neutron and the proton have almost the same mass. Why is it not exactly the same? Why do the carriers of electromagnetism and the strong force have no mass or charge, while the carriers of the weak force are massive and charged? Why *three* families of matter and *four* kinds of forces? Is there a deep reason for these numbers, or would other arrangements do? Why are some but not all particles made of quarks? Nothing in the standard model predicts why the various particles have such a jumble of different masses. In solving problems, researchers must determine the masses from experiment and then plug them into the equations by hand.

This is not what a well-made world should be like. In a perfectly symmetrical universe, all particles would be massless, like the perfect pho-

ton. And so physicists now seek what they call the Higgs field—
the mechanism that caused the symmetries to break, giving rise to all
those ugly, seemingly arbitrary masses. And since all fields must have
particles associated with them, there must be a Higgs boson, what the
American physicist Leon Lederman facetiously calls the God Particle.
Why haven't we seen it yet? Because it is too massive to create in our
existing accelerators.

And so physicists seek even subtler geometries, in which all particles
would appear the same. Just as the strong force doesn't care whether a
nucleon is a proton or a neutron, there is a force deep in the physicists'
minds that doesn't care whether a particle is a quark or a lepton (though
according to one theory, this force would require twelve new "X-
gluons," each so massive as to be undetectable by any conceivable accelerator). Going still further, the inventors of superstring theory seek a
symmetry so deep that fermions (quarks and leptons) and bosons (particles of force) would be united—all would be loops vibrating at different
frequencies in many-dimensional space. Refracted this way, gravity
would finally be united with the three other forces. But the price to be
paid for this seeming economy is a doubling of the number of particles in
the standard model. Each fermion would have a partner boson and vice
versa, giving rise to what are sometimes called "sparticles": squarks, se-
lectrons, photinos, gravitinos, higgsinos, et cetera.

In the midst of all this confusion, particle physicists, like cosmologists,
comfort themselves with a creation myth. Long ago, in the early moments of the big bang, temperatures were so high that all forces, all matter, blended into one. But then the universe cooled and the symmetry
broke, the one superforce splintering, one by one, into four separate
forces, each with its own quirks and idiosyncrasies. From a primordial
gas of quarks, electrons, neutrinos, and photons more solid stuff formed:
quarks congealing into protons and neutrons, which united to form
atomic nuclei; electrons uniting with the nuclei to make hydrogen and
helium atoms; the hydrogen and helium atoms combining to make stars,
the generators of the heavier elements. Then atoms joined with atoms to
make molecules, and finally, at least in this corner of the universe, cells,
organisms, and societies haphazardly evolved—more shattered symmetries, shards from the explosion of a perfect world. How easy physics
would be if only we weren't marooned in this flawed creation, digging
for lost geometries. Like the Gnostics, we feel trapped in matter, longing to ascend to a world where all is symmetrical and all is one.

The consolation is that in a perfect, undifferentiated world we wouldn't exist. We are here because of the broken symmetries. The theorists tell us that in the early moments of a perfectly crystalline creation there would have been an equal amount of matter and antimatter. But it seems that a random fluctuation, a broken symmetry, led to matter's having the slightest edge—a billion and one quarks for every billion antiquarks, perhaps. After most of these particles and antiparticles killed each other off, there was just enough matter left to make a universe.

3

THE HEIGHT OF THE SKY

It seems so clear when you first observe it: Orion the Hunter, one of those rare constellations that actually look like their names. Two stars, Betelgeuse and Bellatrix, mark his shoulders, Saiph and Rigel his legs. And cinching the waist, three stars of approximately equal brightness form his belt. So naturally do these lights seem to link into this hourglass pattern that it is strange to realize that, except from the narrow perspective of our own solar system, Orion's parts are not close to one another at all. Though the stars in Orion's belt are each approximately 1,500 light-years distant, the belt is nowhere near the rest of his body. One shoulder, Bellatrix, is estimated to be about 350 light-years from the earth. Betelgeuse, the other shoulder, is in rough proximity—420 light-years away—if 70 light-years (the distance light travels in seven decades) can be so easily disregarded. One of Orion's knees, Rigel, is 1,000 light-years from us; the other is perhaps twice that far away. At Orion's feet lies his faithful dog, Canis Major; its principal star, Sirius, is a mere 8.6 light-years away.

When we look at Orion, the astronomers tell us, we are seeing stars not only separated by distance but separated in time. In geology, we visit the past by cutting through the layers of sediment beneath us; the deeper we go, the farther we go back in time. In cosmology, the farther we look, the farther we see back in time. According to the picture we have drawn of the heavens, the photons from distant Rigel that strike our retinal cells were emitted a millennium ago. When we gaze through our telescopes, it feels as though we are reaching out into the heavens. But, of course, the photons we are capturing and amplifying are already here in our own atmosphere. All we can really do is sit here on earth and scrutinize the electromagnetic signals that happen to cross our threshold—the light waves, radio waves, x-rays, and gamma rays that bring us news of what we have come to believe are galaxies flying away in all directions, propelled by the ancient explosion we have named the big bang. After estimating the galaxies' velocities and how far away they seem, we can reverse the film in our mind's eye and imagine the whole universe contracting to a point; we can count back and estimate that the explosion must have occurred some ten to twenty billion years ago. But what a curious explosion it is said to have been. We think of something exploding in space over a period of time. But before the big bang there was no time, no space. The explosion created not only matter and energy but the universe in which it expands.

Assuming that nothing can travel faster than light, the big bang theory implies that the farthest we can possibly see is perhaps ten or twenty billion light-years in every direction. That doesn't mean that there is nothing beyond that limit. From our vantage point on earth, we might detect an energy source fifteen billion light-years away, in the direction we quaintly call north, then swing our instruments around and detect another source fifteen billion light-years to the south. These two objects would be separated from each other by thirty billion light-years; they can't be touched by each other's light. Since we are presumably not at the center of the universe, we can assume that there are objects thirty billion light-years from us. But we cannot see them. The light emitted by whatever is beyond our horizon hasn't had time to reach us.

In trying to explain matter, the stuff we can pick up with our hands, science is quickly propelled into abstract realms where we can find pattern only by putting our faith in symmetries that exist in the barely accessible spaces of mathematics. Extrapolating a cosmos from pinpoints of light also takes great ingenuity and imagination. Over the years we have slowly developed a grand picture of the universal scheme—the big bang theory. But inevitably the universe refuses to be squeezed into our formulations. And so, as in particle physics, we end up honing and revising, stacking abstraction on top of abstraction, always striving for a better fit.

Plank by plank, we use theory and observation to elevate ourselves toward the heavens, rising as high as we can above the terrain. But strip away the tissue of concepts and suppositions and the long chains of inferences that get us to these heights and what do we see, standing in the foothills of the Sangre de Cristo Mountains, gazing at the sky? Stars that seem stationary, unless we try to track them with a telescope or leave a camera mounted on a tripod with its shutter propped open. And then what we see is the motion of our planet making it appear that the heavens move around us in a great celestial sphere. The sky shows us only two dimensions, giving no indication that these lights aren't tiny objects all the same distance away. Our only hope is to measure what is near and trust that we can use this data to bootstrap ourselves higher into the sky.

After gazing for a while at all the tiny silent lights, one is sometimes startled from the reverie by a star that suddenly seems to break out of the pack, disturbing the cerebral networks. The light moves too fast to be a planet—or "wanderer," as the Greeks called them, trying to make sense of stars that seemed to come unstuck and meander across the sky. It is too steady to be a meteor. Instinctively laboring to classify this sudden anomaly, the brain throws out a hypothesis—a satellite?—and recalls childhood memories of seeing Echo 1, the astonishment that something made by people had joined the heavenly light show, as though, as in an old Ray Bradbury story, someone had painted the Coca-Cola emblem on the moon.

But we quickly reject that notion, as the ears pick up the sound of metal ripping sky—an airplane flying northeast from Albuquerque. The brain, satisfied, settles back into equilibrium, the wonder dampened—except for a lingering feeling of how eerie it really is, that inside that tiny light are maybe three hundred people, each with a different reason for going to Omaha.

Things shrink as they get farther away, a relationship that seems embedded so deeply into the structure of this fishbowl we call space that we rarely ever think about it. From the window of a car, the fence posts along the highway pass one after another at sixty-five miles per hour, but behind them the telephone poles move slower and the hills beyond more slowly still—tier by tier all the way to the mountains on the horizon, solid earth divided into a continuum of bands by this phenomenon called motion.

Things farther away seem smaller and appear to move at slower speeds. With this simple rule, written into our nervous systems by growing up on this planet, we can calculate how far away the plane is if we compare its true size (we could radio the pilot for information) with the shrunken image we see. Or we could compare its true speed with the speed at which it seems to cut across the sky. Here, though, we need more information. All else being equal, a plane that happens to be flying perpendicular to our line of sight will seem to move faster than one flying off at an angle; a plane flying straight away from us, precisely along our line of vision, will appear not to be moving at all. And so we have to divide the motion into two components: transverse (across the sky) and radial (toward or away from the observer).

It is comforting to think that the same neural circuitry that evolved to help us navigate through woods and mountains can be used to pull a third dimension from the flat canopy overhead; like the Jemez Mountains unfolding in the evening twilight, the sky also reveals hidden peaks and canyons. These days we are confident enough of the behavior of electromagnetic waves that we can gauge the distance to the moon by bouncing radio waves off it and measuring the echo's delay. But there are simpler ways. A triangle has three sides, three angles; if we know the measurements of just one side and two angles, we can calculate the rest. So measure the position of the moon against the distant stars, which seem as steady as the mountains on the horizon, and have someone a known distance away simultaneously take the same measurement. Using the difference between the two apparent positions, the parallax, we can calculate that the moon is 240,000 miles away—a year's drive if we allow ourselves time to eat and rest. Similarly, we can measure the distance of the planets and show that the sun is 93 million miles (eight light-minutes) from us.

But parallax will get us only a tiny fraction of the way up the ladder. Even the closest stars are too distant to permit a detectable parallactic shift from any two points on earth—our planet isn't wide enough. To reach farther into space, we have to use the earth's orbit around the sun as the base of our triangle. Make two observations of Alpha Centauri six months apart, from opposite points in the earth's orbit, and you find a tiny difference in its apparent position against the backdrop of the heavens. How tiny? Look from the horizon to an imaginary point directly overhead; your eyes have traversed a ninety-degree angle. Now imagine one of those degrees and divide it into sixty minutes and each of those minutes into sixty seconds. The parallax of Alpha Centauri is less than a second of a degree. Plugging that number into the trigonometric equations yields a distance of 4.3 light-years. By the same method we can show Sirius to be 8.6 light-years from us; Altair, in the constellation Aquila, is 16.6 light-years. But beyond about 100 light-years, even the parallax from the earth's motion around the sun is too small to detect.

How then can we judge the distance of something like Betelgeuse, and how can we talk of quasars billions of light-years away? The sky is sprayed with lights of all brightnesses—or magnitudes, as the astronomers say. But we can't tell distance from brightness unless we can radio the star, like the pilot of that plane, and ask for its intrinsic luminosity—how bright *it* says it is.

For all but the closest stars around us, our measurements depend on theory and mathematics more abstract than simple trigonometry. We start with a celestial beacon whose distance we have measured, often by the most indirect of methods. Then, using our theories of astrophysics, we analyze the characteristics of its light and guess what its intrinsic brightness might be. The object then becomes what astronomers call a standard candle. If we can find a similar object in another part of the sky, we can hypothesize that it is of the same intrinsic brightness; if it appears a little dimmer, it is arguably farther away than the reference object. Suppose this newly measured light is part of a cluster of stars or a galaxy. Perhaps one of its neighbors can now serve as a standard candle. We now have an estimate of its distance from earth; if we know enough about its physics we can guess at its true intensity and use it to reach out farther still. Layer by layer, we build a house of cards, each resting on a shakier foundation and each testifying to our theoretical bravado.

If we look at the Big Dipper and sight along the pole stars, the points that form the end of the ladle, we see Polaris, the North Star. Like the blinking red lights on the radio towers on the Sandia Mountains, Polaris cycles bright to dim—not in seconds but in four-day cycles. Nearby, in the constellation Cepheus, the star Delta Cephei was observed as long ago as 1784 to vary rhythmically in brightness, going from dim to bright to dim again every six days. In 1912, while studying photographs of the Magellanic Clouds, two galaxies spotted in the skies of the southern hemisphere by Ferdinand Magellan's crew, Henrietta Swann Leavitt of the Harvard College Observatory saw a number of these pulsating stars—the so-called Cepheid variables—and found an interesting correlation: the slower the blinking, the brighter the star. Since all these

Cepheids were in the same formation, the Magellanic Clouds, the stars were presumably of roughly equal distance from earth. Assuming that Leavitt's relationship held throughout the universe, astronomers could now find two Cepheids with the same periods and suppose they were of equal intrinsic brightness. If one appeared dimmer than the other, it would indeed be farther away. Using the inverse square law— an object twice as far away as another is one-fourth as bright—astronomers could calculate their relative distances from earth.

While Cepheids gave clues to relative distance, they said nothing about absolute distance. The problem was that no one knew how far the Magellanic Clouds were from earth. Before scientists could use the Cepheid yardstick, they had to calibrate it, determining as surely as possible the distance of at least one of these blinking stars. It would be comforting to say that astronomers simply found a nearby Cepheid, measured its distance by parallax, and came up with a standard candle. But in fact, Polaris, the closest Cepheid, is far too distant to detect the slightest parallactic shift; it was found through other methods to be some eight hundred light-years away.

To calculate the distances to the nearest Cepheids in the Milky Way, astronomers needed a baseline much longer than the diameter of the earth's orbit. And so they looked to the movement of the sun. The sun, dragging the solar system along with it, circles the center of the galactic spiral of the Milky Way. Wait long enough and the position of closer stars against the backdrop of more distant stars and galaxies should shift ever so slightly. From this parallax, it seemed, one should be able to calculate their distances.

Why believe the sun is moving? In 1783, William Herschel showed that when one looked toward the constellation Hercules it was possible to find stars that seemed to move, as the years went by, as though they were fanning out from an imaginary point, like snowflakes seen through the windshield of a moving car; at an opposite point in the sky, toward the constellation Columba, the stars seemed to be converging in a point, like snowflakes viewed through a rearview mirror. Herschel interpreted this as an optical illusion caused by the sun's motion through the blizzard of stars.

We can calculate the speed of this apparent movement using another phenomenon called the Doppler effect. Moving toward us, the crests and troughs of sound waves are pushed together so the frequency of the signal increases. As the source recedes, the waves are stretched toward the lower end of the scale; the frequency goes down. If we think of light as wavelike, we can take this earthly phenomenon used to explain the rising and falling pitch of train whistles, and apply it to the sky. Stars speeding away from us should have their light waves stretched out, falling in pitch toward the low-frequency red end of the spectrum; stars moving toward us should be shifted toward the blue. Using the Doppler effect, one can estimate the average amount that stars in Hercules are blueshifted and calculate the sun's motion through the galaxy as twenty kilometers per second.

Now, if all of this is correct, we can measure the position of a star and then measure it again years later. Then we use the speed of our sun to calculate the distance between these two vantage points—the base of the triangle. All things being equal, the star's apparent change in position would yield the parallax, and from this we calculate its distance from our solar system.

There is, however, an imposing problem. The stars are not fixed in space. Like the sun, they too are moving; over the eons, the shapes of the constellations subtly change. How can we know how much of a star's shift in position is due to the actual motion of the star and how much to the parallactic illusion?

As with the speed of a distant airplane, we first must distinguish between transverse motion, cutting across our field of vision, and radial motion, traveling along our line of sight. Here, the Doppler effect is our first recourse: by measuring a star's red or blue shift, we can let physics tell us its radial velocity, how fast it is moving toward or away from us. But we can't know the absolute value of the transverse velocity, the motion across the sky, unless we know the star's distance—nearby stars appear to move faster than distant stars, like the fence posts that speed by on the highway.

Of course, it is this distance that we are trying to gauge in the first place. How can we break out of this loop? If we look at a *group* of stars which we have reason to believe are all about the same distance away from us, we can make the simplifying assumption that the directions of their intrinsic motions are random—some are moving this way, some are moving that way. Statistically, the motions should cancel one another out. And so, as we follow our sun through the galaxy, we can conveniently ignore how much of the apparent movement of these *groups* of stars is due to their own motion and attribute their shift in position to parallax. Using this mathematical sleight of hand, Harlow Shapley of

Mount Wilson Observatory calculated the *average* distance of thirteen Cepheids in the Milky Way and used that figure to calibrate Leavitt's period-luminosity relationship, which he hoped would apply to all Cepheids everywhere. According to this yardstick, the Magellanic Clouds were about thirty thousand light-years away. The distance is now believed to be seven times greater, a gauge of the difficulty of the art.

In fact, Cepheids are now calibrated by using other measurements that are at least as indirect. By studying nearby examples of various kinds of stars—there are supergiants, red giants, blue giants, white dwarfs—whose distances can be measured by parallax or some other means, astronomers believe they have found a relationship between a star's temperature, its type, and its intrinsic brightness. A star's temperature is taken by analyzing its spectrum, using the laws of black body radiation that preoccupied Planck. Assuming that these rules hold true outside our neighborhood, we can guess the distances of farther stars: we use the star's spectrum to put it in the proper class, then compare the star's apparent brightness with the intrinsic brightness that theory predicts stars of that variety should have. Finally, we use the inverse square law to calculate how far away it is.

And so we go, from moon to sun to stars. The farther we reach from earth, the deeper our measurements become embedded in our theories of stellar physics, which are based, in turn, on thermodynamics, quantum mechanics, and the nuclear physics we believe energizes stars. Even these methods, with all their uncertainties and assumptions, can take us only so far. Thirty million light-years is about as far as terrestrial telescopes can resolve single stars (though the orbiting Hubble Space Telescope has now extended the range). Beyond that, we use whole galaxies as standard candles. What kind of galaxy is it? How much energy is emanated by similar galaxies, whose distances we are somewhat more certain of? But then we are left with this dilemma: according to the big bang theory, the light of the nearest, most familiar galaxies, the ones we use as our standards, is far younger than the light from the distant galaxies. We do not know whether galaxies in more recent times behave as galaxies did near the beginning of creation. Or whether physical laws were the same. We put our faith in the doctrine of uniformity. Our measurements become even less certain, immersed ever more deeply in theory, as we reach farther out on our limb.

The method we most depend on to measure these barely imaginable stretches is redshift. In the early 1920s, the American astronomer Edwin

Hubble used Cepheids to show that nebulae are huge, distant galaxies and not small nearby hazes of light, as his rival Harlow Shapley thought. Having measured the distances to nearby galaxies, Hubble then used a spectrometer to analyze the color of their light. If it is assumed that stars farther out and farther back in time are made of the same stuff as those closer in, hydrogen and helium, they should show the same patterns of spectral lines. In fact, Hubble showed in the 1920s that the farther a galaxy is from the earth, as reckoned by the pulsations of its Cepheids, the more its spectral rainbow appears shifted toward the red end of the spectrum. Rather than attribute this phenomena to age, concluding that for some reason ancient, more distant galaxies radiate redder light, astronomers quickly realized that, with the proper assumptions, Hubble's observations could be taken as strong support of an expanding universe. According to the big bang, more distant stars should appear to be receding from us more rapidly, the Doppler effect ensuring that their light is shifted toward the red end of the scale. By measuring to what degree redshift is correlated with distance, Hubble invented the yardstick we use to reach to the very edge of the observable universe, measuring the distance of anything whose electromagnetic waves we can detect. The redder an object, the faster it is receding; the faster it is receding, the farther away it is. The theory of Cepheids, supplemented by some sophisticated statistical reasoning, brought us to the nearby galaxies. When we use redshift as a gauge of farther distances, we are assuming the truth of the big bang. Without the theoretical framework, the individual observations would be meaningless.

Hubble used his method to calculate that the observable universe was two billion light-years in radius, and so (according to the big bang) two billion years old. But geologists had used their own measuring stick—the rate at which uranium decays into lead—to calculate that the earth itself was twice that old. Something had to give somewhere. Conveniently for astronomy, it was later decided that there was more than one kind of Cepheid, each of which had to be calibrated differently. Hubble was confusing the two. Once the new numbers were plugged into the equations, the universe doubled in size overnight. It took several more adjustments of the so-called Hubble constant to come up with today's universe: ten to twenty billion light-years in radius. We can expect that the revisions will continue.

In the ensuing decades, the picture of a primal explosion and an expanding universe became so compelling that data stuck to it like iron filings to a magnet, arranging themselves in this wonderful new way. In 1948, George Gamow, Ralph Alpher, and Robert Herman predicted that if the universe had indeed begun with a big bang, space should be permeated with its afterglow, in the form of measurable background radiation. When, in 1964, Arno Penzias and Robert Wilson found that an experimental microwave antenna at Bell Laboratories in New Jersey was plagued with a background hiss no matter in which direction it was pointed, they speculated that the problem was pigeons roosting inside. The birds were evicted and the droppings cleaned out, but the static persisted. A consultation with radioastronomers at Princeton University led to the conclusion that Wilson and Penzias were measuring fossil radiation from the big bang. The mystery of the constant hissing was not only absorbed and explained away—it became one of the most persuasive pieces of evidence for the big bang.

Over the years the celestial net has become thicker and thicker with mutually supporting threads. Whenever possible, distances are gauged by several independent methods. When they converge on similar answers, we can take that as reassurance that the weave is tight, the network robust. The lower levels of our celestial tower have become more solid, we hope, the playing cards replaced by bricks and mortar.

When observation clashes with theory, or with common sense, ways must be found to account for the discrepancy. In 1930 Robert J. Trumpler, an astronomer at Lick Observatory, was studying groups of stars called open clusters when he found a strange correlation between their diameters and their distances from the earth: for no apparent reason, it appeared that nearby clusters had small diameters, distant clusters had large diameters, and that there was a gradation of diameters in between. This would be a strange law indeed, implying that the earth had a special place in the universe: at the center of concentric rings of increasingly wider star clusters. Perhaps, Trumpler thought, there was something wrong with the assumptions behind his calculations. He had gauged how far away a cluster was by comparing what he believed was its intrinsic luminosity with how bright it appeared from earth. Then he measured its apparent diameter and calculated how wide it really should be. But what if there was something in space—interstellar dust—absorbing some of the starlight? Then a cluster would actually be much closer than it appeared to be, the dimness caused partly by dust, not distance. And if it was closer, it would not have to be as large as originally supposed to emit the same amount of light. If we assume instead that all the clusters are of roughly the same diameter, then interstellar dust would create an optical illusion: those farther away would appear larger because their light would have more interstellar dust to traverse.

To gauge how much our distance measurements are thrown off by interstellar debris, we must know how much dust there is between us and the object in question. But how do we measure the amount of dust without knowing the distance, which is what we are trying to determine in the first place? Our assumptions must be expanded, and again we look to our own world for analogies. We assume that starlight is reddened by interstellar dust just as the sun and moon are reddened by dust in the earth's atmosphere. So we predict from our theories of stellar physics the color a star should be and use the reddening to measure the density of dust. We cross-check the estimate with other measurements, taken from other perspectives, adding more threads to the theoretical web.

And so we tinker with our models. From our vantage point on this tiny planet we construct a universe. When, in 1963, objects with redshifts so severe that they had to be billions of light-years away were each found to be emitting the energy of a hundred galaxies, a few astronomers found their faith shaken. What could possibly emit so much light? Some were tempted to conclude that the Hubble method was wrong, that our celestial house of cards, with measurements built on measurements built on measurements, was about to collapse. Perhaps the relationship between distance, velocity, and redshift had been seriously misconstrued. Or perhaps redshifts were caused not by the Doppler effect at all, but by some unknown peculiarity of nuclear physics. If redshifts were not a true measure of distance, then these absurdly powerful beacons might be much, much closer by. The alternative was to accept that these sources—now we call them quasi-stellar objects, or quasars—are indeed fantastically energetic objects at the very edge of the observable universe, and a whole branch of astrophysics has been created to explain what they could possibly be.

When, in the early 1980s, astronomers were faced with the embarrassing problem that—in violation of special relativity—some quasars seemed to be emitting jets of matter that moved faster than light, some were tempted again to assume a much smaller universe; if the quasars were actually nearby stars, then the jets would be moving much more slowly. The tension was relieved when a way was found to dismiss the jets' illegal velocities as an optical illusion caused by the angle at which earthlings happened to be viewing the events.

We can sympathize with Ptolemy and his layers of epicycles. When we can't get the models we build to yield what our instruments detect, we make our own adjustments, hoping that history will show us more prescient than the geocentrists. How deft our brains are at smoothing out the irregularities, absorbing the anomalies, bringing the strange back into the land of the familiar as we search the world for patterns, always trying to extend them to new corners of the universe.

But there must be something at the base of our theoretical towers, a foundation to build upon, something we can take as fundamental. When the superluminal jets appeared, no one seriously suggested overthrowing Einstein's special theory of relativity. Why are we so confident that the speed of light is inviolable?

Until Einstein, the measure of all things was not light but "aether," Aristotle's fifth essence. Aristotle's first four elements—earth, air, fire, and water—had long since been abandoned as fundamental ingredients in the universal recipe. But the ethereal quintessence, said to fill the space within atoms and between stars, was harder to forsake. Something, it was believed, must act as a carrier of the light waves that beam across space. Something must be doing the waving. As Maxwell himself rhapsodized, the universe is "full of this wonderful medium; so full that no human power can remove it from the smallest portion of space, or produce the slightest flaw in its infinite continuity. It extends unbroken from star to star. . . ."

By the time Maxwell made his grand unification of electricity and magnetism, his belief in aether was becoming harder to sustain. In 1887, the Americans Albert Michelson and Edward Morley performed the famous experiment that seemed to show that there is nothing filling the cracks of the universe, no celestial backdrop—just empty space. Using an apparatus of prisms and mirrors, the two scientists split a light beam, sending one part moving in the direction of the earth's orbit around the sun and the other perpendicular to it. They had assumed that the "aether wind," caused by the motion of the earth through the invisible medium, would slow down the first beam (it was struggling upstream). But they found to their surprise that the speed of the two beams was precisely the same.

Interpreting so delicate an experiment is never a straightforward affair. When an experiment fails to confirm a compelling hypothesis— Michelson and Morley had fully expected to find aether—it is a sign that there may be a discrepancy between our picture of nature and the way nature really is. But as philosophers like Willard Quine and Pierre Duhem have pointed out, this misalignment might exist anywhere in the vast web of facts and assumptions that are implicit in the design, execution, and interpretation of the experiment.

In a heroic attempt to preserve the idea of Aristotle's fifth essence, the physicists George Fitzgerald and Hendrik Lorentz suggested that the Michelson-Morley result was an optical illusion caused by a hitherto undiscovered universal truth: things shrink, ever so slightly, in the direction they are traveling. The light beam moving with the earth was indeed retarded by the aether, but the measuring instrument shrank in that direction—the distance the beam had to travel was less. The Fitzgerald-Lorentz contraction conspired to make it appear that there was no aether. This assumption was not as ad hoc as it sounds today. After all, scientists knew from Faraday that a moving charge created a magnetic field. Perhaps as a measuring stick moved, the charges of its atoms generated a field that squeezed the molecules closer together. True, no one seemed to be able to measure the contraction. But perhaps this was part of nature's conspiracy: as the earth moved through space, we and everything else in our world were shrinking in the same direction.

With his special theory of relativity, Einstein suggested that the conspiracy was much subtler than Fitzgerald or Lorentz had supposed. And in doing so he introduced his new absolute, the speed of light, setting a platform on which to erect our towers of observation. As hard as you look you will never see your own instruments shrink. But, he suggested, if you could somehow observe a laboratory in another reference frame—in a spaceship, perhaps, moving by at a fixed velocity—it would appear that those scientists were performing their measurements with shrunken instruments. And they, looking at you, would think that their instruments were just fine; it was your lab that appeared to be moving and your instruments that had contracted. Who was really moving and shrinking? To Einstein that was a meaningless question. The Michelson-Morley experiment should be taken at its face: there is no aether, no privileged reference frame to measure motion by. Motion can only be gauged relative to something else.

No matter how many times we hear them, the implications of Einstein's theory never fail to amaze. An object in relative motion will also increase in mass, and its clocks will slow down. Two events that appear to be simultaneous in one reference frame may appear separate if viewed from another. Space, time, and mass are not any more absolute than the illusory aether; they are just relationships we measure with yardsticks, clocks, and scales—instruments whose readings depend on their relative motion. But the point of all this was not nihilism. The ultimate aim of the conspiracy was a lawfulness more deeply embedded than before. By granting familiar quantities like time and space the freedom to expand and contract, Einstein ensured that from any of the universe's infinite number of moving vantage points, science would be the same. In the universe according to Einstein, all observers would discover the same natural laws. As two laboratories move through space at different velocities, length will contract, time slow, mass increase by just enough to guarantee that the rules that govern creation appear the same in both domains. Looking at the world beyond their own reference frames, the scientists will measure different quantities, see different numbers on their dials, but the relationships *between* the quantities—the structure of the system—will be the same: length contracts, mass increases, and time dilates just enough to even things out.

Of course, there has to be a limit to these dilations and contractions. As you observe a particle speeding faster and faster, at some point its mass will become infinite, its length zero, its clocks frozen still. In Einstein's theory, this limiting speed is the speed of light. Nothing can go faster, and no matter how fast you are moving, the speed of light is the one thing that will always appear the same.

Lorentz had postulated that there must be an aether—that which was doing the vibrating—and he assumed that the logic of classical physics was true. The outcome of the Michelson-Morley experiment—light beams that moved at the same speed no matter how fast the body that emitted them—was the anomaly to explain. Einstein abandoned these assumptions as prejudices and showed that the same facts could be built into an entirely different structure if we turned the anomaly into a postulate: light travels at the same speed regardless of the speed of the observer. He showed how to rebuild the universe in a different way.

In this reconstruction, some things that had been taken as absolutes were now considered relative. But just as important is what became invariant. If some things change—space, time—others, by definition, must remain the same. With everything moving relative to everything else, there must be some kind of glue holding things together, a standard that allows for a sensible world.

There are compelling reasons why it should be light—or, more generally, electromagnetism—that is accorded this dispensation. As far as we can tell, light is the fastest means by which something can make its presence known, by which one event can affect another. In all the thought experiments with laboratories moving by one another at different speeds, the means by which they are aware of one another's existence is electromagnetism: they can observe each other with telescopes, or send radio signals. For that matter, how are we aware of anything beyond our planet, where our atmosphere allows us also to communicate with more parochial signals like sound and smell? We receive signals, information, in the form of light beams, radio waves—electromagnetism.

If there were no upper limit to the speed at which information can be sent, the universe would allow instantaneous action at a distance; or, even worse, a reversal of cause and effect—the future could send signals to the past. It is true that quantum theory presents us with possibilities almost as weird. But for there to be a lawful universe with a strict wall between past and future, cause and effect, there must be an upper limit to the speed of signaling, the speed of causality, the speed of light. Whatever we call it, this constant of the universe is now considered all but sacrosanct—a prerequisite for a rational world. If the house of cards should ever tumble, it will be rebuilt on the same underfooting. There is little in science about which we feel so sure.

Morley and a colleague of Michelson's, Dayton Clarence Miller, later repeated the Michelson-Morley experiment to see if its disappointing results could be explained away by some unappreciated parameter. After all, the original experiment had been performed in a basement. Perhaps if it were repeated on a hilltop . . . But the results were the same. Still, as late as 1925, two decades after Einstein's papers on special relativity, Miller, then the president of the American Physical Society, announced that he had found definite proof that aether existed after all. But by this time, the speed of light had thoroughly dislodged aether as the universal gold standard. Einstein easily dismissed the findings, noting that they might be attributed to temperature variations in the measuring device.

It is natural that in creating an image of the universe, scientists would begin with themselves at the center, humanity as the absolute. Over the centuries, as the illusion became harder and harder to sustain, we could still hope that our most direct experiences—the passage of time, the passage through space—were absolute. If we weren't sitting still at the center of creation, then at least we could measure our motion against the stationary aether.

But in the new view, nothing was at rest. Without aether, we can only measure our motion in relation to something else, which is moving in relation to something else. But for all this relativity, we remain absolutists at heart. In recasting our map of the heavens in Einsteinian terms, we have been careful to maintain the centrality of cause and effect and the existence of universal scientific laws.

But it is a constant struggle to interpret the data we gather so that they obey the laws we believe we have divined. With the platform of special relativity solidly beneath their feet, astronomers have gone on to beguile us with a universe far more bizarre than their predecessors could have imagined. Einstein taught us that light reigns supreme. But in trying to account for an increasingly discordant stream of astronomical observations, scientists have been forced to conclude that most of the universe consists of matter that for unknown reasons seems to emit no light.

The need for dark matter began to insinuate itself into cosmology before World War II when the Dutch astronomer Jan Oort and the Swiss-American astronomer Fritz Zwicky noticed that galaxies behaved as though they were far more massive than they appeared. If our distance measurements can be trusted, galaxies, including our own Milky Way, are spinning faster than the laws of physics would predict—so fast, in fact, that they shouldn't exist, having flown apart long ago. Either we must demote our laws of gravity (as established by Newton and modified by Einstein's general theory of relativity) to the status of local aberrations or we must invent something that is holding the galaxies together—unseen matter that seems to emit no radiation, or emits it too weakly to detect. We can know it only by its secondary effects, phenomena that make no sense in the current theoretical framework unless we come up with more gravity, more mass. Over the years, further measurements have suggested that, from these arguments alone, the ratio of this unseen matter to visible matter is ten to one.

But that is just the beginning of the confusion. For years, cosmologists pointed to the smoothness of the background radiation, the ubiquitous microwaves that we interpret as the afterglow of the primordial explosion, as stunning support for the big bang. Everywhere they pointed their dish antennas, it seemed to rain down with the same temperature. But the more they looked, the more cosmologists worried that the radiation might be *too* smooth. A perfectly smooth afterglow would seem to signal a smooth early universe, which made it difficult to explain why we observe large-scale structures today—the galaxies and galaxies of galaxies that seem to extend as far as we can see.

And so cosmology underwent a shift in emphasis. After marveling again and again at how uniform the background radiation appeared and performing experiments to establish this smoothness over and over, astronomers now needed signs of subtle irregularities—imperfections that might have been magnified by gravity into the lumpiness we see today. Using satellites and ever more sensitive detectors, they scrutinized the radiation at a finer and finer grain. And the closer they looked, the smoother the radiation appeared. Any irregularities must have been very tiny indeed.

At the same time, the structures whose genesis needed to be explained grew larger and larger. In the late 1980s it was discovered that a large number of galaxies are not moving away from us in the uniform manner that the big bang theory predicts. To account for this aberration, theorists were forced to conclude that something rather enormous was pulling them off course. And so they announced the existence of a conglomeration of mass, several hundred million light-years in size, that they called the Great Attractor. Other astronomers found evidence of a giant chain of galaxies: the Great Wall.

It was hard to escape the conclusion that even ten or twenty billion years—the age of the universe predicted by the big bang—wasn't nearly enough time for such enormous configurations of matter to coalesce. Gravity simply wasn't strong enough. A few theorists were inspired to posit the existence of a fifth force to help matter clump together. Others proposed modifications to the laws of gravity, a way to transform it into a more powerful force. A more conservative gambit was to propose that some kind of dark matter provided enough extra gravity to bring about the congealing.

At first physicists were hopeful that this dark matter would prove to be like ordinary matter, that its emanations were simply too feeble for our instruments to detect. Perhaps the universe was filled with very dim stars or black holes, those gravitational whirlpools said to suck in everything around them, including light. Perhaps enough mass was tied up in these vortices to balance the equations.

It is a measure of how rarefied our explanations have become that black holes should be considered conservative candidates for dark matter. Detecting black holes requires immersing oneself deeply into the wells of theory. And even then, the only black holes one can unambiguously see are those within the equations of Einstein's general theory of relativity, which imply that if a collapsing star is massive enough it will go on collapsing forever, tearing a dimensionless pinhole into the space-time fabric.

But to many cosmologists, even this explanation, which once would have seemed daringly exotic, did not go far enough. There is little reason, other than theoretical convenience, to believe that there are enough black holes to make up for the gravitational deficit. And the Hubble Space Telescope has searched in vain for the abundance of dim stars the theorists require. Many cosmologists, in their effort to rectify still other problems with the universal creation story, have been forced to conclude that dark matter is unlike the stuff of planets, people, and stars.

Depending on its total density of matter, a universe can be in one of three states. If the density is low, space will be negatively curved and the universe will be "open," becoming more and more rarefied as it keeps expanding forever. If the density is high enough, space will be positively curved and the universe will be "closed": the expansion will rapidly fizzle out and all of creation will fall back in on itself in a big crunch. Or the universe might have just the density required for space to be "flat," yielding a universe poised between open and closed. Years of observations and calculations have persuaded many theorists that the universe is indeed flat, with the outward Hubble expansion balanced by the inward coalescing pull of gravity. More compellingly, perhaps, it seems that if the universe were not flat, then it either should have collapsed long ago or become so rarefied that there could be no galaxies, no stars, no us.

Here is the problem: If the universe is flat, there doesn't seem to be nearly enough ordinary matter—dark and luminous combined—to provide the necessary gravity to balance the expansion. Luminous stars can account for roughly 1 percent of the needed density. Throwing in ten times as much hypothetical dark matter—the dim stars and black holes that can be inferred but not seen—brings the density up to 10 percent of that required for flatness. So what is the other 90 percent? It seems it would have to be something quite exotic. In fact, according to other arguments, the relative abundance of light elements in the universe—hydrogen, deuterium, helium, and lithium—requires that the primordial fireball that exploded some ten or twenty billion years ago consisted of no more than a tiny fraction of ordinary, "baryonic" matter—the protons and neutrons we once had every reason to believe make up almost everything that has mass. Otherwise nuclear reactions should have produced very different ratios of light elements from those we observe.

We don't know what this nonbaryonic dark matter could be. For a while, it was hoped that it might consist of the elusive neutrinos, with their perverse advantage of being so difficult to detect. Originally invented as accounting fictions, to balance the books of beta decay, neutrinos were now confidently believed to stream through the densest matter as though it were empty space. Neutrinos were thought to be massless, like photons, but perhaps if they had the slightest mass their huge abundance could provide the extra gravity needed to get the big bang theory to cough up the enormous structures we see.

In 1995, scientists at Los Alamos discovered that neutrinos may indeed have a tiny mass. But even if the experiments are confirmed, the particles still appear to be too light to account for more than a small part of the missing gravity. And there are other arguments that may rule out neutrinos as too light and too swift (too "hot") to bring about the congealing of matter. Instead, many cosmologists have concluded, there must be some kind of undiscovered dark matter. Like neutrinos, it would interact only weakly with ordinary matter (or not at all)—explaining why it has gone undetected—but unlike neutrinos, it would be massive and move far more slowly. To play the proper role in the theory, this substance would have to be immune to interference from the high density of photons believed to have existed in the universe's infancy. Thus it would be not only dark but transparent, absorbing or emitting no light. In the early history of the universe, this invisible "cold dark matter" would have clumped together easily, providing the seeds for galactic structure. Candidates for the mysterious glue include monopoles (north and south magnetic poles that have become separated one from the other), particles called axions, or the "sparticles" predicted by supersymmetry.

The big bang theory is impressive in its ability to account for the redshifts of the galaxies that Hubble first measured and the ubiquitous radiation that Penzias and Wilson found. If cold dark matter were enough to explain how the big bang generated the galactic structures we see in the sky, then the search for a satisfying creation myth might have reached an important theoretical milestone. But there was still the problem of the smooth background radiation. Even if one spiked the big bang theory with a hefty dose of cold dark matter, a featureless primordial fireball would produce a featureless universe. There still had to be some kind of measurable irregularities for things as big as the Great Attractor and the Great Wall to have formed. Coming to the rescue in 1992, a satellite called the Cosmic Background Explorer (Cobe) found the tiniest variations—thirty millionths of a degree Kelvin—but only after an enormous amount of computer enhancement to separate what was accepted as signal from what was discarded as noise. Most of the data, by far, was static produced by the instruments or by the Milky Way and other celestial bodies. The experimenters couldn't actually point to any particular variation in the background radiation and say whether it was real or artifact. But taken together, the variations could be interpreted statistically as evidence of irregularities, stirring so much excitement that a press conference was called to herald the results.

Later experiments more firmly established the existence of what came to be called the cosmic ripples. But they are too minuscule to explain, on their own, how primordial lumpiness was magnified into the huge structures we find. An enormous amount of dark matter—or something extra—is still required. And we still don't know what it is. Using computers, cosmologists simulate the early conditions of the universe, tinkering with various mixes of cold and hot dark matter, trying to bring about the structure we see. But the right recipe continues to elude them.

The big bang theory remains very much a work in progress. The Hubble Space Telescope allows us to measure distant Cepheids far more reliably than before. Recent observations suggest that the constant first calculated by the telescope's namesake must be recalibrated again, making the universe only eight to twelve billion years old. This would leave even less time for galaxies to form. And it would create another kink in the fabric of knowledge: there are stars that, according to laws of stellar physics, are more than fifteen billion years old.

There are still many cosmological mysteries to explain. Despite the variations measured by Cobe, the background radiation still appears smoother than it has any right to be. Unless the initial flash of the big bang just happened to be extremely uniform, a coincidence or divine decree that few scientists are willing to suppose, then there must have been a way for the hotter parts of the early universe to radiate heat to the cooler parts, evening things out. But one of the most striking implications of the big bang is that most regions of the universe have never been in contact. There is no way they could have exchanged heat. An observer anywhere in the universe should find itself surrounded by an imaginary sphere perhaps ten billion light-years in radius—the distance light has traveled since the big bang. Objects that are farther apart, say twenty billion light-years, can never have been in contact. When the universe was one minute old they would have been two light-minutes apart, twice as far as light would have time to travel. No mixing of background radiation could possibly have occurred.

In the early 1980s, Alan Guth, of the Massachusetts Institute of Technology, began developing a mechanism to account for the smoothness: Suppose the universe we live in is only a tiny part of the initial creation that began with the big bang. According to the so-called inflationary universe version of the big bang theory, an infinitesimal bit of this primordial megaverse, small enough to have a uniform temperature, was pinched off in the earliest microseconds; then, propelled by the brief existence of a kind of antigravity, arising from quantum effects, this tiny region was inflated at enormous speed, expanding instantaneously, perhaps 10^{50} times, into the seed of the universe we live in today. Guth's scenario also provides an explanation for why the universe is flat. Any positive or negative curvature that existed would have been flattened out, and tiny quantum ripples amplified into the irregularities that later became the seeds for galaxies. Other pieces from the big bang may also have been inflated into other universes, perhaps with their own unique laws, but they are beyond our horizons and can never be detected.

Many cosmologists hold out hope that a way will be found to account for galaxy formation without having to declare that most of the universe is invisible. But if science is ultimately compelled to conclude that as much as 99 percent of matter indeed consists of particles that emit or absorb no light, then we are confronted with a startling reversal of figure and ground. Most of what we know about the universe comes to us through photons filtering through our atmosphere or striking the receptors of our occasional space probes. We are creatures of light. We depend on sunlight to power the earth's biochemical reactions and electromagnetism to learn of the universe and, in turn, to make our existence known. How strange that in mapping a universe whose one seeming certainty is the speed of light, we are faced with the possibility that only an insignificant portion can be known this way.

What we have taken for creation—the stars and galaxies we see and imagine around us—may be only a froth on a wave made of mysterious dark matter, an essence that can be, at best, only obliquely detected. We and the universe we know may be no more than a bit of static, noise in the cosmic signal, as central to the universal scheme as the pigeons Penzias and Wilson expelled from their microwave antenna.

Few deny the magnificence of the big bang, its power to explain so much of what we see. Perhaps this is as good as we have a right to expect a theory of the universe to be. But we cannot escape the fact that the one explanation it seems a creation story should include—how *our* universe of stars and galaxies and galactic clusters came to be—continues to elude us. What the big bang theory strains to explain is the very platform on which we make our observations and construct our towers of abstraction—the very platform on which we, the builders of the big bang theory, stand. So we make adjustments to our vision of the celestial fireworks. We rearrange and embellish until we come up with a version of the big bang theory that can explain flatness, smoothness, and, most important, the origin of structure—but only if as much as 99 percent of the universe is essentially invisible. Should we of the 1 percent congratulate ourselves on being clever enough to discern the rest? Or should we worry that we have been backed into a theoretical cul-de-sac where the only way out is to make such extravagant assumptions? We evolved on the earth with a marvelous ability to find patterns. But our brains were not selected for their ability to understand cosmology or particle physics. There is the constant danger that we are being too clever, too good at absorbing discrepancies to our theoretical inventions. We cannot always know when, like Ptolemy, we are adding epicycles, elaborations to keep our theories standing.

And yet we maintain the conviction—at least as a working hypothesis—that we can comprehend the whole. So vivid is the picture we have drawn of the heavens, with its stunningly brilliant quasars, its infinitely deep black holes, that an alien reading our literature might think we had traveled great distances. But, in fact, we have sent space probes no farther than just beyond the solar system, we have stepped no farther than the moon. The rest of the picture is built from the photons that happen to come our way—magnified by telescopes, sifted for patterns.

With all this effort, the cosmological model we have constructed has become so firmly lodged in the brain that mere humans can be heard to speculate confidently about the very origin of the universe. What caused the big bang? That is where science once left off and religion began. But who could be satisfied with a science that could say no more than Genesis, "Let there be light"? And so in recent years, cosmologists have looked to their colleagues in particle physics, wondering if quantum theory, the tool that has proved so powerful in explaining the universe inside atoms, could be stretched to achieve understanding of the most fundamental particle of all: the infinitely dense pinpoint whose mysterious explosion is said to have given rise to all we see and imagine. How do you get something—in this case, everything—from nothing?

It seems that nature not only abhors a vacuum, it doesn't allow one to exist. It is not just that space is filled with cosmic dust. According to quantum theory, the vacuum we once thought was empty actually seethes with energy, constantly creating pairs of "virtual particles"—matter and antimatter—that jump out and flash their tails for an instant before annihilating one another and returning to the void. Heisenberg's uncertainty principle guarantees that this energy is untappable: a particle's timespan and energy, like its position and momentum, are reciprocally related. So any particles that the vacuum creates will be either so low in energy or so short in duration that they will escape our grasp.

Apply relativity and quantum theory to the primordial mass and you get a cosmic loophole big enough for all creation to jump through. Suppose that a tiny amount of energy—a virtual particle—randomly arose from the vacuum. According to Einstein's special theory of relativity, this energy will be associated with mass ($E = mc^2$), and according to Einstein's general theory of relativity, the mass will bend space-time, giving rise to gravity. Thus all of the mass-energy created spontaneously from the void could conceivably be offset by the resulting gravitational pull—the net value would be zero, so no conservation laws would be broken. Heisenberg's principle tells us that an infinitesimal amount of energy can exist for an infinite amount of time.

The universe then would be a quantum fluctuation, and, like all things in quantum theory, it could be described as either a particle or a wave. Just as scientists talk about the wave function of the electron, they talk about the wave function of the universe, assuming that a set of mathematical tools devised to explain the nucleus can be applied to the whole of creation.

But then we are left to wonder: How can one have a quantum fluctuation before there was space and time in which anything could fluctuate? And where were the laws and the mathematics this fluctuation obeyed before there was a universe to contain them? Pushed to the edge, we come up against the limits of our terrestrial notions. But who would have thought that they would take us this far?

When we look upon the grand architectures of cosmology and particle physics with the advantage of hindsight, developments take on an illusory sense of inevitability, as though the picture that has emerged, with all its strengths and weaknesses, is the only one that could have been. When we read in the history books how the great Harlow Shapley believed that the Milky Way *was* the universe, we want to grab his hand and say, "No. Don't you see? Those nebulae that look so small and nearby are really huge distant galaxies. If you calibrate your Cepheids correctly and allow for all that cosmic dust, you'll see that our own galaxy is really much smaller than you believe." We struggle in vain to imagine what it would have been like to be in his shoes—or in Thomson's or Rutherford's—observing the phenomena without the filters that would later channel everything we see.

But what a false sense this gives us of the scientific quest, of the never-ending effort to map our inner and outer worlds. Looking back, knowing what we know, it is hard to shake the conviction that the astronomers and the particle physicists are uncovering a preexisting order, converging on the way the universe really is. If we could go back in time and see the enterprise through their eyes, we would have a stronger sense of science as a glorious human construction, an artful fitting of the data into a carefully crafted mental framework, a construction of towers that just possibly might have been built another way. And so we must turn from science past to science future, to ventures so new that there is no way to know how the story will come out.

Just as New Mexico's stars and stark geology set the mind to wondering about the pictures we have drawn of the world, its laboratories invite us to see through the eyes of contemporary scientists, those whose work is so pregnant with potential upheaval that it puts them up against the very edge of the aquarium, theorizing into empty conceptual space. Northern New Mexico happens to be a haven for some of these explorers. Faced with a universe inevitably more complex than the brains trying to capture it in their mesh, these scientists, like their predecessors, struggle to sort order from randomness. And, in the process, some of them go even further, examining more closely just what we mean by randomness and order, and why it is possible to have a science at all.

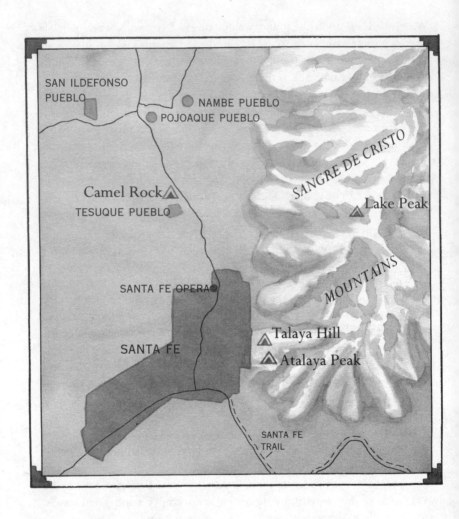

SAN ILDEFONSO
PUEBLO

NAMBE PUEBLO
POJOAQUE PUEBLO

SANGRE DE CRISTO

Camel Rock
TESUQUE PUEBLO

Lake Peak

MOUNTAINS

SANTA FE OPERA

Talaya Hill

SANTA FE
Atalaya Peak

SANTA FE
TRAIL

THE RIDDLE OF THE CAMEL

The cosmopolitan world of Santa Fe is separated from the land of the Tewa by a large alluvial hump of sand washed down from the Sangre de Cristos in the last million years or so, in what geologists call Pleistocene time. Just past the top of the hump, where Santa Fe's open-air opera looks out toward the Jemez Mountains, one can spot the random, twisted features of the Española badlands: volcanic ash, sandstone, and other deposits carved by the wind into a whimsical landscape like something out of a Dr. Seuss book. Farther north on Highway 285, just past the turnoff to Tesuque pueblo, is a roadside monument that might charitably be described as northern New Mexico's answer to the Sphinx, a naturally carved rock formation that looks like a large, impassive camel resting by the side of the road.

The camel's stony gaze is trained along the highway, where, late every afternoon, cars and pickup trucks driving from both directions pull into a large parking lot in front of an enormous prefabricated steel building on the Tesuque reservation. In one of the closest things multicultural New Mexico may have to an interdenominational ritual, the passengers of the vehicles stream in through the doors and line up at the cashiers' desks to buy packets of cards randomly arrayed with numbers. There are Tewa Indians who on another day might be at a corn dance or deer dance at Tesuque, San Ildefonso, Nambe, Santa Clara, or San Juan, sharing the centuries-old hope that by dancing with just the right rhythm in just the right frame of mind they can fend off life's uncertainties. There are Hispanics, from the town of Española on the Rio Grande or the little villages huddled in their niches in the Sangre de Cristos, some of whom might, at another time of day, be reciting the rosary or perhaps listening to the preacher Pat Robertson on cable TV. And there are a handful of Anglos, many of them retired people seeking diversion and a chance to supplement their pension checks.

Many of the visitors look less than prosperous, as though they might have reason to believe that they, more than others, have been singled out as victims of life's unpredictability. But tonight they have come to celebrate randomness, not fight it. They have come here hoping that this time Fortuna's wheel will spin in their direction. Not a wheel really, but a large mechanical generator of randomness that sits on a platform at the focal point of the room. Though its metallic cabinet resembles that of an old mainframe computer, this is a machine ruled by chance, not determinism. In a transparent chamber, seventy-five numbered balls bounce up and down like popcorn, colliding with each other until their trajectories become so thoroughly scrambled that when the operator pushes a button, opening a door to the outside world, no one can know which number will come popping out: O-66, I-23, G-57, O-75, I-22, B-15.

As the various combinations are called out, the players, hundreds now, expertly mark off their cards, each consisting of five-by-five grids of numbers. If the beast spits out just the right numbers to fill a row or create the sought-after constellations—Four Corners, Small Picture Frame, Crazy Kite, Six Pack, Crazy T—the winners shout the word that appears outside in huge red letters against the sky: BINGO. They collect the prize, a couple hundred dollars perhaps—or, if they are extraordinarily lucky, a new GMC pickup truck. At the end of the evening, winners join losers as they walk out into the night, beneath the eyes of the camel and the scattered lights of the starry sky. And the biggest winner of all, the pueblo of Tesuque, the owner of the concession, is richer than it was the night before.

So the riddle of the camel is this: What forces conspire to make a few people winners while returning the rest to the well-worn trajectories of their previous lives? Many simply blame it on Fortuna, the ancient goddess of chance: there are reasons behind her whimsy, but ones we cannot fathom. Perhaps we can importune her with offerings and prayers. Those who cannot convince themselves of a correlation between the faithful and the favored might simply attribute the outcome of the game to some blind, mysterious force called luck. But what do we mean by that? Where, in a lawful universe, is there room for things that happen without cause? Or would we find reasons if we looked hard enough?

During the bingo marathons at Tesuque, there is one game in which players are called upon to choose their own fate. Before the round begins, you are given a blank card of five columns, labeled B, I, N, G, and O, and asked to fill in what you perceive somehow to be your lucky numbers. If there were a way to predict which balls were most likely to pop out of the bingo machine, you could get an edge on randomness. Suppose there were a way to take a three-dimensional snapshot of the chamber of bouncing balls and convert it into a string of numbers giving the position and velocity of every one. To be as precise as possible, we might also factor in the Coriolis effect of the spinning earth, the gravitational pull of the stars. Feeding these data into a computer, already programmed with the mass and elasticity of the balls, we might use Newton's laws to calculate what the arrangement inside the chamber will look like one second later. Then, from that starting point, we could calculate what will happen during the next second and the next—until the very moment when the operator pushes the button and asks the mechanical oracle for the next ball. We would not be surprised by which one popped out. By gathering and processing enough information, we would have beaten the vagaries of chance.

Could it be, then, that randomness is just another name for ignorance? The configuration of the balls unfolds according to the laws of physics; it's just that we cannot possibly process all of the data necessary to make a good guess. Even if we could imagine gathering all the information, it might take too long to make the calculation. If it takes three days, or hours, or even three seconds to calculate the configuration of the bingo balls one second later, then we cannot claim to have seen into the future. But suppose there were creatures with brains surpassing ours. Suppose we endowed the camel, sitting stoically at the side of the road, with a brain far more powerful than the world's best supercomputer. Could it gaze into the canister of bouncing balls and see order where we see randomness? Is chance in the eye of the beholder?

While the physics of bingo seems intractable, at least to humans, roulette is another matter. Instead of taking a snapshot of seventy-five interacting balls, we need only worry about two objects: the ball orbiting the circular track of the roulette wheel and the spinning rotor. Doyne Farmer and Norman Packard, two physicists who are part of the invisible college that has grown up around the Santa Fe Institute, spent the early years of their careers trying to use Newton's laws of motion to analyze the game and give a bettor a sporting chance against the casino. Like many of their colleagues, they have come to believe that understanding the subtle interplay between chance and determinism will help cast light not only on gambling but on the very nature of the scientific enterprise. Growing up together in Silver City, New Mexico, Farmer and Packard were members of an Explorer Scout post that specialized in science. A childhood spent disemboweling radios and engines seems to have imbued them with the belief that all systems, no matter how complex, can be penetrated by an inquiring mind.

Scientists have tried for centuries to find subtle orders hidden within games of chance. The French mathematicians Blaise Pascal and Pierre de Fermat are said to have invented the science of probability, in the mid-seventeenth century, for the express purpose of calculating gambling odds. Some of the greatest mathematicians, including the Bernoulli brothers, Pierre Simon Laplace, and Henri Poincaré, have tried to devise systems for beating roulette. They would analyze the string of numbers generated by the wheel and look for statistical regularities, departures from randomness. What they lacked was good computers, corrective lenses for the brain.

A round of roulette begins when the croupier spins the rotor in one direction, then throws the ball so that it circles around the track in the opposite direction. Dragged by friction and wind resistance, the ball moves slower and slower until the centrifugal force holding it to the track falls so low that it is overcome by gravity and drops onto the spinning wheel. By clocking the relative velocities of the ball and rotor and calculating their rates of deceleration, one could try to guess the outcome of the game. In the early 1960s, Claude Shannon, the inventor of information theory, and Edward Thorp, a mathematician with a passion for gambling, built a small analog computer to perform the calculations. After squeezing it down to the size of a cigarette pack, which broadcast signals to a hearing aid, they tried their luck at Las Vegas. Their crude technology was powerful enough to give them a slight advantage over the house, but ultimately they were defeated by unreliable circuitry and inadequate computer power.

By the mid-1970s, when Farmer and Packard were graduate students in physics at the University of California's Santa Cruz campus, the wonders of digital technology held out the promise of succeeding where Shannon and Thorp had failed. Using little more than Newton's laws, they found that they could often predict which octant of the wheel a roulette ball would land in. Sharing a house near the beach with some other aspiring scientists and computer hackers, they dreamed of using reason to defeat chance, siphoning some of Las Vegas' riches into utopian ventures of their own. Pooling their money, they bought a regulation roulette wheel and scrutinized it with strobe lights, cameras, and photo detectors. Gathering data and analyzing it on a university computer, they devised some equations that seemed to govern roulette.

The next step was to capture these rules in an algorithm and program it into a miniature machine. A data taker would click a switch when the roulette ball reached a fixed point on the rim of the wheel, then click again when it came around a second time. The data would be fed to the computer, which would calculate the ball's velocity and rate of deceleration. The rotor would be clocked in a similar manner, with the click of a switch each time the green double-zero went flying by.

In an ideal world this would be all the information needed to predict where the ball would land. Inside its circuitry, the computer would play a simplified, high-speed version of the game, revealing the outcome in time for a player to put chips on the right range of numbers. But roulette wheels in real life don't match those idealized in the equations. Before any bets were made, the data taker would have to fine-tune the machine, adjusting for discrepancies caused by different kinds of roulette balls or the tilt of the wheel. Even atmospheric pressure might have an effect. Not all of these real-world factors were an impedance. The experimenters quickly found, as had their illustrious predecessors, that a tilted wheel was much easier to predict than a level one: the ball would tend to fall off the high side of the track.

After several years, Farmer, Packard, and their friends—including Jim Crutchfield, an expert programmer on the verge of a career in physics and complex systems—managed to fit all the necessary software and circuitry into a device small enough to hide inside a shoe, or actually three shoes that communicated by electromagnetic waves. Beating roulette was a two-person effort. The data taker's equipment was divided between two modules, one for each foot. By clicking a micro-switch with his right big toe, he would clock the ball and the wheel; with his left and right big toes he would calibrate the various parameters. He would do this while sitting at the roulette table, pretending to be just another greenhorn gambler, gazing wide-eyed at the wheel and (to avoid suspicion) occasionally placing a bet. When he was satisfied with the fit between the computer and the wheel, he would place a bet on even; this was a signal to his accomplice, the bettor, that it was time to begin playing. Each time the croupier spun the wheel and threw the ball, the data taker would clock the coordinates with his toe and the computer would calculate the octant of the wheel in which the ball would likely land. The information would be received by the bettor's shoe, which was equipped with three solenoids that vibrated at three different frequencies. It was these patterns that told the player how to bet. *High, middle, low* might be a signal to place a bet on octant two, *high, low, low* on octant seven.

Living for days at a time in run-down motels in Las Vegas, the Santa Cruz players won enough money to show that some of the randomness could be squeezed out of roulette. But they never earned enough to justify becoming full-time gamblers. Though they believed their system was the best yet developed, they found, like Shannon, that it was not a straightforward matter to go from laboratory to casino. Loose wires and software bugs caused the miniature computers to crash. Shoe-to-shoe signals were scrambled by the electromagnetic glitter of Las Vegas's lights. And then there were the vagaries of the wheel itself. Over the long run, random fluctuations in the behavior of ball and rotor could be counted on to even themselves out. But how long would the long run be? Flip a perfect coin a hundred times and you will get half heads, but along the way there might be a stretch of tails that never seems to end. In roulette, chance deviations like these might burn up your bankroll, leaving you with no capital to keep playing, no way to win your losses back.

There were also problems that had nothing to do with physics and mathematics. The statisticians for the casinos calculate odds very carefully. Every game is designed to give the house an edge of several percentage points—not enough to discourage players but enough to ensure a one-way flow of wealth. The opulence of Caesars Palace and every casino on the strip testifies to the fact. A player who is acting suspiciously or enjoying an unusually long winning streak attracts the interest of the powers that be and can expect to be treated in a less than friendly manner. Crutchfield calls this "the big, burly man effect." Eventually the co-ercion was made official: roulette computers were banned from Las Vegas casinos.

The veterans of the roulette project still speculate on what one might do with today's technology — their miniature computers used only four or five kilobytes of memory compared with the megabytes now readily available. But before they were done with college, they found themselves preoccupied with other kinds of randomness. In trying to beat roulette, they concentrated on what might be called the randomness of ignorance, the hope of getting an edge on chance by gathering information. But the universe also seems to be imbued with randomness that is inherent, that cannot be expelled by any amount of knowledge.

We can imagine our all-powerful intelligence, the camel, keeping track of all the balls bouncing around in the bingo machine and calculat-ing which will pop up through the chute. If the balls were ejected me-chanically, say every forty seconds, then this interval could be factored into the calculations. But the machine is not automatic. A human decides when to press the button. Could the camel conceivably predict when each decision will be made? It is clear that the emcee picking the balls and calling out the numbers operates within certain constraints: he is under pressure not to pick numbers so quickly that players will not have time to mark their cards, nor can he move too slowly. He works for the house, and the house wants to keep winnings at a respectable minimum— enough to encourage people to keep coming back but not enough to erode profits. The house too is buffeted by chance, though the game wouldn't exist unless the odds were stacked in its favor. One way to pre-vent too many bingos is to call numbers so fast that people's information-processing skills are overwhelmed. This is also a reason for changing the pattern that counts as a bingo from game to game—it adds variety, alle-viates boredom, but also confuses people. Hidden within the cards are bingos that are never called.

One could probably come up with an ideal rhythm that would guar-antee, on average, a certain return for the house. But no matter how closely the emcee watches the clock, he cannot issue a uniform stream of numbers. As Farmer and Packard discovered with roulette, the best-honed human reflexes can only be so accurate. The human nervous system is plagued with noise. No matter how hard an emcee tried to push the button at the right instant, a random twitching of a neuron might throw off his timing.

But couldn't this randomness be expunged? Suppose the camel monitored the emcee's nervous system and used the laws of physics to predict the flow of ionic currents and neurotransmitters that govern the behavior of every neuron. While the camel was tracking bouncing balls in the chamber, it would also track bouncing electrons in the brain. And that is where the laws of quantum mechanics come in, declaring that the behavior of a single subatomic particle is essentially undecidable. Given a large number of electrons, we can predict their average behavior. But the twitching of a single electron is essentially random. The problem is not the clumsiness of the experimenter. No matter how precise and delicate the instruments we measure with, the mathematics of quantum theory dictates that there will always be a residual amount of randomness that no measurement can expel. If reading the mind of the bingo operator requires monitoring his electrons with precision, then quantum mechanics will defeat the most exacting clairvoyant.

Up here in the macroscopic world of bingo and roulette, we can usually ignore quantum uncertainty. With large objects, the random fluctuations average out so that reality can be approximated as a game of cause and effect. But at about the time Farmer and Packard were strolling into casinos wearing electronic shoes, they and two of their fellow graduate students, Robert Shaw and Jim Crutchfield, were becoming entranced by what seemed to be yet another kind of randomness. Chaos, as it has come to be called, manifests itself even in seemingly simple equations describing familiar events in the macroscopic world. Unlike the randomness of ignorance scientists had been accustomed to, chaos was not caused by the interaction of too many variables for a human mind to monitor. Under the right conditions, it seemed, the simplest systems could generate randomness that was inherent and inexpungable.

At their root, chaotic systems marched to the laws of Newtonian physics, not quantum mechanics. But unlike other deterministic systems, chaotic systems exhibited what physicists called "extreme sensitivity to initial conditions." Encountering a rock, a stream of smoothly running water suddenly shatters into turbulence. Place a cork upstream and it will travel what appears to be a random path. Repeat the experiment, placing the cork as close as humanly possible to the same starting position, and it will follow a completely different trajectory. The slightest uncertainty in the initial conditions is amplified so rapidly that prediction becomes impossible. As a mathematician would put it, the trajectories diverge exponentially. In a linear relationship, $10x = y$, the output rises slowly as x increases from 0 to 6: 0, 10, 20, 30, 40, 50, 60. But consider what happens when we feed the same input to an exponential equation: $10^x = y$. The value of y starts at 1, then, as x increases, jumps to 10, 100, 1,000, 10,000, 100,000, 1,000,000. Faced with a chaotic system, in which tiny perturbations are amplified exponentially, even the perspicacious camel would have to know the initial conditions with infinite accuracy—to the nth decimal point—to foresee what it will do.

Even more striking was the fact that a system need not be as complex as a turbulent stream to exhibit chaos. In fact, mathematicians showed that equations simple enough to solve on a pocket calculator could be chaotic: change the numbers you plug in by the tiniest amount and the output will be wildly different.

Chaos is an example of what physicists call nonlinear phenomena. In a linear equation, a change in the input causes a proportional change in the output. Think of a seesaw. Push the left-hand side of the board down an inch and, depending on where we place the fulcrum, the right-hand side will move up half an inch, one inch, two inches. Input and output are locked into a linear relationship. But suppose the seesaw is made of stiff rubber. Push down slightly on the left-hand side and the board begins to bend, absorbing the motion; the right-hand side doesn't move at all. Push harder and the right side will begin to move, but not in the straightforward manner we saw before. Vibrations might cause the output to wiggle up and down, the frequency depending on how hard and how quickly we push, and on the stiffness of the rubber. If we happen to hit the system's resonant frequency, vibrations from the output might feed back to the input, completing a mechanical circuit that causes the board to vibrate harder and harder. In a nonlinear system there are many, many factors to take into account.

Like a seesaw, an electronic amplifier is a kind of lever. The input, a feeble sine wave generated by the vibrating string of an electric guitar, is leveraged into a sine wave large enough to drive a loudspeaker. If the circuit is adjusted correctly, the relationship will be linear: the shape of the output is the same as the input, only bigger. But if we drive the amplifier too hard, feeding the input with too strong a signal, the resemblance breaks down. Instead of smooth undulations we get waves with their tops and bottoms clipped off; our once pure sine wave is entangled in secondary waves, called harmonics, that beat at frequencies that are multiples of the one we fed in. The result is cacophony. From our speaker we hear a wild buzzing sound. The amplifier has gone nonlinear.

In the abstract world of mathematics, an equation can be purely linear or nonlinear. In the real world, linearity is always an idealization. The rigid wooden seesaw also bends and vibrates, but the nonlinearities are dampened enough that we can ignore them. In most of life, however, it is nonlinearity that dominates.

The Santa Cruz gambling system worked as well as it did because roulette is a fairly linear game. One can never click one's big toe at the precise instant that the ball or the rotor passes a mark on the base of the wheel. The ball is not perfectly spherical, the mark not perfectly even. If roulette were chaotic, then these imprecisions would be amplified exponentially. Within seconds the ball in the computer and the ball in the real world would wildly diverge. The Santa Cruz researchers did uncover some nonlinear effects: When a ball leaves its track, it might strike one of several small diamond-shaped reflectors on the rotor. Tiny differences in the position and velocity with which the ball strikes a diamond will be exponentially amplified. But these nonlinear fluctuations were not enough to swamp predictability.

The game of pool, or pocket billiards, is a very different story. Each time two balls collide, their spherical surfaces amplify uncertainty. Suppose we rack up the balls into the customary triangle on the felt and hit them with the cue ball. The balls bounce off the edges of the table, against one another, and we record the configuration in which they land. Now we repeat the process. But no matter how hard we try to position the balls in exactly the same location and to strike the triangle at the same place with the same force from the same angle, the outcome will be completely different. With each collision, the tiniest differences in the initial conditions will be amplified and amplified again. Two balls that collided in the first round might miss each other entirely; a ball that previously went on to collide with two other balls might be taken out of play when it falls into a side pocket. In fact, as they began to explore the intricacies of chaos, Crutchfield calculated that one minute after the opening shot, a game of pool will be significantly altered by a discrepancy in the initial conditions as small as that caused by the gravitational pull of an electron at the far edge of the Milky Way. Suppose that the electron shifted position, randomly, because of a quantum fluctuation. This seemingly negligible effect would be amplified until it had an impact on the game. Chaos, it seemed, was an amplifier of uncertainty, providing a way that quantum fluctuations could rise from the subatomic substrate and manifest themselves in the Newtonian cosmos.

Chaos makes perfect predictability impossible. Even our all-knowing camel would be defeated by uncertainty—quantum and otherwise—amplified again and again by the chaotic collisions of the bingo balls. But for those compelled to seek order in the world around them, chaos also held out cause for hope. Even very simple systems could be chaotic. So if you looked hard enough, behavior that seemed random might turn out to be generated by a few simple equations. Unlike the pure randomness of quantum mechanics, chaos displayed a hidden harmony.

As members of what came to be known as the dynamical systems collective, or the chaos cabal, the four aspiring physicists began to look for chaos in the world around them, finding it in things as simple as the flapping of a flag or the dripping of a water faucet. Turn on a tap slowly and the water drips rhythmically. Open it a little more and you might hear what mathematicians call period doubling: drip-drip, drip-drip, drip-drip. Open the tap a little more and the period doubles again and again until at some point all semblance of pattern disappears into a random, never repeating arrhythmia. Using strobe lights, cameras, microphones, and photocells, the members of the group studied the dripping faucet as assiduously as they had studied the roulette wheel. Once they found a simple set of equations that produced rhythms resembling those they had recorded from the tap, they began to analyze them.

They were looking for strange attractors. Mathematicians had shown that for all their apparent randomness, chaotic systems were steered by these mathematical objects. Acting like a magnet, a strange attractor imposed a rough shape on the randomness. A system might be disorderly, but only certain kinds of disorder were allowed.

In simple systems, attractors are easy to visualize. The classic example is the pendulum. If we graph its motion, plotting position on the horizontal axis and velocity on the vertical axis, the bob will trace out a circle as it swings back and forth. As friction slows the pendulum, the circle will get smaller and smaller, spiraling into a point. The point is said to be the attractor of the system. It exists in what mathematicians call phase space. More complex systems are pulled by more complicated attractors. A doughnut-shaped torus, for example, will set a system spiraling in two directions: circling around the circumference of the doughnut and winding along its inside.

The more structure and pattern to the attractor, the more interesting the behavior of the system becomes. The most complex of attractors, the strange attractors, spiral and fold in on themselves, over and over, leaving a scribble of loops that is infinitely dense with structure. They are what mathematicians call fractals. If we could take a small piece of the attractor and magnify it, we would find the scribbling repeated on a smaller scale. If we took a piece of that pattern and magnified it, we would find tinier patterns, and so on, ad infinitum. A system ruled by a strange attractor will dart this way and that, probably never repeating itself in the lifetime of the observer, or perhaps the universe.

The members of the Santa Cruz collective found that if they took the simulated drips generated by their model and graphed them in just the right way, a strange attractor emerged like a face peering from behind clouds. Mathematicians had long been adept at constructing strange attractors on paper and showing that they generated chaos. By the time they had earned their doctorates, all in the early 1980s, the members of the chaos cabal had developed a technique for taking seemingly pattern-less data and reconstructing the attractor. Lawful chaos could sometimes be sifted from what otherwise appeared to be random noise. From the complexity, or dimensionality, of the attractors—measured using fractional dimensions, or fractals—one could gauge just how haphazard a chaotic system would be. For systems with low-dimensional, relatively simple attractors there would be hope of rough prediction. Failing that, one could at least admire the beauty of the geometry underlying what had seemed on the surface to be pure happenstance. One could satisfy the human longing for buried symmetries shattered giving birth to the real world.

From contemplating bingo, roulette, and dripping water faucets, one might conclude that there are three kinds of randomness in the world: randomness due to complexity (there are just too many factors for our poor brains to comprehend), randomness due to chaos (the underlying system is simple but is ruled by a strange attractor), and randomness due to quantum uncertainty, which can never be overcome. Faced with something we do not understand, we can try to supplement our brains with computers, as the roulette players did, and search for relationships evolution has not made us acute enough to see. Or we can search for hidden order, in the form of a strange attractor. But most of the time when we look at the world we cannot distinguish between randomness and order that is simply too deep for us to fathom. How deep should we dig for an attractor? If we scrutinize a little longer, will a pattern emerge? Much of what seems orderly to most of us must seem random and arbitrary to a dog, or for that matter to a child or a retarded person. In this sense randomness really is in the eye of the beholder. As Doyne Farmer once put it, "Roulette appears random unless you happen to have a computer in your shoe." But we can imagine creatures born with the neurological equivalent of computers in their shoes, with brains so powerful that they can see orders invisible to us.

Suppose we receive electromagnetic waves from space and translate the fluctuations into numbers: 1111317739. Signal or noise? If, try as we might, we can't find a rule for generating this string, we would conclude that the sequence is apparently random and not an alien transmission after all. But are we missing a hidden pattern? Perhaps the sender is much more intelligent than we are. A mathematically inclined sixth-grader, studying circles and radii, might recognize this signal from geometry class: 314159. Perhaps beings with more powerful brains would recognize the earlier string as part of a familiar landscape way out in the hinterlands among the endless digits of pi.

The difference between the digits of pi and what we take as a truly random number—generated perhaps by the flipping of a perfectly balanced coin or the neutrons shooting quantum-mechanically from decaying uranium atoms—is that pi can be produced by a simple process: dividing a circle's circumference by its diameter. The infinite information in the decimal expansion of pi can be compressed into a compact algorithm, a program that will generate the number. Likewise the sequence 1,2,3,4,5,6,7,8,9,10,11,12 . . . can be compressed into a simple rule or computer program, as can the sequence 121212121212 . . .

Suppose we are given digits caused by the random rolls of a die: 2663165611 . . . There is an infinitesimal chance of rolling a compressible number. But generally there will be no way we can express the number we generate in shorthand. The smallest algorithm that will spit it out of a computer is one that says "Print 2663165611 . . ." The program for generating pi is infinitely smaller than pi itself. The program for generating a random number is longer than the number itself: it contains every single digit plus the instruction to print them. A random number, in other words, is its own shortest description.

In this sense, a random string can be said to be more complex than pi;

it takes vastly more information to express it. In fact, the American mathematician Gregory Chaitin and the Russian Andrei Kolmogorov define the complexity of a number as the length of the shortest computer program that will spit it out. (While in many ways this is an intuitively satisfying definition, it leaves us with the curious conclusion that random systems contain more information—and are more complex—than orderly ones. As we will later see, this algorithmic information is often contrasted with Shannon's information theory, in which orderly, unlikely systems are said to be higher in information content than disorderly ones.)

Given Chaitin and Kolmogorov's definition, the question of whether bingo is at all predictable comes down to this: How much can we compress the string of numbers that emerges from the bingo machine? Examining the data from a typical game, we would find some patterns immediately evident: the numbers 1 to 15 are always associated with the letter B, 16 to 30 with I. The letters are just a convenience, so one need not scan the entire bingo matrix looking for number 44—it will always be in the N row. The letters are redundant information. Eliminating them, we can immediately compress the string by half. Whenever we wish, we can take the compressed string and restore it with a simple rule.

But this won't help us much. Are there less trivial compressions in the numbers, significant departures from randomness? Perhaps some balls are the slightest bit heavier than others. It takes less paint to make the letter I than the letter O. O's are also accompanied by heavier, inkier numbers: 60s and 70s. Suppose that factors like these conspired to make some letters more likely to appear than others, just as the tilt of the roulette wheel favored certain octants. By analyzing the output of the bingo machine we might detect these regularities. A tiny bit of randomness would be squeezed out of the game. Most of the time we look in vain for such compressions. But there is always the hope that if only we can squint our eyes and focus a little harder we can see what no one has before.

At some point our brains will bump up against unsurpassable limits. Chaitin proved that no program can generate a number more complex than itself, "any more than a one-hundred-pound pregnant woman can birth a two-hundred-pound child." Conversely, he showed that it is impossible for a program to prove that a number more complex than itself is random. If the mind is a kind of computer, then we are left to conclude that there is complexity so deep that we can never penetrate it; order that will always look to us like randomness. If we could somehow prove that the phenomenon was random, then we could accept our limits and give up the quest. But even that is impossible. There is no way to know for sure if we are overlooking a subtle compression. So we are doomed to keep searching, never knowing if there is anything to find.

Chaitin's discovery is not just of interest to gamblers and mathematicians. For what is science if not the search for compressions, for simple rules? We gather data—the motion of the lights through the heavens—and we make compressions: Ptolemy's epicycles, Kepler's laws. But more often we stare at the data, casting it this way and that way, searching in vain for underlying simplicity. Is the system random or simply more complex than we can fathom, no matter how powerful the computers we put in our shoes? Whether the signals we scrutinize come from an atom, a star, or a bingo machine, we can never know when we are up against the limits of our power to compress.

In fact, from the point of view of the mathematician, we are far more likely to find randomness around us than order. All but a tiny, insignificant fraction of the real numbers are random, consisting of an infinite sequence of digits that never repeat. They are incompressible, their own shortest descriptions. So, we're left in a curious world where most numbers we encounter are random, but where we can't count on proving whether an individual number is random or not.

The randomness seems to run even deeper than that. The mathematician Kurt Gödel proved that any mathematical system of sufficient power contains statements that it cannot prove true or false. Chaitin went further, showing that there are an infinite number of statements one can make about arithmetic that are incompressible—i.e., they cannot be understood by reducing them to simpler truths. There is no method to prove them true or false; the best we can do is flip a coin. As Chaitin sees it, this is tantamount to saying that the structure of arithmetic is random. "God not only plays dice in quantum mechanics," he said, "but even with the whole numbers."

Gambling and science: what it all comes down to is finding compressions. Sometimes the compressions elude us and we see randomness. But sometimes we imagine compressions that aren't really there. In the bingo hall, we might notice that the last three times we won a small stake we were using a pink marker to cross off the numbers on our cards. So, from now on, we make a point of marking our cards with the magic color. After weeks with no luck, we might switch to green, which the winner next to us seems to have prospered from. Or we might look deeper and infer that pink works best on Saturdays, green on Sundays. Or does the position of the moon have an effect? When pigeons are put in Skinner boxes and given random reinforcement in the form of pellets of food, they develop ritualistic bobbing and weaving motions, as though they were trying to will the gods to deliver good fortune.

Our own false compressions—we call them superstitions—can be as simple as carrying a rabbit's foot or wearing a lucky fishing hat. Some of the New Age followers in Santa Fe insist that wearing a crystal with just the right resonant frequency will divert spiritual and material riches your way. Adorn your life accordingly and you will be in harmony with the heavens. With the proper crystal you can tune out chaos and listen to the music of the spheres.

Once its members had earned their degrees, the Santa Cruz collective broke up. In the early 1980s, Packard moved on to the Institute for Advanced Study in Princeton, New Jersey, and then to the Center for Complex Systems Research at the University of Illinois. Farmer and his future wife, Letty Belin, an environmental lawyer, moved to a house on the Tesuque River, not far from the camel and the Tesuque bingo hall. Commuting to Los Alamos' Center for Nonlinear Studies, Farmer explored systems as complex as the immune system and as seemingly random as the stock market. Feeling uncomfortable at times about working for what began strictly as a weapons lab, he customarily ended his papers with a disclaimer: "I urge the reader to use these results for peaceful purposes."

In 1991, Farmer's and Packard's paths converged again when they formed the Prediction Company in Santa Fe, hoping to apply their skills at analyzing complex systems to the financial markets. Most economists hold that the stock market is what statisticians call a random walk. Millions of buyers and sellers, influenced by factors rational and irrational, interact in ways too numerous and intertwined for us or our computers to track. If a company issues a surprisingly bad earnings report, it is likely that the stock's price will fall (though almost as often, it seems, the price perversely goes up). But such simple causal linkages are undependable and rare, feeble signals swamped by a background of noise. In general, a stock's day-to-day fluctuations are said to be as independent as the flips of a coin. We can take the price record for the last year and graph it this way and that, looking for patterns. But nothing in the zigs and zags will foreshadow the future any more than past configurations of heads and tails will predict what the next fall of the coin will be. It is assumed that even if there were reliable signals, they would be self-correcting: if three zigs followed by a zag meant an uptick tomorrow, then so many people would notice this pattern and buy the stock today that the information would already be reflected in the price. In the jargon of the trade, the information would be discounted by the market.

In this idealization, the market is a perfectly efficient processor of information. In the short term, the theory goes, you cannot even profit by following the business news: if a company is about to announce a layoff or a new contract you can be sure that the market has already absorbed the fact; so many people (with better connections than you) have bought and sold on the information that it is reflected in the price.

A curious implication of the random walk theory is that most people must not believe it is true. For if all available information is almost instantaneously reflected in the price of a stock, then someone must be gathering it. No amount of studies from the Sloan or Wharton business schools have kept people from trying to outwit the market, from trying to find compressions in the zigs and zags of the financial charts. If the market isn't perfectly efficient—if there is a lag between the moment when some bit of information is discovered and when everyone knows it— then a limited amount of predictability may be possible. After all, roulette was supposed to be random too.

While most prognosticators, the so-called fundamentalists, study economic data and company financial reports for clues, others study the price fluctuations as diligently as radio astronomers scrutinizing invisible signals from a distant star, searching for the faintest of patterns. To these technicians, or chartists as they are sometimes called, a firm's fundamentals and the economic news are only distractions. They prefer not to know even the name of a company, just the shape of the line plotting the stock's swings in price. Some are so adamant that they try to seal themselves off from the world, like a permanently sequestered jury. Just as all the bits of information about the bouncing of the bingo balls and the construction of the bingo machine, and perhaps even the laws of mechanics, both quantum and Newtonian, are implicit in the stream of numbers called out by the emcee, so, they believe, the crucial information about a stock is reflected in its fluctuations. All that is left is to analyze the record and predict how it will unfold.

Some of the shapes the chartists see may be simple reflections of human nature. A stock may oscillate like a sawtooth wave, rarely dropping below 40 or above 50. In the mental pictures we make, the price is like an elastic ball bouncing off the ceiling and floor of a room that is 10 points high. If the price breaks through 50, some people might take it as a signal to buy—better times ahead. Others, though, may use it as an excuse to cash in their profits. Thus a rough equilibrium is struck.

The most imaginative of the chartists see constellations that have names like those the bingo players look for on their cards: the Head and Shoulders formation, the Rounded Bottom, the Pennant, the Island, the Breakaway Gap. Some swear they see what are called Elliot waves: vibrations and vibrations within vibrations undulating lawfully, carrying riches in their foam. Are these compressions true or false, subtle correlations or superstitions? Sometimes it is impossible to tell. If many investors become convinced that a certain pattern means buy or sell, then it might become a self-fulfilling prophecy: enough people will respond to the signal to move the market. But if the belief becomes rampant, then it will become a self-defeating prophecy: hordes of people buying on Tuesday on the expectation of a stock's rising on Wednesday will even the ripple out.

But what if one used sophisticated mathematical analysis to find patterns too subtle and esoteric for most people to see? Could a strange attractor, for example, be governing a stock's seemingly random fluctuations? If so, the information could be hoarded and exploited. The mathematical elite could skim off the riches before the knowledge disseminated and the window of opportunity slammed shut.

With their studies of the dripping faucet, the Santa Cruz collective found ways to test for chaos and ways to tease out strange attractors from a stream of data. Throughout the 1980s, scientists had polished these mathematical lenses, and a few economists were using them to peer at the stock market. Maybe some of the randomness of the market is chaotic, generated by simple nonlinear equations. If one could see through the noise, there might be orders to exploit, attractors that would give at least some shape to the data.

There are reasons to believe that markets might be analyzed with nonlinear mathematics. If millions of nervous investors are waiting to bail out of a stratospherically priced stock, then the slightest fluctuation can cause an avalanche: sensitivity to initial conditions. Prices are also subject to feedback effects, another source of nonlinearity: a stock goes up and more people buy it, so it goes up faster, attracting still more buyers. It is as though we put the pickup of an electric guitar too close to the loud-speaker: output becomes input becomes output, like images ricocheting between mirrors. The sound becomes louder and louder, driving the amplifier into a dizzying, nonlinear scream.

In the late 1980s, two economists, José Scheinkman of the University of Chicago and Blake LeBaron, who divides his time between the University of Wisconsin and the Santa Fe Institute, found evidence of a strange attractor in stock price data. In 1987, when physicists, mathematicians, and economists gathered at the institute for a conference called "The Economy as an Evolving Complex System," several speakers, including Farmer and Packard, discussed the possibility of detecting underlying rhythms. They might never succeed in finding the generator of the beat, but there was evidence that it at least made sense to try.

Just southwest of Santa Fe's plaza, the spirit of the roulette project and the Santa Cruz chaos collective has been reborn at the Prediction Company. Staked by investors in Santa Fe and Chicago, the company began using computers to try to predict the currency exchange market, which Farmer and Packard have reason to believe is less noisy than the stock market. They automatically track hundreds of financial variables—the Dow Jones Industrial Average, the Standard & Poor 500, the price of the dollar, the price of the yen, the price of pork belly futures and stock options. Each can be thought of as an axis on a graph: plotting the points yields strange shapes in multidimensional space. If there are any low-dimensional attractors lurking within these hyperlandscapes, they have remained elusive, but Farmer and Packard hope there are other patterns too evanescent for the chartists to see—and that they can sift them out by supplementing their brains with different brands of artificial intelligence. Neural nets, programs that simulate simple webs of neurons, have been shown to be adept at recognizing some patterns. At the Santa Fe Institute, followers of John Holland, a University of Michigan computer scientist, are studying ways to use "genetic algorithms" to evolve software that would be too complex for a human to design. Borrowing from the tenets of Darwinian evolution, the scientists write programs in which strings of computer code are randomly mutated and made to compete against one another until a winner emerges: the program best able to perform the task. Set loose a horde of genetic algorithms on market data and maybe some will evolve the ability to find pockets of predictability. Using these and other techniques, Farmer and Packard hope to realize the dream they had in Santa Cruz: to make enough money to separate themselves from the establishment and pursue knowledge on their own. If the patterns they find are real and not statistical illusions, they would only have to be right a small percentage of the time. With enough money to bet, even the slightest edge could make one wealthy.

In time, others similarly equipped can be expected to find these hidden orders. The window of predictability will close and other, more subtle orders will have to be found. One can imagine a financial arms race as more and more powerful computational lenses are trained on the data, gazing for a hidden message that might generate profits for a week or two. "The question," Farmer said, "is whether we can evolve and stay ahead of our competitors."

In the end, all of our searches for order are fated to bottom out in randomness. No matter how far we go in compressing the signals pouring in around us, there will always be some irreducible uncertainty. With chaos, we might at least get a loose grip on the disorder by finding a low-dimensional attractor. With quantum randomness, not even that much control is possible.

Something about the mind, wired to find patterns both real and imaginary, rebels at this notion of fundamental disorder. A few scientists, like Farmer and Crutchfield, have come to question whether quantum randomness is truly impenetrable. After all, Chaitin showed that we can never know if we have achieved the ultimate compression: there may be hidden redundancies to squeeze out. How then can we know that quantum randomness is inherent? Neutrons may seem to fly out of decaying nuclei with no pattern. But maybe we just can't see the compressions. To most physicists, this is the worst kind of heresy, like trying to show that the earth is really at the center of the universe.

Still, against all odds, Farmer can't help but wonder. If we "peel back the next layer of the scientific onion," he ventures, perhaps we will see that quantum indeterminacy arises somehow from chaos, that there is a method to the madness. Like Einstein, he doesn't want to believe that God plays dice.

PART TWO

"THE COLD, GRAY CAVE OF ABSTRACTION"

Why is it that showers and even storms
seem to come by chance, so that many people
think it quite natural to pray for rain or fine weather,
though they would consider it ridiculous to ask
for an eclipse by prayer?

—*Henri Poincaré,* SCIENCE AND METHOD

4

THE DEMONOLOGY OF INFORMATION

In the beginning, the main route connecting Santa Fe to the rest of the known universe was the Camino Real, the royal highway that ran up from Mexico City, meeting the Rio Grande at El Paso and following it northward through Albuquerque, Santa Fe, and on to the hinterlands of New Spain. Today the American portion of the Camino Real has been replaced by Interstate 25, but the scenery along the route remains pretty much the same. Those who fly into Albuquerque International Airport and drive north for a scientific conference in Santa Fe or Los Alamos pass through a stark landscape very much like that the Spanish conquistadores saw.

To the east, as one leaves the suburban sprawl of the Albuquerque metropolitan area, the Sandia Mountains rise nearly six thousand feet above the already mile-high terrain, exposing a rocky facade so fractured and so sheer it looks as though half the mountain has been sliced away. In a sense that is what happened. The Sandias are an example of what geologists call a fault-block mountain. Like the Sangre de Cristos they were squeezed from the earth when two continental plates collided, but in the case of the Sandias, one side collapsed; instead of a slope, the western face of the mountain is a bare, almost vertical expanse of steep granite walls. The most prominent of these is the Shield, so formidible, the guidebooks say, that some of its more onerous ascents can take days of hard climbing, the nights spent roped to the cliff like a tent worm, trying to fall asleep on vertical ground.

To the west, beyond a line of dormant volcanoes, one can barely see Mount Taylor, a jagged blue bump on the horizon that was named after General Zachary Taylor, after he took this land from the Mexicans in the War of 1846. The Mexicans, and the Spanish before them, called the mountain Cebolleta, "Little Onion." They took it from the Navajos, who still call it Turquoise Mountain and consider it the southern border of their universe and the home of Monster Slayer, one of the legendary Hero Twins who fought against the evils of the earth. Drive west from Albuquerque on Interstate 40, old Route 66, and just before Grants, a mining town that in better times billed itself as the Uranium Capital of the United States, you cross over the petrified bubbles of the Malpais ("Bad Land") lava flow; the Navajos say it is the dried blood of Ye-itsa, one of Monster Slayer's victims. Ye-itsa's head can be found to the north in the form of an old volcanic plug with sloping shoulders that the Spanish named Cabezón Peak. (*Cabeza* means head, and a *cabezón* is one that is particularly big and ugly.) Ye-itsa's bones (the geologists say they are petrified trees) lie as far east as Albuquerque. Though Ye-itsa was killed and turned to stone, some of the other monsters survived, the legend goes. Demons called hunger, greed, filth, and old age still stalk the land.

The pueblo Indians included the Navajos among the monsters and still remember the stories of their raids on the adobe villages that lie between Albuquerque and Santa Fe along the Rio Grande—Sandia Pueblo, Santa Ana, Santo Domingo, San Felipe, Cochiti, little worlds with their own languages and, like the Tewa pueblos to the north, their own quartets of magic mountains marking off their personal universes. The landscape on this part of the journey is like nothing else on earth. Far to the west the Jemez Mountains reach toward the river with fingers of lava, hardened into the black, flat mesas that, to use another metaphor, look like frozen breakers of stone. The turnoff to San Felipe, a hive of adobe houses shaded with cottonwoods hunched against the base of one of the larger mesas, marks the halfway point of the drive to Santa Fe. A few miles later, just after the highway crosses the dusty arroyo known as the Galisteo River, a steep volcanic wall looms into view. The Spanish called it La

Bajada, "The Descent," though when one is driving up from Albuquerque it is quite the opposite, an eight-hundred-foot rise that divides the lower country of southern New Mexico from the highlands of the north. Until this point the highway has been cutting across what the Spanish cartographers called Río Abajo, "Lower River," the part of the northern kingdom that lay closest to Mexico City. In those days of horses and wagons, La Bajada was known for the treachery of its hairpin turns—the price one paid for entering another realm: Río Arriba, "Upper River," the vast, barely explored region that extended north of La Bajada and then off the top of the maps.

It is fitting that La Bajada was named from the Río Arribans' point of view. Sitting in their perch seven thousand feet above sea level, the people of Santa Fe and beyond literally and sometimes figuratively looked down on their neighbors in Río Abajo. Except for a gradual rise to reach the top of the La Bajada hump, it was a two-thousand-foot slide from Santa Fe to Albuquerque. The change wasn't simply one of geography. La Bajada was, and is, a psychological and a cultural divide. Though southern New Mexico has its share of mountains, it is largely a flat, desert land whose subtle beauty requires the heart and eye of a connoisseur. There is nothing subtle about the topography of northern New Mexico. Once you ascend La Bajada, with the Sangre de Cristos looming straight in front of you and the carved symmetrical volcanoes rising from either side of the highway, Río Arriba opens up all around. You know you are in another country, where even the light seems changed.

It wasn't easy, going between the land of the familiar and the land of the strange. Wagon drivers coming down La Bajada often had to brace their wheels with rocks to keep from succumbing to the force called gravity. Cars heading the other way sometimes had to back up the hill, reverse gear providing them with more leverage, as their boiling radiators protested against the heat. Today the endless turns have been straightened into a more gradual ascent; cars and trucks barely slow down as they surmount the divide. But they are still bound by the same laws of physics that held sway in the conquistadores' time. Then and now, it takes energy to cross the divide.

In May 1989, some three dozen scientists, mostly from the United States but a few from as far away as Germany, Britain, France, Israel, and Japan, flew into Albuquerque and boarded rental cars and shuttle buses for the journey up the Camino Real. Skirting the edges of the pueblo universes, they ascended La Bajada, arriving in Santa Fe for a conference sponsored by the Santa Fe Institute and held in the spectacular setting provided by St. John's College, which sits at a confluence of arroyos that cut through the foothills of the Sangre de Cristo Mountains. Hike three miles up the canyons from St. John's and you reach Atalaya Peak. *Atalaya* means "watchtower," and if you stand on its heights and look down on Santa Fe and across to Los Alamos you will be seeing what may be the world's largest concentration of scientists (granted, there aren't many) working in a new field called the physics of information, which sits at the boundary where mind and nature, subject and object, seem to collide.

In some ways, St. John's seemed an incongruous setting for a conference on so revolutionary a subject as information and physics. The school is known for its classical curriculum: students learn physics by starting with the pre-Socratics, then moving on to the more recent ideas of Plato and Aristotle. The physicists and mathematicians were coming to St. John's to discuss ideas at the very edge of twentieth-century science. They were responding to a manifesto with the intriguing title "Complexity, Entropy, and the Physics of Information," which had been dispatched by Wojciech H. Zurek, a Polish-born physicist who works at the Los Alamos National Laboratory's Theoretical Astrophysics Group.

In building a tower of abstraction, one must start with a foundation, those things that are taken as given: mass, energy, space, time. Everything else can then be defined in terms of these fundamentals. But gradually over the last half century some scientists—and Zurek was among the most adamant—had come to believe that another basic ingredient was necessary to make sense of the universe: information. "The specter of information is haunting the sciences," his manifesto began. There is a "border territory," he believed, where information, physics, complexity, quantum theory, and computation meet. So in another way, St. John's wasn't so strange a setting for the conference after all. What Zurek and his colleagues had in mind was a return to basics, a rethinking of reality's pillars as thorough as any undertaken by Thales, who thought all was made of water, or Heraclitus, who thought all was made of fire.

Most of us are used to thinking of information as secondary, not fundamental, something that is made from matter and energy. Whether we are thinking of petroglyphs carved in a cliff or the electromagnetic waves beaming from the transmitters on Sandia Crest, information seems like an artifact, a human invention. We impose pattern on matter and energy

and use it to signal our fellow humans. Though information is used to de
scribe the universe, it is not commonly thought of as being part of
the universe itself. But to many of those at the Santa Fe conference, the
world just didn't make sense unless information was admitted into
the pantheon, on an equal footing with mass and energy. A few went
so far as to argue that information may be the most fundamental of all;
that mass and energy could somehow be derived from information.

There was, first of all, the mysterious connection that seemed to exist
between information, energy, and entropy, the amount of disorder in a
system. We learn in school that, left on its own, any closed system becomes more and more disorderly; its entropy increases. It is because of
this fact, embodied in the second law of thermodynamics, that neat geological strata become gnarled into formless Precambrian rock. The planar geometry of an adobe village melts until it is barely distinguishable
from the surrounding hills. Along the way, pattern is washed away; information is lost. Information can be thought of as a measurement of distinctions, the simplest being 1 or 0, the presence or absence of a certain
quality. By this measure, there is more information in something that is
orderly than in a homogeneous, undifferentiated mess.

On the other hand, by gathering and processing information, we can
create order—we can take the matter and energy of our world and
arrange it into songs, civilizations, fragile eddies in the entropic tide.
Using our powers as information processors, we can find unlikely structures that already exist—water trapped in a mountain lake, carbon molecules strung in a volatile chain, protons and neutrons stacked into a
precarious nuclear sphere. And then we simply let them follow the path
of least resistance. As they topple and move down the hill from order to
disorder, we can extract work by harnessing the entropic flow. The nucleus disintegrates, the bonds of the carbon atoms break, the water flows
from its pool to the formless sea. Entropy increases, information is lost,
but the energy released in the process can be tapped to build new structures, to create information, though all our creations must eventually
succumb to the second law.

No wonder the mind craves patterns. It is the ability to find order in
the world that allows us to make use of its resources. For many scientists
this would be reason enough to believe that information is fundamental.
But, going beyond the laws of thermodynamics, some believe information plays an even deeper role. According to some interpretations of
quantum theory put together by Zurek and his circle, without informa-

tion there would be no resources to exploit and no one to exploit them; there would be nothing that resembled what we call the real world. The mathematics used to describe the subatomic realm tells us that, left to its own devices, an electron lacks the very attributes that we, on our macroscopic plateau, consider the very hallmark of existence—a definite position in time and space. It exists, we are told, as a probability wave, a superposition of all the possible trajectories that takes on substance only when it is measured, when, as it is often put, an observer collapses the probability wave. How this transformation occurs is one of the deepest mysteries of physics, the so-called measurement problem: How does the rock-solid classical world, in which things occupy definite positions in space and time, crystallize from the quantum haze? In the past, quantum theory has often been embraced by those who would elevate subjectivity over objectivity, championing a mystical world view in which consciousness brings the universe into being. By making information fundamental, Zurek and some of his colleagues hoped to demystify quantum theory. For what is an observation but a gathering of information? And if information is fundamental, it exists as surely as does matter and energy, without the need of conscious beings. The quantum wave might collapse not because it was beheld by a mind but simply because information flowed from one place to another in the subatomic realm.

Of course, it is easy to be fooled by our own metaphors, becoming so dazzled by the concepts we invent that we can see the world only through their glare. In the nineteenth century, entropy and the laws of thermodynamics were invented to deepen our understanding of the steam engine and make it as efficient as nature would allow. Any closed system, sealed off from its environment, would inevitably march from order to disorder. Soon scientists and philosophers were applying these new mental tools to the universe itself, declaring that, as the most closed of closed systems—what could possibly be outside of it?—the universe was marching inevitably toward thermodynamic death, a state of equilibrium, lifeless, unstructured, random. In the twentieth century, information theory was invented to help engineers make electronic communications channels as efficient as possible. And before we knew it, people were speaking of information as real, a few going so far as to imagine that we live in a universe of computation, created from the shuffling of bits.

One of the challenges implicit in Zurek's manifesto was to find new ways to think about whether computation—and therefore information—is natural or artificial. The computers we have built over the years have been crafted from macroscopic parts: first gears, then vacuum tubes, then transistors, and now chips inscribed with thousands of transistors that get smaller and more densely packed every year. We stamp our designs on nature's designs; circuitry onto silicon lattices. But the finer the blueprints of our artifices, the more they begin to clash with the physics underneath. Quantum randomness scrambles our neat choreography of 1s and 0s. But perhaps as engineers reach tinier and tinier scales they can somehow exploit the natural behavior of atoms to make their machines more efficient, bridging the divide between the circuitry we design and the "circuitry" of nature. An atom with an electron that could be in one of two states might naturally be thought of as a register containing a 1 or a 0. How thin can we make this gap between the laws of computation and the laws of physics? Where will the shrinking bottom out? If computation can take place only down to a certain scale, requiring components made up of many, many molecules, then perhaps information is simply an artifice, secondary to the laws of physics, a pattern imposed by people as they struggle to describe the world. But if single molecules or even atoms can be said to somehow process information, then maybe computation is as fundamental as what we think of as the laws of physics. Like mass and energy, information would be irreducible, at the roots of creation.

For many of the people who gathered in Santa Fe to talk about information, thermodynamics, and quantum theory, this would be the first of many visits to northern New Mexico. Another conference followed a year later, this one at the Santa Fe Institute, which was then housed in an old convent among the galleries and adobe houses on Canyon Road. In a way, though, the first conference never really ended. Over the years, the physics of information group Zurek started at the Santa Fe Institute has attracted a changing cast of visitors. Rolf Landauer and Charles Bennett, two of the first people to make a connection between physics and information, visit often from the IBM Thomas J. Watson Research Center in New York. At his retreat in Tesuque, a rural village that provides refuge for those who find even Santa Fe's slow pace too frenetic, Murray Gell-Mann and his frequent guest James Hartle, of the University of California at Santa Barbara, try to use information to make sense of quantum cosmology, in which the whole universe can be thought of as a quantum probability wave.

As one listened to these scientists' lectures, read their papers, and spoke to them privately, at dinner or in hikes through the mountains, it was hard not to be struck by hints of an even deeper purpose to their travails. The physicists at Santa Fe were not simply doing science. In this land where so many people see the universe in so many different ways, they were examining the very nature of the scientific enterprise, this curious drive we have for gathering bits and weaving them into pictures of the world.

In fact, to some of the visitors making the drive up La Bajada to discuss their ideas with colleagues in Los Alamos and Santa Fe, it has become natural to think of information as the fuel that, quite literally, takes them over the divide. During one of the Santa Fe Institute conferences, Charles Bennett of IBM declared that given a long enough memory tape—a blank string to be filled with 1s and 0s—he would have all the energy he needed to drive from Albuquerque to Santa Fe. Several years later, at a conference in Dallas called "The Symbiosis of Physics and Information," Bennett said he couldn't recall making the statement, but that it was not one with which he would disagree. "It's what I believe," he said. "It definitely sounds like something I would say."

To all but the handful of initiates, it sounds impenetrably mysterious, this notion that information and energy could be somehow intertwined. To understand what Zurek, Bennett, Landauer, and their colleagues have in mind, one must become submerged in a way of thinking and carving up the world that has its origins in the late nineteenth century, when James Maxwell tried to pick open a loophole in what was thought to be an unassailable universal law. In 1871, several years after inventing the equations braiding together electricity and magnetism, Maxwell publicly introduced, in his book *Theory of Heat,* an imaginary imp, later to be dubbed Maxwell's demon, that seemed to have the ability to outthink the second law of thermodynamics.

In the age of the computer it is hard to imagine how something as prosaic as the steam engine could have done so much to shape nineteenth-century thought. Someday, perhaps, our own preoccupation with the digital computer will seem just as quaint. In contemplating how to get Robert Fulton's engine to mesh as closely as possible with the laws of nature, squeezing the maximum amount of work from the steam, Sadi Carnot, a French army engineer, concluded that even with his utmost efforts, he could never hope to reach an efficiency of 100 percent. In transforming the energy of the steam into the energy needed to turn a wheel, some would inevitably, irreversibly, leak away. This truth was expressed in the form of the two laws of thermodynamics. The first law can be taken as the good news: it declares that energy is indeed conserved, that it can be neither created nor destroyed, but simply changed from one form to another. The second law, however, tells us that whenever energy is put to use it is degraded: the potential energy of water stored behind a dam turns to kinetic energy and then to electricity as it rushes down the spillway and turns the turbine blades of a generator. In the end the accounts must balance: the energy coming out must equal the energy that went in. But not all of the energy of the water can be converted into electricity. Some is dissipated in the form of heat—the friction of water molecules bumping into air molecules and into each other, the friction of the imperfect bearings on the turbine blades, the resistance of the electricity in the wires. The energy of the wasted heat is still somewhere in the environment, in the form of randomly vibrating molecules. We can imagine ways to recapture some of this random motion and channel it back into the system. But it can never be completely recovered. If it weren't for this loss, we could use a generator to power a motor and then use the motor to turn the generator and have a perpetual motion machine.

Rudolph Clausius, in Berlin, was so struck by this inevitable change from useful to useless energy that he gave it a name: entropy. Water above a dam, steam compressed in a chamber, a spring wound tight, a battery with its negative charges sequestered from its positive charges— all are in highly structured states and are said to have low entropy. As they do work they become randomized. Viewed this way, entropy is a measure of disorder, and what the second law is telling us is that the march toward randomness is inevitable. One can reduce entropy (water can be pumped back uphill; a dead battery can be recharged, its homogenized negative and positive ions divided between the two poles again), but only by expending energy. And this produces more entropy. Our refrigerators freeze shapeless water into the crystalline lattices called ice, but as they do so, heat, the random vibration of molecules, is exported into the room. In the long run, entropy always wins. Pockets of order must be paid for with larger pockets of disorder, and the system as a whole—the universe—increases in entropy. We are fortunate in finding around us huge stores of potential energy, clocksprings already wound— food, fossil fuels, rivers, uranium. By letting them flow down the energy hill, we can run our civilization. For now, the whole system is continuously recharged by the sun. But eventually it too must run down.

There is also a third law of thermodynamics, which insists that it is impossible to reach absolute zero, the temperature at which all molecular motion would cease. Thus there will always be heat in the world, the energy of these randomly moving molecules. But according to the second law, it always takes more work to harness this scattered, ubiquitous motion than we can possibly gain from the attempt. Otherwise our cars and our appliances—for that matter, trees, animals, anything that requires power—could run by themselves, fueled by nothing more than this bottomless sea of vibrations.

It was quite a radical move when, in a thought experiment, Maxwell tried to devise a way to break the second law, to show that if a creature were clever enough it could create energy out of thin air. He began by imagining a vessel divided into two chambers, connected by a small tubular passage. Suppose you place a barrier in the passage and then fill one chamber with a hot gas. Remove the barrier and this initially ordered system, with all the heat on one side and none on the other, will quickly move to a state of equilibrium, with both sides filled with a gas at a lower, uniform temperature, a homogeneous expanse of randomly moving molecules. Place a paddle wheel or a piston in the passageway and this rush to disorder lets us do work. But once the system is in thermal equilibrium, with entropy at a maximum, the second law tells us that there is no way to extract any more work from the gas. We would have to pump it back into one chamber, and that would require energy.

But, Maxwell wondered, why couldn't you instead place a small, intelligent being in the middle of the contraption to observe the movements of the molecules and manipulate the valve so that the faster ones congregated in chamber A, while the slower ones stayed in chamber B?

When a molecule came speeding from chamber B toward chamber A, this "very observant and neat-fingered" being would open the valve and let it through. It would close the valve if it saw a fast molecule about to escape from A to B. Merely by the exercise of its wits, Maxwell contended, the demon would cause the temperature in A to exceed that in B; it could build up a potential and use it to do work. With intelligence, it seemed, one could overcome entropy.

Maxwell wasn't seriously interested in building a perpetual motion machine. Where, after all, was one going to find one of these uncomplaining little slaves? His purpose was to show that, unlike the laws science had proposed in the past, the second law is not absolute but statistical; the best we can say is that it works the overwhelming majority of the time. Even in a system without a demon and a valve, there is a tiny but real chance that the fast molecules would just happen to congregate on one side and the slow molecules on the other. But the vast likelihood is that the temperature would even out: for each chance fluctuation that put a fast molecule in the left chamber, another fluctuation could be expected to put a fast molecule in the right chamber.

Still, it *could* happen. Individual gas molecules don't know about the second law and the one-way flow toward entropy; they simply obey the laws of mechanics. Nothing in Newton's writ bans the possibility that after the valve was opened and the gas flowed from chamber A to chamber B, a majority of the molecules might reverse course and flow back into the first chamber. The system would be restored to its original state, providing us with work for free. But the chance of this happening is so remote that we would be better off waiting around for a swarm of fireflies to spell out messages in the sky.

The moral, Maxwell once wrote, is that the second law of thermodynamics "has the same degree of truth as the statement that if you throw a tumblerful of water into the sea, you cannot get the same tumblerful of water out again." But again, it could happen, either by chance or the good graces of an observant, nimble-fingered demon gathering the molecules up again and putting them back into the glass. Entropy, it seemed, was a measure of ignorance; it depended on the observer.

The demon was a fantasy. But between its perfect perception and nimbleness and our myopic clumsiness, one can imagine a continuum of creatures endowed with different powers. Nature is like a text and these various beings will vary in their ability to decipher its code, or even to suspect that there are patterns there to divine. In an article he wrote for the 1878 edition of the *Encyclopaedia Britannica,* Maxwell compared the situation with trying to read a notebook written in the owner's personal shorthand: "A memorandum-book does not, provided it is neatly written, appear confused to an illiterate person, or to the owner who understands it thoroughly, but to any other person able to read it appears to be inextricably confused. Similarly the notion of dissipated energy would not occur to a being who could not turn any of the energies of nature to his own account, or to one who could trace the motion of every molecule and seize it at the right moment. It is only to a being in the intermediate stage, who can lay hold of some forms of energy while others elude his grasp, that energy appears to be passing inevitably from the available to the dissipated state." The implication was that entropy existed for moderately intelligent creatures like people but not for demons or dogs—that order and disorder were in the eye of the beholder.

If the second law was indeed statistical, then the best way to treat it was with the mathematics of probability. Seen this way, systems tend to move from ordered (unlikely) to disordered (likely) states because there are vastly more disordered ones. Try to imagine the countless ways in which gas molecules could arrange themselves in a closed vessel. In a tiny number of these configurations, the molecules will appear bunched into one corner or another, or sequestered in various-shaped blobs; in a precious few cases, they might arrange themselves in spheres or cubes. According to what statisticians call the ergodic hypothesis, the gas will eventually visit every one of its possible arrangements as its molecules wander randomly through the chamber; one is no more likely than the other. But in the vast, vast majority of possible arrangements, the molecules will form what appears to our myopic eyes as a featureless mix uniformly occupying the container.

In Maxwell's engine, we begin with the molecules forced into an unlikely arrangement, all occupying chamber A. When we open the door they rush to assume one of the vastly more likely arrangements in which the gas is uniformly distributed throughout both chambers. Another way to say it is that by opening the door we give the gas more "degrees of freedom," twice as much room to roam.

Probability could also be used to explain the inevitable sucking away of energy through friction and other forms of dissipation. The environment, after all, represents a huge, essentially infinite number of degrees of freedom. If we shatter Maxwell's vessel, the molecules will escape, fanning out through a labyrinth so vast and convoluted that they can never find their way back again. And so it is with the heat produced by an engine or any kind of machine. When these vibrating molecules are allowed to bump up against the molecules in the open wilderness of the air, the energy follows the path of least resistance, flowing irretrievably into the great beyond.

Among Victorian intellectuals, the demon elicited two extreme reactions, both reaching far beyond anything Maxwell seems to have intended. To those who took comfort in the objectivist creed, that it was possible to stand outside creation and see it whole, the notion that entropy was subjective was seriously disturbing. Could the second law really be no more than an anthropomorphic effect caused by our myopia and clumsiness and the fact that we are so much larger than molecules? Others found Maxwell's thought experiment liberating and declared that intelligence was a force that could somehow overcome the constraints of physical law, a solace against the gloomy idea of a universe doomed to increasing entropy. By precisely monitoring and manipulating molecules, a creature could (theoretically at least) outwit the second law. There was something special about life and mind that eluded the cold equations of the physicists. Or so some people wanted to believe.

In an attempt to dispel such wishful thinking, some scientists tried to lobotomize the demon by showing that Maxwell's paradox would arise even when mind was removed from the mix. One didn't need to actively sort the molecules. The second law could be overcome, they argued, with nothing more than a one-way door; it would passively swing open when a molecule traveling in one direction collided with it but would stay closed when it was struck by a molecule coming the other way. Eventually more molecules would accumulate on one side of the door than the other: energy for free. Maxwell himself found this convincing— after all, his only intention was to show that the second law was statistical, not to elevate intelligence to the supernatural realm. "I do not see why even intelligence might not be dispensed with and the thing made self-acting," Maxwell wrote, adding later: "This reduces the demon to a valve. As such value him. Call him no more a demon but a valve."

But as it turned out, Maxwell conceded the point too easily. In 1912 the Polish physicist Maryan Smoluchowski showed that a trapdoor tiny enough to serve as an automatic demon would absorb heat and vibrate so wildly that it would be completely ineffective. But, he allowed, "such a device might, perhaps, function regularly if it were appropriately operated by intelligent beings." Could thought overcome entropy after all?

The argument lay in this murky realm until 1929, when it was taken up by Leo Szilard, the Hungarian-born physicist who would later be so instrumental in the founding of the Manhattan Project. The title of his paper, "On the Decrease of Entropy in a Thermodynamic System by the Intervention of Intelligent Beings," sounds like another attempt to elevate mind over matter. But actually Szilard's intent was to demystify the demon by replacing the ethereal notion of mind with the more concrete notion of information processing. By doing so he set off a chain of arguments and counterarguments that can be traced sixty years later to Charles Bennett's pronouncement about using information to get up and over La Bajada.

To crystallize his argument, Szilard reduced Maxwell's apparatus to its simplest possible form: a chamber with a single gas molecule wandering randomly inside. First the demon would insert a movable partition in the middle of the chamber. Then it would determine which side the molecule was on, left or right. By hooking up a rope and pulley to the proper side of the partition, the demon could use it as a piston. As the molecule pushed against the barrier, it would pull the rope, turn the pulley, and lift a weight. Potential energy would now be stored in the weight hovering above the ground. By dropping it on a piezoelectric crystal, which generates electricity when it is squeezed, or by using it to pull a belt attached to the armature of a generator, the demon could do work. Then it could decouple the weight and remove the piston from the chamber. With the system back in its original state, the demon could repeat the process, seemingly creating work from nothing more than its ability to perceive which side of the partition the molecule was on.

So far this sounds like just another version of Maxwell's tale. But Szilard reached quite a different conclusion. His breakthrough was to realize that the demon's measurement, determining whether the molecule was on the left or right side, entailed making a binary record, recording what we now would call a bit of information, 1 or 0, left side or right. And making this measurement, Szilard suggested, inevitably consumed a certain amount of energy—enough to ensure that the second law was not violated. In rigging a Maxwellian demon one was traditionally allowed to assume things like perfectly frictionless pistons, the justification being that there is nothing in the laws of physics that would keep one from approaching this ideal as closely as technology and cleverness allowed. But Szilard proposed what amounted to an underlying limit. The very act of gathering information, he implied, must always dissipate at least enough energy to offset any gain in work and ensure that perpetual motion was impossible.

In the old demon arguments, mind had been looked upon as something separate from the material world; it was an essence with powers of its own. Long before the beginning of information theory and computer science, Szilard focused the argument by showing that, in this simple case at least, intelligence could be thought of as processing bits. And processing bits expended energy. In building our intellectual cathedrals, we might think of ourselves as detached observers, but our powers are finite, our observations rooted in the physical world.

As some would later put it, Szilard showed that there is no such thing as an immaculate perception. This idea was further explored in a paper published in 1951 by the Frenchman Léon Brillouin, who proposed that there would be no way for a demon to sort molecules without seeing them—using a flashlight to bounce photons from the molecules to its eyes. One could imagine making the beam weaker and weaker, but eventually, Brillouin argued, you would bottom out at a minimum intensity. The chamber is, after all, filled with vibrating molecules. A signal that was too weak would be indistinguishable from the surrounding noise. In the same year, the physicist Dennis Gabor calculated that as the light beam was made weaker and weaker, it would, for reasons of quantum mechanical uncertainty, become harder and harder to focus. Again, the implication was that processing information required a minimum amount of energy.

Though this idea was fated to undergo an important modification, in a fundamental sense Szilard, Brillouin, and Gabor were on the right track. Information was becoming less ethereal, a choice between two states of a physical system: molecule on the left or molecule on the right. In retrospect we can see that Szilard showed it was indeed true that the demon could be replaced by a machine, but only if the machine was an information processor, a computer of some kind. And computers must be plugged into the wall. The work the electronic demon gained by processing information and lowering entropy would be offset by the kilowatt-hours it consumed; the entropy is not eliminated but rather exported into the environment—the heat produced by the generating station and the resistance in the wires.

What Szilard had intuited about information and entropy Claude

Shannon made more solid and precise. Shannon, who worked for Bell Laboratories, was studying the best way to encode signals so that they could be transmitted without becoming hopelessly garbled by the random molecular vibrations called noise. Though Shannon was dealing with telephone lines, not heat engines, the general issues were the same as those for thermodynamics—how, in a universe ruled by an inexorable tendency toward disorder, do you preserve structure amid randomness? The fruit of Shannon's investigation was a pair of papers, published in 1948, in which he derived a mathematical expression for the amount of information in a signal. As it turned out, the expression was essentially the same as the one derived in the previous century for entropy.

In retrospect, this connection is not so surprising. Chaitin and Kolmogorov later showed that random, incompressible numbers have a higher algorithmic information content than orderly compressible numbers—it takes longer computer programs to spit out the random numbers. However, many of Shannon's followers found it more intuitively satisfying to put a minus sign in front of the expression for information, making it the opposite of entropy. A highly ordered, low-entropy system can be said to contain a high level of information—distinctions that can be encoded with bits. All the gas is confined in chamber A, none in chamber B. But open the valve and the information content decreases as the entropy rises. What can one say about a random, homogeneous mix of molecules? It is featureless, with no distinctions to be made. Since orderly systems are less probable than disorderly ones, Shannon's measure of information is sometimes called statistical information (as opposed to Kolmogorov and Chaitin's algorithmic information). It is also sometimes called negentropy, the opposite of disorder.

Since, in the general scheme of things, an orderly, high-information system is a rare device, information is also sometimes called a measurement of the degree of surprise. Hiking up a mountain, we are startled to look down and find a perfect arrowhead at our feet: it has a higher statistical information content than a rough piece of granite. If we lose the trail, we look for a marker—stones piled up to form a cairn. The more stones in the pile, the less likely it is that they fell that way by chance. Of course, the amount of structure in a system lies in the eye of the beholder—remember Maxwell's story about the memo book, or think of the labyrinthine molecular world open to those who could see past the surface of a piece of rock. Thus Shannon's new information theory reinforced the notion that there was something subjective about entropy and order.

Though many scientists were intrigued by the resemblance between the mathematical expression for entropy and the one Shannon derived for information, not everyone liked the idea of introducing this slippery concept as one of the atoms of creation. Shannon himself was skeptical of the interpretations. It was one thing for engineers to introduce a concept called information for analyzing man-made systems, but quite another to claim that it was an important part of the physical world. If the cost of saving the second law required accepting that there was a subjective element to our perceptions of randomness and order, the very basis by which we carve up the universe, then many nonbelievers wanted no part of the trade-off.

Their challenge was to show that there wasn't really any information floating around in Szilard's engine, that the reason it was incapable of generating perpetual motion was not the cost of processing information but more mundane considerations, like the thermal vibrations that had defeated Smoluchowski's trapdoor. By replacing the demon with ingenious arrangements of sensors and electromechanical devices to engage the gears and pulleys, they tried to design an automated version of Szilard's single-molecule engine in which no binary decision—left or right, 1 or 0—need be made. But like debunkers of magic acts, their opponents were able to show time after time that there was something hiding behind the curtain; information was lurking in the cracks of the machines. Once one had tried on Szilard's newly ground eyeglasses it was hard not to see bits everywhere. If, at the end of a cycle of one of the automated machines, a weight was left dangling on either the left or right side of the piston, or if a lever was flipped one way or the other, this was considered information, 1 or 0. The machine, no matter how crude or lumbering, had a memory—it stored a bit that represented the state that the molecule had previously assumed. To repeat the cycle and keep the engine turning, someone or something would have to reset the machine. And how this was done depended on which of two states it was in, which weight was suspended, which way the lever leaned. Information would have to be gathered, a decision made. Implicit in this was an idea that would not fully emerge until Charles Bennett entered the picture in 1973: that it was not the actual gathering of information but its erasure—resetting the apparatus—that necessarily dissipated energy and saved the second law. And that, at last, is where the notion of driving up La Bajada fueled by a memory tape comes into the story.

In 1961, Rolf Landauer of IBM set out to do for the digital computer what Carnot had done for the steam engine: plumb its thermodynamic depths. The second law showed that nature sets limits on how efficient a heat engine can be. Steam engines could never convert 100 percent of their heat into energy because some was dissipated irrevocably into the environment as heat. When work is performed, a minimum amount of energy must always be irreversibly lost. Everything in the saga of Maxwell's demon suggests that the same might be true for the labor we call computation. It was left for Landauer to clinch the argument.

Punch 2 + 2 into a calculator, press the "equals" button, and the display says 4. But if you find a calculator someone has left on a desk and *it* says 4, you have no way of knowing where the number came from. Did someone punch in 2 + 2, 3 + 1, 1 + 1 + 1 + 1, or perhaps 9 − 5, or 1,239,477 − 1,239,473? There are an infinite number of calculations that can yield this same answer. Such a computation is irreversible. You can't go from 4 back to 2 + 2. The expression 2 + 2 contains more information than the expression 4—a surplus that is lost when you complete the computation.

Where does the information go? Landauer showed that it is dissipated into the environment as heat, and is as difficult to gather up again as the friction generated by a turbine or the molecules in a glass of water dumped into the sea. His argument went something like this: Recall the apparatus in Maxwell's original thought experiment, with two chambers connected by a valve. Once the gas has been loosed from the confines of the left-hand chamber, so it is free to fill the whole vessel, thermodynamics tells us that it takes energy to squeeze it back into the first chamber again. We are taking a system that now has many more degrees of freedom—all the ways the molecules might be arranged throughout the entire container—and squeezing it back into a system with many fewer degrees. The same is true for the calculator. Information has to be represented by physical states, whether voltages in a wire or positions of beads on an abacus. In an electronic calculator, 2 + 2 is represented by a string of 1s and 0s each held by a transistor that is either on or off. Each of these memory cells then has two degrees of freedom; it can represent either a 1 or a 0. To erase it, Landauer figured, the two degrees would have to be squeezed back into one: a memory cell that could only be empty. And, as with the gas, that would require a minimum amount of energy. As a digital computer churns through a long series of calculations, clearing registers so they can be filled again and again, the machine throws away information, shedding heat into the environment.

In the computers we build, the energy lost from clearing memory registers is insignificant compared with the energy consumed by resistance in the connections, the filament that lights the video display, or the motor that turns the disk drive. Still, these losses are dependent on the technology used; in theory they can be made as slight as we wish. But nature seems to put a limit on how cheaply we can erase bits. Below a certain level, the loss cannot be reduced. Information, Landauer argued, indeed is physical.

But that was not the end of the story. A little more than a decade later, Landauer's colleague Charles Bennett was struck by one of those questions that seem both simple and profound: What if you don't erase? Imagine that each time a computer made a calculation, it saved the intermediate result. As the machine ran through a chain of computations, it would accumulate a tape of its history. It would come up with an answer without having thrown away information. Aha, you might think. This is where it must pay the thermodynamic cost: the tape of all those intermediate steps must be erased to make room for more. But no. Bennett showed that the machine could be reset simply by running the tape backward, retracing its computational history until it was in its original state. Computation is merely the converting of an input (the question) to an output (the answer) according to a set of rules. Usually this is a one-way flow—given 4, we can't uniquely infer 2 + 2. But with a reversible computer, we have the extra information needed to convert the output back into the input. And why should going from output to input require any more energy than going from input to output? Of course, a huge disadvantage would be that in reversing the computation the answer would be lost. But, Bennett pointed out, before we kicked the machine into reverse, we could copy the answer onto a blank tape. And copying, unlike erasure, does not incur a minimum energy cost.

While such a machine would dissipate energy through electrical resistance, whirring disk drives, and glowing video screens, the actual act of computation could be done with no minimum energy cost. Though erasing information requires an amount of energy below which it is impossible to go, computation can otherwise unfold using an arbitrarily tiny amount of work. In fact, Bennett designed a hypothetical computer powered by nothing more than Brownian motion, the natural thermal vibration of molecules. While this boundless thermal reservoir could not be used to power a perpetual motion machine, as some fans of Maxwell's demon had hoped, it apparently could be tapped to perform computations, as long as you were provided with one of these carefully designed, reversible computers.

A few scientists, such as Edward Fredkin of Boston University, believe that the possibility of reversible computation implies that information is more fundamental than matter and energy, unconstrained by the second law. He envisions a hidden layer beneath what is currently taken as the laws of physics, where the shuffling of bits somehow gives rise to the world we see. The implication, of course, is that reality is some kind of simulation. The question of what is running the simulation or why is left as an exercise for the reader. Fredkin has called for an effort to recast the laws of physics in the form of algorithms for this hypothetical machine, carving up the world in an entirely different manner. But little work has been done in this direction.

Most scientists in the small world of information physics take Bennett's work as an amplification of Landauer's principle rather than a contradiction. Bennett strengthened the notion that it is not the gathering but the erasure of information that necessarily dissipates energy. The demon could make each measurement expending an arbitrarily small amount of energy. But before it acted—opening or closing the trapdoor—it would have to store the result of the measurement in its memory. The second law would exact its toll when the bits were erased, for that would require at least as much work as was generated by the engine. Another way to look at it is that as the demon is lowering the entropy of the gas, creating a more orderly arrangement of molecules, it is funneling all that randomness into its memory, scrambling its brain. Or, if you subdivide the system a little differently, the memory can be considered part of the environment. So, once again, creating order in one place requires exporting entropy to another.

But again—and here is what Bennett was thinking of that day in Santa Fe—what if you don't erase? A memory can be made as gigantic as you like. Imagine it as a long tape. The demon could just keep filling it with bits and postpone erasure indefinitely. Hook the demon and its engine to a set of wheels and you could drive down I-25, up and over La Bajada, and back to the Albuquerque airport, spewing out an exhaust of 1s and 0s all the way.

There is nothing mystical about this subtly different way of carving up the physical realm. Like anything, Bennett's engine is mired in the laws of thermodynamics. One doesn't find blank memory tape sitting around in the world waiting to be used as fuel. Its state, all 0s, is highly improbable, as unlikely as a gas that sits only on the left side of a container. It takes work to create this order. Then this work can be exploited by letting the tape run down the information hill, from orderliness to randomness, just as the energy behind a dam or stored between the poles of a battery runs down the entropy hill.

Viewed this way, there is no reason why a battery cannot be thought of as a memory tape. It begins in a blank, orderly state (positive charges at one pole, negative charges at the other) and is randomized as it runs down. Once the battery is spent, the rearrangement of its molecules is a memory—a history of its use. This record is erased by recharging. The positive and negative charges are sequestered again, the order is restored, the system reset. But the battery charger dissipates heat, transferring the randomness to the environment, filling it with jumbled bits. In fact, we can think of the universe as a memory tape, blank and structured. As all this order turns into entropy, the Universal Memory Tape is filled with random bits. But it can never be erased. There is nothing to erase it, nowhere to export the randomness. You cannot reset the universe. The randomness just keeps on accumulating. And that is the information version of the second law.

Maxwell saw entropy as purely subjective and concluded that the more intelligent a creature, the more work it could extract from a source of fuel. Szilard took the first step toward demystifying this notion. But the trade-off was that he had to elevate information—something most of us think of as subjective and man-made—to the objective realm.

Though intelligence doesn't allow us to overcome the second law, it remains true that creatures with more acute senses and more powerful brains will see pattern where others see randomness. How can we have a science if every observer, depending on its abilities, looks at the same system and perceives a different entropy, a different order? Zurek showed that we can get around this problem if we take up our ontological scalpel and slice the world yet another way, into two kinds of entropy.

Zurek's division is based on the notion that there are two kinds of information. First there is Shannon's information, measuring how improbable a structure is: an intricately patterned system is highly unlikely, and so it can be said to contain a lot of information. But there is also Chaitin and Kolmogorov's way to look at the situation, using algorithmic information. From this opposite perspective, a highly patterned, compressible system takes fewer bits to describe than a random one. If all the gas molecules are in chamber A, our demon's memory tape might say 1111111111111111111111111 . . . And this could be reduced to a simple algorithm: "repeat '1' ten billion times." On the other hand, once we open the valve and let all the molecules into the whole vessel so that they are distributed at random, the only way to describe the system's state would be to specify where each molecule is. By definition, this random string of 1s and 0s cannot be reduced to a shorter algorithm. It is incompressible.

Zurek proposed that we define what he calls physical entropy as consisting of two reciprocal quantities: the ignorance of the observer, measured by Shannon's statistical entropy, and the randomness of the object being observed, measured by its algorithmic entropy—the smallest number of bits it would take to record it in memory. The beauty of this explanation is that, during measurements, the ignorance of the measurer decreases as its memory tape gets longer—so their sum, the physical entropy, remains the same.

Different demons with different acuities may disagree over how much order or entropy there is in a system. The more precisely something is measured, the less random and entropic it might seem. But the more precise the measurements, the more scrambled the demon's brain will become. Taken together, the two kinds of entropy balance out, so that from the point of view of an outsider looking in, the physical entropy of the whole system remains the same.

The important lesson to take away from all this is that the measurer must always be included as part of the system. The result is nothing less than a law of conservation of information, which Zurek would like to see stand alongside the conservation of energy and the conservation of momentum as the pillars on which science stands.

There is a final coda to the story of the demon. Looking back at the march of ideas that began with Maxwell, we can see that the key to exorcising the demon was to replace it with an information processor, a juggler of bits. But then measuring molecules and deciding when to open trapdoors doesn't require much intelligence. In all the thought experiments, the demons could be made from a few dollars' worth of electronic parts.

But, Zurek began to wonder, what if the demon was *really* smart? Suppose, for example, that in observing the position of the molecules, its memory tape recorded a string that included this sequence: 100100100100100100100100100. Now with some simple pattern-recognition algorithms, the demon could notice the regularities. The first string could be replaced by an algorithm that said "repeat '100' nine times." Or suppose part of the string included the sequence 11111111111111111, which could be replaced by "repeat '1' seventeen times." The algorithms are much shorter than the strings. After it made its measurements, a smart enough demon (using a reversible computer so that no energy was consumed) could compress its memory. Then it would have fewer characters to erase. Less energy would be dissipated. The demon would pay less than the full thermodynamic price. Purely random sequences couldn't be compressed. But if there were pockets of regularity, the string could be squeezed shorter. (This, in essence, is what a program like DiskDoubler does when it compresses the information on a computer's hard disk so that it takes up less room.) Some sequences that appeared random might in fact contain hidden order. If the demon was really smart—and here we are encroaching on the realm of artificial intelligence—it might notice that part of the memory string represented the decimal expansion of the square root of 2 or of pi.

Was there something transcendental about intelligence after all? Could a demon this intelligent get a slight edge on the second law, where dumber demons had failed? Zurek suspected not.

Chaitin showed that it is impossible to prove whether or not a particular number is random, whether a compression is the most concise—it is always possible that there is more order that can be squeezed out. The demon could never know if it had made its memory tape as compact as possible. But the demon wouldn't have to find the most concise description. Anything that significantly reduced the number of bits it had to erase would save energy. The question was whether compression could save so many bits that a very intelligent demon got more work out of the system than it put in. But once again, the second law was rescued. Zurek showed that Shannon's information theory and the laws of computation put a limit on compressibility. The most concise description of a message still must contain at least as much information as was in the original. The demon's memory tape can be thought of as the receiver of a message whose source is the pattern of the gas molecules. It is this very information that allows the demon to do work, manipulating the molecules to lower entropy. But while the "message" might be compressible, whatever is squeezed out is redundant information that does not contribute to the demon's success. In the end, when the demon erases its memory, it cannot throw away any less information than was required to make the engine run. Compressing the memory tape will make the demon more efficient, and if one was lucky enough to stumble upon the most compact description, then the engine would reach maximum efficiency, expending just as much energy as it generated. But Zurek showed that the best it can ever do is break even. It can generate only as much energy as it takes to erase its memory.

It may strain our intuition to think of batteries as information stores and memory tapes as fuel. We think of matter and energy as fundamentals—we can feel the heft of a rock or the jolt of electricity. Information seems subjective. Yet why should what we know through our bodies be more fundamental than what we know through our brains? In the end, we only know about matter and energy through the signals sent by our senses—our eyes, ears, noses, the receptors in our skin. It all comes down to information. And yet what is this information but matter and energy—charged ions carrying electrochemical signals through our nervous systems. Landauer and Bennett showed the limitations that physics puts on computation; Zurek showed that laws of computation—the limits of compressibility—have implications for physics. And so the circle turns.

As one of the world's premier demonologists, Zurek sometimes finds himself identifying with Maxwell's little creature. To the demon, the gas in the chamber is the universe; its quest to find hidden orders is like the scientific quest to find universal laws. We decrease our ignorance by measuring, but only at the cost of this informational exhaust. If there were little order in the universe, if it were in equilibrium like the gas, then we would simply be funneling the randomness intact to our memory tape—the library of scientific knowledge. We would be no better off than the demon, measuring and measuring but never getting ahead. We could gather bits and bits of data, but we couldn't compress them into more compact forms, the succinct statements we call universal laws. Science would be reduced to cataloguing every fact about every particle. The universe, like a random string, would be its own shortest description.

Of course, a completely random universe wouldn't have information gatherers at all. There would be no structure. Our very existence stands as proof that the universe we live in is far from being in equilibrium. There is order to exploit, compressions to be made. And so, as Zurek says, it pays to measure.

5

THE UNDETERMINED WORLD

Struck by the way sunlight glancing off granite can seem to set a mountain on fire, Río Arriba's Spanish occupiers named the Sangre de Cristo Mountains after their sanguine glow, a light so pure and uniform that a nineteenth-century physicist might have been tempted to think of black body radiation, as though the rocks themselves were hot and radiating from within. Albuquerque's mountains, the Sandias, were named for the Spanish word for watermelon, again because of the way they redden as the night falls.

Photons from the sun scattered by electrons in granite. For all the talk about the menagerie of particles locked inside the nuclei of atoms, the universe we directly experience is generated almost entirely through the play of electrons and photons, a dance whose steps are laid out in one of the supreme accomplishments of twentieth-century science, a theory called quantum electrodynamics, or QED. The theory ignores gravity and stops short of the nuclear frontier, but it is stunning how much is still encompassed within its grasp. We are electromagnetic creatures in an electromagnetic world, existing at the intersection between light and electricity.

Moonlight reflecting off a lake becomes, in the language of QED, solar photons bounced by the electron shells of silicon and oxygen atoms in lunar rock and bounced again by the electrons in the hydrogen and oxygen atoms that clasp together to make water. And the photons ricocheting from the water interact with electrons again—the charged haze surrounding the carboniferous chains of protein molecules in our retinas. Almost everything we experience comes to us as reflected light, and so QED gives us a theory of how we know the world.

But electromagnetism is more than a carrier of signals, or a beacon to illuminate matter. Even for creatures that have no eyes, light and the way it plays with electrons is as fundamental as anything can be; it is the very reason atoms stick together to form matter. And they do so according to the rulebook of QED. In the way twentieth-century physics has carved up the world, quantum electrodynamics lies at the foundation of chemistry. Whenever two atoms pull together or push apart, the force arises from photons bouncing back and forth between their electron shells. Most of these interactions are invisible, but sometimes a chemical reaction will shed such an excess of photons that they light up the night: oxygen rapidly binds with carbon to make a forest fire; a firefly phosphoresces with a dull green glow. But even when the light is too weak or vibrates at frequencies our nerves cannot register, it is there providing the medium through which electrons communicate, through which atoms become objects and objects disintegrate into atoms again. With every step we take, it is electrons exchanging photons that generates the repulsive force that stops our feet from going through the sidewalk, that creates the illusion of solidity in a world that, we have come to believe, is mostly the empty space inside electron shells.

There is another, parallel science, quantum chromodynamics, that explains how colored gluons play with colored quarks to generate the stuff inside the nucleus. But that dance takes place on a hidden realm. Atoms and all that is made from atoms "know" one another through their electron shells, the charged facades they present to the world. To them, the nucleus is a black box, obeying rules they need not fathom. In the world where we find ourselves, it is QED that provides the rules of the game.

We live in a world orderly enough that it pays to measure. And we make our measurements using electrons and light. We feel the tug of gravity as we walk the earth, but once we train our sights beyond the planet we measure gravitational forces indirectly, by the effect they have on celestial objects: the pull of one planet on another, the centripetal effect of dark matter on a rotating galaxy. And how does news of these effects reach us? From signals of light that are registered only when their photons scatter off the electron shells of the atoms that make up our instruments and our eyes. The same is true when we try to see within the nuclei of atoms. The hypothesized chain reactions of particles creating particles creating particles ultimately must end with photons scattering off electrons, leaving their mark on our brains. The maps we make, the patterns we find, are rooted in this most basic interaction.

Explorations of Maxwell's demon showed the complications that arise when we contemplate the subtleties of how we find order in the world. There is no such thing as an immaculate perception; we are inevitably part of the world we are trying to measure. And everywhere we look, we come face-to-face with randomness. We build our orders, but only at the expense of creating randomness elsewhere.

QED tells us that this thing we call measuring is even stranger and subtler than thermodynamics suggests. The Q at the beginning of the name is a sign that the mathematical choreography of our subatomic messengers will be nothing like that which reigns up here in the macroscopic world. Like all subatomic particles, photons and electrons obey quantum logic. In building a theory of the most ordinary phenomena, we are led into abstract realms as remote as anything in the scientific imagination, into mathematics that defies our mental imagery. Electrons and photons do not ricochet off one another like billiard balls, we are told. When we say that electrons repel one another by bouncing photons back and forth, what we really mean is this: the first electron creates a photon, which is absorbed by the second electron rapidly enough to avoid violating conservation of energy. (Once again, Heisenberg's uncertainty principle provides the loophole that allows energy to be seemingly created out of thin air.) But the process is even more counterintuitive than that. The path that a single electron or a single photon follows is said to be random. This is not the randomness of human ignorance, which we find with thermodynamics; it cannot be reduced by gathering more information. Since this is quantum mechanics, we must deal with *inherent* uncertainty, even when calculating something as seemingly straightforward as the route a photon takes when it is reflected from the face of a meter and into our eyes.

At Santa Fe, Los Alamos, and elsewhere, the scientists exploring the physics of information are trying to understand the implications that this inherent randomness holds for our attempts to measure and find structure in the world. In trying to cast quantum theory in a new light, they have been led again to a bedrock in which information seems irreducible and fundamental. But before turning, in the next chapter, to these attempts to recast quantum theory, we will consider anew the strange map bequeathed to us by twentieth-century physics in its attempts to chart the subatomic world.

We are used to thinking of quantum randomness as something hidden away on a realm too small to see. But through the theoretical lenses of QED, quantum effects become magnified until they seem to manifest themselves as a familiar part of our world. As you look out the window at a streetlight illuminating the snow, you might be momentarily startled to catch an image of flames from the fireplace behind you hovering in the nighttime air, or a ghost of your own reflection. A fraction of the photons passing through the window is bouncing back into your eyes. Why do most of the photons emanating from the room go right through the glass, unimpeded, while some are reflected? It seems that a single photon arriving at the surface of the glass must be faced with two possibilities, sail through or bounce back. How does it "decide" the trajectory it will take?

In the real world, of course, not all photons are the same. They come in different colors—frequencies—and arrive at different angles. But we can take steps to reduce the complications caused by these variables. We find in the laboratory that even if we illuminate a piece of glass with a uniform source of monochromatic light, a beam in which all photons are as similar as we can make them, we still get this phenomenon called partial reflection. Using a photoelectric cell, which counts individual photons, experimenters find that a certain percentage of these particles of light pass through the glass, while a certain percentage are reflected. If the device produced a stream of bits, 1 for a photon that passed through the glass, 0 for a photon that reflected, we would find a fixed ratio of 1s and 0s. Each time we ran the experiment, however, we would get a different arrangement of these bits. We might find that time after time we get eight 1s for every 0, but one time the pattern might look like this: 101111111, and another time like this: 111111011. We can no more predict the identity of an individual bit than we can the roll of a die. While we can measure the average behavior of swarms of photons, a single photon's "decision" to go through the mirror or bounce back seems to be random.

What determines the ratio of 1s and 0s? By tinkering further, we find that we can change the percentage of reflected photons by varying the thickness of the glass. As we increase its thickness, the fraction of reflected photons rises from 0 percent to a maximum of 16 percent. But before we conclude that thicker glass means more reflection, we make the glass still thicker, and find that the number drops back toward 0 again. If we could rig a dial to adjust the glass's thickness, we would find that as we turned it higher and higher, the percentage of light reflected would rise and fall, rise and fall, in smooth sinusoidal undulations.

If we are content to think of light as waves instead of photons, partial reflection is not quite so mysterious. In the classical view, a wave passing through a piece of glass divides in two: part of it passes through, part of it is reflected. But if we carve up the world so that light is entirely wavelike, then we are hard-pressed to explain why our photodetector seems to register the beam as photons; if we hooked it to an amplifier and a loudspeaker we would hear a steady stream of clicks. And, more telling still, if we raise the brightness of the light the clicks remain just as loud, but they come at a faster pace. (If we want louder clicks—more energetic electrons dislodged from the photoelectric cell—we will have to increase the frequency of the beam. This is none other than Einstein's photoelectric effect.) In a lucid series of lectures, published as *QED: The Strange Theory of Light and Matter,* Richard Feynman, one of the principal architects of quantum electrodynamics, estimated that it takes five or six photons to fire the receptors in our retinas. If our eyes could be made only a bit more sensitive, he declared, we would be startled to see very dim monochromatic light as pulses.

So the dilemma remains: If light consists of these starbursts called photons, then how does a single particle "make up its mind," as Feynman put it, whether to bounce off the glass or go on through? And how does it adjust its behavior for different thicknesses of glass? Arriving at the front surface, it seems, the photon would have to somehow send out feelers and gauge how thick the glass was, then calculate the odds so it could decide which path to take. But even if a photon could sound the vitreous depths by sending out some kind of probe, we would be left with an insurmountable problem: for the signal to travel through the glass and back to the photon in time for it to adjust its course, it would have to move faster than the photon, faster than light.

Since superluminal signaling is supposed to be against the law, violating special relativity, our only recourse for explaining why seemingly identical photons are not treated identically by a windowpane is to blame the phenomenon on randomness—a randomness that, so far as we can tell, is inherent, not based on our ignorance of some of the facts. QED makes no attempt to offer a mechanism for how partial reflection, or any other optical phenomenon, works. It describes but it does not explain. The photon, apparently for no reason whatsoever, just goes one way or the other, and the best we can do is calculate the odds.

For us, the information gatherers, this is a curious situation. In trying to describe the one phenomenon most fundamental to our world, electromagnetism, science gives us powerful tools to make statistical predictions, but it is incapable of offering an explanation we can picture in our heads. Since Planck's experiments forced the quantum on the world, physicists have honed the mathematics diamond-sharp. But the possibility of describing some kind of machinery behind the equations becomes more and more remote. Those whose intuition tells them that events in all worlds, invisible or not, should be linked in a tightly drawn web of cause and effect are left in the same state of confusion as Professor Jakob, the subject of Russell McCormmach's novel *Night Thoughts of a Classical Physicist*. "Physicists used to seek picturable mechanisms for understanding the world," he lamented, "but now many of them had pretty well given it up." They had retreated into a "cold, gray cave of abstraction."

But it is hard not to be beguiled by how the utterly counterintuitive mathematics of quantum electrodynamics so neatly predicts the ways electrons and photons interact. We learn in high school physics that the angle at which light strikes a mirror (the angle of incidence) is equal to the angle at which it is reflected because of the "least time principle" put forth by Fermat in the seventeenth century: light takes the fastest path from A to B. But forget what we see in the macroscopic world—a beam of light reflecting off an area in the center of the mirror. In *QED,* Feynman shows that to reconcile the least time principle with quantum theory, we must assume that, behind the scenes, photons are behaving in ways that seem impossible, that they are bouncing off every single spot of the mirror, trying out even the unlikeliest of paths. Using the rules of QED, we assign an "amplitude" to each of the paths (the square root of the probability that a photon will go that way). When we add the amplitudes of these multiple routes, or "histories," we find that almost all of them cancel one another out, leaving the path in which angle of incidence equals angle of reflection, the one that takes the least time to traverse.

Feynman's "sum over histories" method can also be used to explain partial reflection or mirages on a desert highway or why your leg seems to bend when you step into a pool of water. In each case we are asked to imagine every possible way the photons can travel, then add them together. To be as precise as possible, we can even allow for hypothetical trajectories in which the photon is moving slower or faster than light. Some possibilities reinforce one another, others cancel out, and we are left with the trajectory that we see in the classical, macroscopic world.

Feynman's method does not simply apply to photons traveling en masse. We can almost imagine one photon from a light beam bouncing here, another one bouncing there. But to address the original question— how a single photon "knows" where to go—we must assume that *each single particle* tries out every possible path simultaneously, and that they cancel one another out, leaving the classical trajectory.

What is true for photons also applies to electrons. If we want to consider the seemingly simple case of an electron moving from point A to point B, we must consider every conceivable route and entertain the possibility that along the way the electron might emit and absorb any number of photons. According to the rules of subatomic physics, when an electron collides with a positron, the two self-annihilate in a flash of light; conversely, a flash of light—a photon—can give birth to an electron-positron pair. And so we must also allow for the possibility that every photon emitted by a traveling electron becomes an electron and a positron, which might collide to form a photon again. Making matters weirder still, in Feynman's formulation of QED, a positron is equivalent to an electron moving backward in time.

As a calculating tool, QED is as good as they come, yielding numbers that agree so closely with those measured by experiment that, as Feynman put it, it is as though one could gauge the distance from New York to Los Angeles to within the width of a human hair. But is QED just a mathematical device, like the cumbersome algorithms once taught in high school for extracting square roots and cube roots by hand? Or does it really describe an underlying reality? In *QED*, Feynman insists that many optical phenomena such as diffraction (the cause of the rainbow you see when you hold a phonograph record or a compact disc to the light) can only be understood if we assume that each photon glancing off a surface really is taking every possible path. It is not reality's fault that our brains are incapable of imagining this. It seems that this level beneath the classical, Newtonian world operates according to different principles. And from these quantum rules arise the familiar rules of our realm.

In the old days, scientists would study a system and imagine a mechanism that could explain how it worked, then they would discover or invent some mathematics to make it precise. With quantum theory we have the mathematics but we don't know what it means. In contemplating this strange situation, we might sympathize with the fictional Professor Jakob, left out in the cold by the new quantum theory:

"Over his lifetime," McCormmach wrote, "physics had taken a turn toward increasingly advanced mathematical conceptions of nature. Fifty years had not proved long enough for him to see into the depths of the equations of classical physics, certainly not into the final revelations of Maxwell's equations, which Hertz correctly saw were wiser than their creator and his followers. And it was unrealistic of him to expect to see into the depths of the equations of physics that came after Maxwell, if they had depths and were not a mathematical trick in the end. (For weeks he had been struggling with a paper on atoms by Sommerfeld, only to conclude that he was not doing physics but conjuring with numbers.)"

Feynman's "sum over histories" method is only one of the formalisms used to solve problems in quantum mechanics. Quantum phenomena can also be described using mathematical devices called Heisenberg matrices, Dirac state vectors, and, the most familiar, Schrödinger waves. All are mathematically equivalent and equally strange. The surprising predictions of quantum theory have become such a staple of popular culture that, by now, we all know the drill. While in transit a particle cannot be said to have a definite position or momentum. Until we measure it, it hovers in a limbo in which all of its possible positions or all of its possible momenta somehow exist simultaneously, represented by the Schrödinger wave function. Only when the particle collides with a detector, when we make a measurement, does it assume an actual value. The wave describes the likelihood that it will end up in one state or another, but the outcome is uncertain until it occurs. Whichever formalism we use, the mathematics dictates that the more precisely we know one of these values, position or momentum, the less precisely we can know the other. Once we determine precisely where a particle is, we can know nothing about how fast it is moving. This, of course, is the Heisenberg uncertainty principle, which also holds that time and energy and other pairs of attributes are complementary. We the observers must decide which to measure. Likewise, light can seem like particles or waves, depending on our experimental point of view.

How can the world we live in be so different from the world that lies underneath? Up here in the classical world, things look different from different perspectives, but like the blind men investigating the elephant we can reconcile our differences and agree on the shape of what is before us. But quantum theory takes this subjectivity to a strange extreme: There is no elephant, only blind men.

To the pueblo Indians, Sandia Crest is known as Turtle Mountain, a name that seems inappropriate to anyone who grew up in Albuquerque. From that vantage point, at the foot of the eastern escarpment, the mountain looks nothing like a turtle. But head north, up the Rio Grande, and it becomes stretched and distorted until by the time you get to Santo Domingo pueblo and look back, its animal form has snapped into view. Most people know the mountain from its Albuquerque angle. But just as there is no privileged reference frame in the universe where we can step outside the system and see it whole, neither is there a canonical view of Sandia. Settlers coming in wagons from the east would have seen a long gentle slope rising toward the granite knife edge where the ground suddenly gives way. Still, all observers will agree that there *is* a mountain there and that if we all got together and rose directly above it in a hot air balloon we would see the same geography.

With quantum theory there is no mountain, only the views. If Sandia obeyed the rules that each of its particles does, it would be wrong to grant it an independent existence. Until observed, it would hang in a quantum limbo, as a superposition of all the possible ways it could be.

How do we get from this level of quantum mushiness to our world of objects that have definite positions, not superpositions, that have both positions and momenta at the same time? Built atop the quantum rules, it seems, is a higher level called Newtonian mechanics. How do we make the jump? That is what QED does not even attempt to explain: why objects in our world do not seem to obey the same rules as the stuff they are made of.

The notion of rules on one level giving rise to completely different rules on another level is not so mysterious in itself. Our world is a whole wedding cake of layers. Subatomic particles obeying laws of quantum electrodynamics and quantum chromodynamics give rise to atoms and molecules obeying the laws of chemistry, which give rise to cells obeying the laws of biology and creatures obeying, to some extent, laws of psychology, sociology, and economics. Without straying far from the classical world, we can find such emergent phenomena everywhere. Water with all its properties emerges from hydrogen atoms joined to oxygen atoms, but a single molecule of H_2O cannot be said to be wet. In the brain, components called neurons, each obeying simple rules, send signals back and forth and properties we call perception, intelligence, and consciousness arise. Each of these levels is, in a sense, sealed off from the one below it. Gas molecules jostling about in a container give rise to emergent qualities called temperature and pressure. But it is meaningless to speak of a single particle having a temperature or pressure just as it would be meaningless to say that a neuron is conscious or a water molecule is wet; these are ensemble properties that exist only on a higher level.

Burrowing through the levels of a digital computer, we find a hierarchy of languages. At the bottom of the ladder, microprocessors and memory chips communicate in a binary tongue, in which everything—numbers, letters, images—consists of strings of 1s and 0s. The rules that reign are those of binary logic. Riding on top of this machine code is a higher-level, more abstract language whose tokens are not 1s and 0s but simple commands like ADD and MOVE. By harnessing these tokens, one can devise still higher-level languages like BASIC, FORTRAN, and C and use their more powerful commands to write word processors, painting programs, and video games. Between each of the levels is a program—called an assembler or a compiler—that translates from one set of rules to another. The rules in a video game bear no resemblance to those of FORTRAN, which bear no resemblance to the machine code dictating how the 1s and 0s must move. But if we had the patience, we could translate the succinct higher-level rules—"When you destroy an asteroid, you get 10 points"—into a long binary string.

The difficulties arise when we try to go one level deeper, to the murky bottom where the silicon and other atoms that make up the logic and memory chips are mired in the world of quantum mechanics. As different as the upper levels are from one another, they all obey the same deterministic logic; everything happens for a reason and is woven into a tight skein of cause and effect. On the quantum level, the language is indeterministic. We can only speak of the probability of an electron's moving this way or that. Yet somehow this gnarled quantum bedrock, like the formless Precambrian gneiss at the bottom of the Grand Canyon, supports all the neatly arrayed layers above it. To return to our computer metaphor: What is the nature of the compiler that allows us to make this mysterious transition? How does classicality emerge? Why do we agree that there is one mountain?

One of the windows we are given to glimpse the hard truths of quantum theory is the infamous two-slit experiment. Shine a light beam at a photosensitive screen, and between the source and the screen place a barrier with two holes in it. The image cast is called an interference pattern, which is just what we would expect if light consisted of waves: light and dark bands indicate the regions where the waves, passing through the two holes, are in phase, reinforcing each other, or where they are out of phase, canceling each other out. But repeat the experiment using a beam of electrons and they too leave a striped interference pattern, the classic signature of a wave. We can turn down the intensity of the beam and see evidence of particles: individual flashes each time an electron or photon strikes the target. But if we wait long enough, the individual collisions will trace out, point by point, the same dark and light bands.

We expect a wave to leave an interference pattern: it passes through both holes simultaneously and is split into two waves that interfere with each other. But why would a steady stream of particles behave this way? We are faced with the same problem that arises with partial reflection. Explaining why some particles go through the first hole, while other particles, presumably emitted with the same initial conditions, go through the second hole, is difficult enough. The choice seems to be made at random. But why is the result of this strange behavior an interference pattern? How do the particles "know" to arrange themselves this way?

One recourse is to suppose that our particle source isn't as uniform as we believed, that the particles are actually emitted with slightly different trajectories, causing them to glance off the edges of the holes at different angles. Or perhaps they are affected in transit by other particles or waves. Then we might conclude that the wave pattern is an illusion, a statistical distribution of particles which land in different places because of randomness caused by ignorance—factors we did not account for. If we knew all the information we could predict exactly where each electron would land.

If things were only so simple there would be no need for quantum theory; the two-slit experiment or partial reflection could be explained with Newtonian mechanics and ordinary statistics. But as we look closer, this interpretation breaks down. Close one hole and repeat the experiment with the electrons. We simply get a circular spot where the electrons or photons go through the open hole and hit the screen. They are acting like particles again. It is easy to think of this pattern as a probability distribution; slight fluctuations in the initial conditions cause the particles to strike the target in various positions clustering around the open slit. Close that hole and open the other and we get a similar distribution. Put shutters on the holes and close one and then the other, alternating back and forth, and we get two of these blurs side by side. So far nothing has happened that our classical intuitions would find objectionable. But when we open both holes at once we get the wavelike interference pattern again—dark regions where most of the electrons strike the screen and light regions where few or no electrons land. But what is interfering with what? How can it be that simultaneously opening both holes *prevents* an electron from landing in places where it was previously free to go? As the physicist John Bell wrote in 1985, "It is as if the mere possibility of passing through the other hole influences its motion and prevents it going in certain directions."

Are we dealing with waves or particles? Suppose we zero in on the holes themselves, replacing the shutters with detectors. Now when an electron arrives at the barrier we can see if it acts like a wave, passing through both openings, or like a particle, passing through one or the other. What we find is this: when one detector clicks, the other is inevitably silent; they never go off at once. That seems to resolve the matter in favor of particles going through one hole or the other. But then we look at the target and see that the interference pattern has disappeared. We are back to two circular spots again. When we test for particles we get particles, when we test for waves we get waves. As with partial reflection, the demand for a mechanism reduces us to imagining that the particles must somehow sense the nature of the experimental apparatus by sending out superluminal probes, so they can adjust their behavior accordingly.

The compromise adopted by most physicists is to give up hope of a mechanical interpretation of quantum phenomena and think of each electron in the two-slit experiment as a mathematical abstraction, a wave that can be thought of as representing every possible path, or history, the particle can take on its way to the target. When the particle collides with the target, this wave function "collapses" or is "reduced" and the particle randomly assumes one of the possible positions. In this interpretation, the probability wave itself goes through both slits, splitting into two probability waves that interfere with each other, producing the striped pattern. When we put detectors at each of the slits, we are simply collapsing the mathematical wave sooner, before it has the opportunity to divide in two and create the interference bands.

Ever since Max Born proposed this odd view in the summer of 1926, physicists and philosophers have been arguing over what it could possibly mean. It is easy to think of probability waves involving many electrons, but how does one picture a single electron as a probability wave? On a day-to-day basis, most physicists content themselves with the fact that quantum theory yields such precise predictions. But when pressed for an interpretation, many will insist that the probability wave really does propagate through space. Forced to settle for mathematics without mechanism, they reify the mathematics, treating this abstraction as we would a wave of water, allowing it to engage in refraction, reflection, interference.

Some might argue that this elevation of mathematical devices to the status of real stuff is done all the time in classical physics; witness the electromagnetic field, an abstraction that we have little trouble granting substance to. But quantum wave functions seem far more ethereal than electromagnetic waves. For one thing, an electromagnetic field exists in physical space; we can walk about inside one, measuring its intensity from point to point. The wave function exists in a mathematical domain called configuration space. It does its waving in an imaginary region we can no more experience directly than we can the space in which a nucleon exhibits isospin.

And so, it is all the more amazing that these waves of probability can be manipulated as confidently as if they were musical waves. The French mathematician Joseph Fourier showed some two centuries ago that a wave of any shape—a flat-topped square wave, a jagged sawtooth wave—can be expressed as a sum of sine waves, those smooth, symmetrical undulations we associate with a pure musical tone. We have also learned how to express a sine wave as a sum of square waves, or sawtooth waves. In fact, any wave can be broken into a sum of any other kind of wave. These components can be manipulated and juxtaposed so that those in phase will reinforce each other, those out of phase will cancel out, leaving us with any waveform that we like. Some of these compositions are more natural than others. Sine waves and square waves are so similar that it takes relatively few of one to make the other. But other waves are complementary, occupying opposite corners in the space of waveforms. An impulse wave, the sharp spike you would get if you hit a stick suddenly on a piece of wood—all the amplitude concentrated in an infinitesimal time, a temporal point—is so different from the pure tone of a tuning fork that it takes a vast sum of impulse waves to make a sine wave, and vice versa. In fact, to duplicate these tones with perfect precision would require an infinite number of complementary waves. Abstract as this seems, these wave conversions are done all the time in digital recording. An impulse wave is either on or off, 1 or 0, and so they form the two letters in a binary alphabet. A digital compact disc recording of a symphony can be thought of as sine waves from the instruments—vibrating columns of air—refracted into impulse waves by the recording studio and turned back into sine waves as they emerge from the loudspeakers on your stereo.

Sound waves are made of stuff: vibrating strings, vibrating air, vibrating eardrums. In quantum theory, physicists apply the same rules to waves of probability: refract a wave function representing a particle through the proper mathematical prism and it becomes a sum—a superposition—of sine waves. Each sine wave represents the possibility that the particle will have a certain momentum. (Square the wave's amplitude—the measure of its "loudness"—and you get the probability.) Refract the wave function another way and you get impulse waves, which represent the particle's possible positions. And the Heisenberg uncertainty principle follows from the fact that sine waves and impulse waves are complementary. A sine wave representing a particle's momentum is a superposition of an infinite number of impulse waves, a haze of possible positions. Pinning down one value with perfect precision smears the other value all over the mathematical map.

We can see, then, what some theorists mean when they say that a quantum particle is pure potentiality, that it doesn't have coordinates until it is observed. Sine waves aren't *really* made of square waves, but we, as observers, are free to break them down that way. We could take our favorite aria and refract each tone into a superposition of square waves, but that doesn't mean that there really are square waves emanating from the soprano's mouth. Likewise, a wave function representing a particle isn't really made of sine waves or impulse waves; these stand-ins for momentum and position exhibit themselves only because we choose to refract the wave function a certain way. The position or the momentum is created by the measurement.

In this interpretation of quantum theory, every attribute of a particle can be thought of as a different prism through which we pass the wave function. Sine waves mean momentum, impulse waves mean position. Other waveforms stand for energy, time, spin, and so forth. But where do we draw the line? Mathematically, a wave can be refracted into a limitless variety of shapes. There is nothing to stop us from breaking up the wave function into sawtooth waves, square waves, tulip-shaped waves, or a superposition of any waveform we randomly scribble. Implicit in the wave function is an infinity of attributes, most of which seem to have no counterpart in the way earthlings carve up the world. A few we have given names to—isospin, charm, strangeness—even though they have no clear physical interpretation. But the vast majority must remain nameless as well as unimaginable.

Why are position and momentum so fundamental to us? Can it be neurological, an accident of evolution? Or a happenstance of the way the universe itself evolved? If we had ended up on a different twig of the evolutionary tree, or in a different universe, perhaps position and momentum would be meaningless while qualities we can only think of as Xness and Yness were second nature. But it is easy to get carried away, taking our symbols for reality instead of as mere tools of description. When are we doing physics? When are we just conjuring with numbers? We build these systems to represent the world, then we are left to wonder what they mean. What is map, what is territory? Is there really any difference at all?

Niels Bohr believed the distinction was meaningless, that all we can hope for is good maps. The problem, he believed, is that the languages, both verbal and mathematical, that have evolved to aid our survival on earth are simply not equipped for navigation in the subatomic realm. "We must be clear that, when it comes to atoms, language can be used only as in poetry," he told Heisenberg one day as they trekked through the German woods. "The poet, too, is not nearly so concerned with describing facts as with creating images and establishing mental connections." Then, Heisenberg asked, "how can we ever hope to understand atoms?"

"I think we may yet be able to do so," Bohr replied. "But in the process we may have to learn what the word 'understanding' really means."

In building an interpretation of quantum theory, the reality behind the mathematics, we have to decide how to lay the foundation. We have two rough choices: either we can take our own world for granted and explain the quantum world in classical terms, or we can take quantum theory as fundamental and try to explain classicality in its terms.

In what has come to be called the Copenhagen interpretation, Bohr and Heisenberg took the first course. The world of phenomena, that which we can observe, is the only reality we can know. Since trying to describe the quantum world in classical terms leads to contradictions and absurdities, we can only conclude that beneath the surface we see, reality operates in ways that are inscrutable to us. Quantum theory is a kind of triangulation in which complementary pairs of imperfect concepts—wave/particle, position/momentum—are used to home in, as best we can, on phenomena beyond the pale of human nervous systems. The mathematics is not a picture of an underlying reality but simply a tool to describe the baffling interaction between the quantum and classical realms. An observer makes a measurement, the possibilities of the wave function collapse like an accordion, and we get a definite but unpredictable result. To Bohr and Heisenberg it was meaningless to speculate on whether the wave itself is somehow real. At the doorway into the atom, we have reached the limits of our powers. As Heisenberg put it: "What we learn about is not nature itself, but nature exposed to our methods of questioning."

Bohr was fond of saying that there is no deep reality. The confusion that arises when we contemplate the nature of the subatomic world is rooted in a conceptual mistake: our insistence that there is something behind the reflections in our experimental mirrors. All we have is our observations. There is no deeper realm. And even if there were, there is little reason to suppose that our brains would be tuned to understand it. With quantum theory we may have taken the mind as far as it can go.

But there is more than one way of looking at these things. The mathematician John Von Neumann found it arbitrary and anthropocentric to divide the universe into two separate realms, as Bohr and Heisenberg did, giving privileged status to our own. Einstein showed that within the neoclassical world of four-dimensional space-time there is no privileged observatory. Why should we abandon this democracy when we try to enter the subatomic realm? In building his interpretation of quantum theory, Von Neumann assumed it was the quantum world that was fundamental. Classicality was the mystery to explain. One could almost think of our classical world as an aberration, a snapshot taken when an observer makes a measurement and collapses the probability wave.

If quantum theory is indeed universal, then the macroscopic world, like the subatomic world, must be represented by wave functions. When we perform an experiment, the waves of the measuring device interact with the waves of what we are measuring. And the waves of the scientist interact with the waves of the measuring device. Somewhere along the way, possibility becomes actuality. But at what stage does the collapse occur, and what causes it?

Picture a scientist sticking an imaginary probe into an atomic shell, getting some kind of reading—an image on a dial, a number on a computer screen. This information then registers on the scientist's retina—a network of receptors that send electrochemical signals up the chain of neurons that form the optic nerve. At each step, neurons are measuring neurons. Where does the probability wave collapse to give a specific answer? At what point do we go from a superposition of every possible state to a single outcome? In the measuring instrument, in the eye, somewhere along the optic nerve? On what basis would we choose one synapse over another? Which link in the chain will we call the "observer"?

Von Neumann suggested that this seemingly infinite regress of measurements, measurements of measurements, and measurements of measurements of measurements bottomed out when it reached the mind, the only place along the chain where anything special seems to occur. We might call this the Budapest interpretation. For it was the Hungarians Von Neumann and his colleague Eugene Wigner who laid out the sequence of arguments that, if you accept the premises, leads to this conclusion. For them, the lesson of quantum theory was not that there are limits to the mind's comprehension, as the Copenhagenists believed, but that it is the mind itself that makes it possible for the world to exist—by performing the final measurement that collapses the quantum wave.

But this simply shifts the mystery to a new locale, from something inscrutable happening on the laboratory bench to something inscrutable happening in the brain. And isn't giving a special status to consciousness at least as anthropocentric as giving special status to the classical world? In the Copenhagen interpretation, any classical measuring device can collapse the wave, a photographic plate as easily as a brain. It almost seems as though consciousness were chosen by the Hungarians simply because it is one of the two phenomena—the other being quantum theory—whose reality we least understand. In his recent book *The Emperor's New Mind,* the mathematician Roger Penrose takes the opposite tack, trying to explain consciousness by supposing that the brain operates according to quantum principles, that when we are thinking we hold myriad possibilities in quantum superposition, a wave that collapses when the answer to a problem strikes us like a bolt out of the blue. So take your pick. We can use the inscrutability of consciousness to make sense of quantum theory, or the inscrutability of quantum theory to make sense of consciousness.

In cosmology, the speculations of Von Neumann and Wigner are the root of what is called the strong version of the anthropic principle, which holds that through a dizzying symbiosis, the universe creates observers who create the universe by collapsing probability waves. Heisenberg said that all we can ever know is "nature exposed to our methods of questioning." Going further, the physicist John Wheeler put it like this: "No question, no answer."

Science once took a perverse glee in dethroning mankind from the center of the cosmos. Galileo removed us from the eye of the solar system; Darwin denied us even a terrestrial throne. With the strong anthropic principle and the Budapest interpretation of quantum theory, we see a departure from that progression. Humanity, or at least consciousness, moves back to center stage. Though it might be comforting to imagine that the universe somehow depends on our measurements, sometimes it seems that the strong anthropic principle should rather be called the strong anthropomorphic principle. How reasonable is it to assume that neurons—these tiny sacs of seawater that evolved the power to strike sparks of electricity when disturbed by the shadow of an approaching predator or the force of a wave—can not only divine the harmony of the spheres but somehow participate in bringing about the universe's very existence? Rightfully proud of our ability to measure, we come to identify our measurements with reality. Then, going even further, we entertain the possibility that it is the measurements that create reality in the first place.

In another corner of the scientific universe, neuroscientists have been trying to close the gap between brain and mind, to show that consciousness is simply an emergent property arising from brain cells, whose behavior can be explained with chemistry, the grammar of molecules and atoms. The mind arises from the laws of matter. So while some scientists are trying to reduce matter to consciousness, others are trying to reduce consciousness to matter. We seem to be caught in one big loop, like the fish with their noses against the aquarium fascinated by their own reflections.

For those who find either the Copenhagen or the Budapest interpretation too incredible, there is an alternative, first put forth by the American physicist Hugh Everett III in 1957: suppose that the wave function never collapses—not in the brain, not in the measuring instrument, not anywhere along the way. In a sense, the Everett interpretation of quantum theory is the most economical. There is no need to come up with what both Bohr and Von Neumann lacked: an explanation for what causes the reduction of the probability wave. But this economy comes at a price. The good news is that we don't have to grant a special, God-given status to either measuring devices or consciousness. The bad news is that we have to accept the notion of what seems to be an infinity of parallel worlds.

Suppose we want to measure an electron's spin, determining whether it is "up" (counterclockwise) or "down" (clockwise). Like Von Neumann, Everett assumed that quantum theory was universal, that it applied with equal vigor to the classical realm. So when the wave function of our measuring instrument interacts with the wave function of the particle, the instrument will also be in a superposition of states, spin up and spin down. And when we interact with the instrument, then we will also be in superposition: in one component of the wave, we will discover that the instrument has registered an up spin; in the other component, we will discover that it has registered a down spin. And, for that matter, anything that interacts with us from then on will also be put into superposition, and so on, ad infinitum. The wave function never collapses; instead the quantum limbo radiates forever, giving rise to one universe in which the electron's spin is up and another universe in which it is down.

The big problem here, of course, is to explain why we perceive just one of these outcomes. The Copenhagen and Budapest interpretations cannot explain how we collapse the wave function; the Everett interpretation gets rid of the collapse, but how do we picture such a world? One interpretation, that of the physicist Bryce S. Dewitt, is to imagine that, invisible to us, there really is an identical observer who perceives the other outcome. One of us sees an up-spinning particle, the other a down-spinning particle. If we are measuring something, like position, that can take a range of values, then we split into a range of different observers, each in a different universe. So confident are some physicists of humanity's mathematical powers that they are willing to reify even this most counterintuitive of formalisms. If there are parallel states in the mathematics, then there must be parallel worlds.

But even many who find Everett's formalism the most attractive alternative resist its weirder implications. Zurek and his colleague Seth Lloyd, who frequently works at Los Alamos and the Santa Fe Institute, tried to persuade a cosmologist who subscribed to Everett's "many worlds" interpretation that he should be willing to play Russian roulette for a million dollars. Ultimately, where the pistol's cylinder stopped spinning could be traced to a quantum fluctuation. In one world the cosmologist would die, in five other worlds he would be alive and richer. The cosmologist finally conceded that he couldn't imagine taking such a risk. "I wouldn't want to cause my wife such suffering in any of the worlds," he said.

To Lloyd there are compelling reasons to consider only the branch we live in to be real. Samuel Johnson is said to have reacted to Bishop Berkeley's contention that there is no real world, that reality is all in the mind, by kicking a rock: "I refute him thus!" To Johnson, something wasn't real unless we could get information about it. The parallel branches in the Everett formalism are sealed off from one another; no information can flow between them. They are, Lloyd says, the best examples he can imagine of what *isn't* real. They are better thought of not as parallel universes but as possible histories. We can pretend what it would be like if we were in another of these possible worlds. But that is no more metaphysical a problem than the one posed by the movie *It's a Wonderful Life,* where we are asked to imagine what the village of Bedford Falls would be like without George Bailey and the savings-and-loan.

Ever since Einstein said that he found it hard to believe that God plays dice with the universe, some physicists have tried to argue that quantum theory must be incomplete, that the randomness it finds cannot be inherent. Like thermodynamic randomness, it must be subjective, arising from ignorance.

When a photon "decides" whether to bounce off a piece of glass or go straight through, when an electron chooses one slit and not the other, when a nucleus suddenly decays, shooting out a neutron, is the event really indeterminate? Or could there be a hidden mechanism? It is our fate never to know whether a string is random or whether we just aren't clever enough to compress it. As with the numbers emerging from the bingo machine at Tesuque, might there be hidden biases, departures from randomness, orders we aren't acute enough to see?

In fact, such a "hidden variable" interpretation of quantum theory is always an option. If cosmologists can posit transparent dark matter, why not posit that the particle in the two-slit experiment is indeed sending out feelers, judging whether one hole is open or two and adjusting its trajectory accordingly? The physicist David Bohm called these hypothetical feelers "pilot waves." With that time-honored method of proposing an entity that is in principle unobservable, he recast quantum theory in a way we can almost picture in our heads. In this view, first proposed in 1925 by the French physicist Louis de Broglie, electrons, photons, and the like are indeed both particle and wave. Traveling along with the particle, an undetectable pilot wave scouts out the territory, sending back information. The catch to Bohm's interpretation is that it requires us to believe in superluminal communication between the pilot wave and the particle. Again we are faced with the problem that a photon is traveling, by definition, at the speed of light, so any quantum radar beam that it used to probe the environment would have to travel faster.

In fact, a celebrated finding known as Bell's theorem concludes that all hidden variable theories necessarily imply some kind of instantaneous, superluminal contact. Bell reached this conclusion after carefully examining the implications of the notorious Einstein-Podolsky-Rosen paradox. Einstein and his colleagues Nathan Rosen and Boris Podolsky tried to undermine quantum theory by imagining a particle decaying into two particles which, by the laws of physics, must be spinning in opposite directions. According to quantum theory, the pair of particles forms a single system which is in a superposition of possible states. After letting the particles fly apart for miles or even light-years, an observer measures one of them, collapsing its probability wave so that it randomly assumes either an up spin or a down spin. Thereupon, the celebrated punch line: Once we have made this measurement, the second particle will by definition have to be spinning in the opposite direction. Somehow the effect of the first measurement seems to propagate instantaneously across space (or, alternately, backward in time). Since this "spooky" action at a distance seems to violate special relativity, Einstein naturally concluded that it is quantum theory that must be wrong: like classical particles, the two quantum particles must have had definite spins all along, not just when one of them was measured. The uncertainty wasn't inherent; it was simply due to our ignorance.

Bell showed, however, that Einstein was succumbing to wishful thinking. His theorem (later supported by experiments at the University of Paris under Alain Aspect) showed that no purely deterministic theory can explain the behavior of subatomic particles, unless it invokes some kind of "nonlocal," instantaneous connection. And the only way to get rid of this obnoxious notion is to deny the existence of hidden variables. In choosing which way to carve up the quantum realm, we are backed into a corner where we must either accept inherent uncertainty or believe that particles on the opposite side of the universe can be, in some un-fathomable sense, intertwined, that locality is an illusion.

Even if we choose the latter interpretation, Bell's theorem shows that the superluminal connections cannot possibly be used as channels to transmit information instantaneously within the classical realm. It seems that quantum information and classical information are two very different things. Classical information can be read without disturbing it and copied as many times as we wish, but it cannot travel faster than the speed of light. Quantum information—consisting of qubits, as some call them—is capable of these strange EPR effects, but it cannot be read or copied without being fundamentally altered.

In a collaboration with five other scientists, Charles Bennett has proposed that qubits and classical bits might be used for what he calls quantum teleportation. In this scheme, the sender (Alice, she is usually called) wants to transport a quantum particle to the receiver, Bob. Using two channels, one transmitting classical bits, the other qubits, she can send Bob two signals, which he can combine to create a perfect replica of the particle (the original is destroyed in the process).

First they create an EPR pair of quantum-correlated particles. Alice keeps one and Bob takes the other with him to another part of the world. Then Alice performs an experiment, allowing her EPR particle and the particle she wants to transport to come into contact. Since Alice's EPR particle is correlated with Bob's, the disturbance will be instantaneously transmitted to him as qubits. Then, using a classical form of communication—a telephone call, a radio broadcast, a classified newspaper ad—she tells Bob the outcome of the experiment. Then he can use this information to re-create the original particle's state in his EPR particle.

A few scientists, driven by a hunger for a deterministic universe, still hold out hope that the randomness of quantum theory will turn out to be an illusion, that there is a deeper order lingering underneath. At the end of a lecture sponsored by the Santa Fe Institute in 1992, Jim Crutchfield, who moved on from the Santa Cruz roulette collective to become a mathematical physicist at Berkeley, made an offhand remark about the problems arising from quantum theory. "*What* problems with quantum theory?" Murray Gell-Mann interrupted. "Well," Crutchfield replied, "I guess I am just a diehard determinist." Crutchfield was quick to admit that one can hardly quibble with the overwhelming success of quantum theory in explaining the outcome of particle experiments. But he hopes there is some way to show that the act of measurement causes some kind of chaotic effect. If that is true, then the random outcome of the collapse of the wave function might be governed by a strange attractor, the hallmark of deterministic chaos. Quantum randomness would be complexity too deep for us to fathom.

So far, though, no one seems to have made any progress in finding chaos beneath quantum randomness. A huge problem is Bell's theorem. Replacing quantum randomness with deterministic chaos would constitute a hidden variable theory, and so, once again, superluminal signaling would sneak in the back door. Some entertain the possibility that perhaps there is a hidden flaw lurking within Bell's network of beliefs and assumptions. Bell assumes, obviously, the ability of mathematics to accurately mirror the physical world. Some doubters have suggested that quantum randomness might somehow arise from the inherent randomness Chaitin found in arithmetic. But so far no one has begun to unravel what all agree is a beautifully knit argument.

In fact Lloyd, Gell-Mann, Zurek, and others believe that those trying to attribute quantum randomness to chaos are looking through the wrong end of the telescope. Recall the calculation in which the position of an electron at the edge of the Milky Way is amplified by nonlinear interactions until it affects the outcome of a billiards game. From this perspective it appears that the randomness of chaos comes from quantum indeterminacy, not the other way around.

Of course, none of these interpretive contortions would be necessary if we were willing to accept that maps are not territory, that there are limits to our mental powers and our mathematics, that we are stuck with our classical conceptions because of our evolution, that the best we can do is talk about how our classical world and our language interact with the hidden subatomic world. Perhaps such a phenomenon as wave/particle duality simply shows that there can be two internally consistent but mutually exclusive models—human mental constructs—that let us make predictions about the world. We can use the one that works best under the circumstances.

But there is something about the human mind that rebels against limits the way an animal rebels against the bars of its cage. Instead of conceding that maybe our mathematics is not universal and omnipotent, that it is simply a fallible human invention, some prefer to accept its conclusion that causality doesn't exist outside our own domain. How foolish and parochial we were to think it would. With quantum theory, our brains and our mathematics seem strained to the breaking point, yet it is part of human nature that we keep straining. Never really believing that it is impossible for us to know the ultimate, we seize on quantum theory not as a tool for interpreting experiments but as a statement about how the world really is. In a kind of mathematical transubstantiation, our numbers, like the Word, take on substance and become flesh.

6

THE DEMOCRACY OF MEASUREMENT

In the annals of demonology, Maxwell's imaginary critter, with the ability to keep track of the position of every molecule of gas in a chamber, has an even more venerable ancestor, a being we might call Laplace's demon. While we humans can identify with Maxwell's little imp, frantically gathering bits, scrutinizing the sequences for patterns, trying to squeeze what order it can from the deluge, Laplace's demon sits back calmly, studying the world with the all-seeing eye of God. In his celebration of a completely knowable Newtonian world, Laplace imagined, in the eighteenth century, an intelligence so powerful that, given the position and velocity of every particle in the universe, it could calculate the unfolding of history. Laplace's conclusion has been quoted many times. "For such an intellect," he wrote, "nothing could be uncertain; and the future, just like the past, would be present before its eyes."

These days it is tempting to think of the Laplacian demon as a great computer, an automated version of the Tesuque camel. If the universe were classical and we fed the demon the position and momentum of every particle at a certain instant, it would know all that could be known. Everything would be implicit in the data stream, including the events that came together to form Laplace and the experiences that marbled his brain, leading ultimately to his proposition that the universe is Laplacian. For this ultimate intelligence, there would be no point in doing science: every experiment ever conducted—including who did it, when and where, how it came out—would be as trivial as the tickings of a clock.

To an all-seeing eye, nothing would ever really happen in this static, Tralfamadorean world. Only its inhabitants, these Maxwellian demons with their tiny brains and blindered senses, would find life interesting. The compensation for ignorance is surprise. Even the most diehard determinist—sure that history is encoded in advance, in the Book of Revelation or (as some Santa Feans would have it) in cosmic pictographs carved somewhere on the etheric plane—would be blessed with never knowing nearly enough to make life a bore.

Even before the emergence of quantum theory, the rise of mathematical chaos spelled the downfall of the Laplacian god. We have learned that even simple, deterministic systems can be ruled by strange attractors. The tiniest change in the initial conditions, the numbers we plug into these seemingly innocent equations, can cause huge, essentially unpredictable swings in the output. Our ignorance is amplified exponentially until within moments it overwhelms all. A tiny imprecision in the measurement of the position of a single particle—an 8 instead of a 9 in the millionth decimal place—would be rapidly compounded until the course of cosmic history was changed. In theory the behavior of such a system is completely determined—it is all there in the initial conditions—but only if they are known with infinite precision. We can still imagine a Laplacian demon, but we have to grant it infinite powers. By definition, an infinitely powerful being can do anything, so Laplace's powerful vision is reduced to a tautology. The universe is completely predictable, but only if you happen to be God.

It may be surprising to realize that for all the inherent unpredictability of quantum theory, the wave function itself is quite well behaved. It is linear with a vengeance. Left unmeasured, it is deterministic, and not in the strange sense of chaotic equations, which are predictable in theory but not in practice. Given the shape of a probability wave at one point in time, we can predict what it will look like later, or we can extrapolate backward and see how it must have looked in the past. The complex of possibilities, all in superposition, evolves in an orderly manner. It is only when a measurement is made that randomness enters into the picture, as the wave collapses unpredictably into one of the possible states.

The primordial mass of the big bang can be thought of as a particle, and so, the cosmologists tell us, nothing in quantum theory prevents us from imagining a wave equation for the whole universe, with every possible history superposed. Again, we are left with the vision of a world of perfect information, where there doesn't seem to be room for science. Every possible outcome of every experiment that could be done, including the circumstances in which it was conducted, would all be implicit in the quantum wave function of the universe—a universe that, from the outside, would be as tedious as Laplace's machine.

But again, to those of us on the inside, blessed with ignorance, such a world is not boring at all. In our modest attempts to illumine the darkness, to gather information, we shatter the beautifully symmetrical world of pure potentiality—where all unrealized possibilities somehow coexist—into our messy world of specific objects and specific events. By measuring, we break the symmetry, and the outcomes of these quantum experiments inevitably contain a residual bit of uncertainty. To the Laplacian demon, on the outside looking in, our measurements may simply be nodes in the ever-forking maze of Everett space, as our observations cause the universe to split again and again, once for each of the possible outcomes. Behold all the branches simultaneously and nothing appears to happen. In one universe the electron has an up spin, in another universe it has a down spin. But, stuck as we are on our own branch, sealed off from the rest, we see a universe of constant surprise. We look at our instrument and there is a fifty-fifty chance of seeing a down spin or an up spin. We cannot predict which it will be. From our blindered point of view, the quantum dice seem to fall at random.

When we contemplate the mystery of how this quantum limbo gives rise to a real world of measurers and measurements, we run into many of the same problems raised by thermodynamics. The supposedly magical effect of the observer—Maxwell's demon seemed to be able to generate perpetual motion simply by making measurements—was brought back to earth by taking a harder look at what happens when one gathers and processes bits. The key was to realize that there is nothing ethereal about intelligence, that information is physical, that as finite creatures we pay a price for its processing—the energy that it takes to erase our limited short-term memories and start over again. The hope of Wojciech Zurek, Murray Gell-Mann, and others drawn to northern New Mexico's scientific frontier is that quantum theory and the role of the observer can be demystified in a similar way, by looking at them in terms of information. Working with others around the world, they are crafting a new interpretation of quantum theory and quantum cosmology. Both are built on a scaffold of information.

With both thermodynamics and quantum theory, there is a mismatch between the very small—or at least our conceptual picture of the very small—and what we actually experience. Rules on one level give rise to completely different rules on another level. Viewed individually, molecules conform approximately to the laws of classical mechanics, which are perfectly reversible: a particle can just as readily move from A to B as from B to A. But en masse the particles give rise to a breaking of the Newtonian symmetry: the one-way flow from order to entropy that seems to generate an irreversible arrow of time. If we see a film in which shards of hardened clay fly together to make an Anasazi pot, we can be sure that the projector is running backward.

To those of us in the macroscopic world, the entropy arrow points one way because information is dissipated into the bottomless pit of the environment. Breaking a pot is irreversible because we cannot keep track of the pieces and gather them together again. We have lost the information needed to reassemble the original object. Quantities like pressure and temperature also arise from this inherent ignorance: they are statistical averagings of the individual behavior of more gas molecules than our brains could ever hope to track.

Like the equations of classical physics, the quantum wave function is reversible; it contains no arrow of time. But once the wave function is measured and unpredictably assumes a specific value, we cannot reconstruct it from the way the data happen to break any more than we can confidently go from 4 to 2 + 2 rather than 2 + 1 + 1. Like breaking a pot, breaking a wave function is irreversible. Information is inevitably lost, as we go from a symmetrical world of pure potentiality to an asymmetrical one of actuality, creating randomness along the way. Whether we look through quantum or Newtonian eyeglasses, it is the finite nature of us, the observers, that gives rise to a macroscopic world where things happen, one after another, where the search for pattern inevitably creates randomness—where there is an arrow of time, so that our actions, like our measurements, can never be undone.

How far can we take these parallels? Could classicality itself have some rough equivalence to temperature and pressure, arising from our ignorance of submicroscopic details? In a letter to Wolfgang Pauli, Heisenberg proposed that "time and space are really only statistical concepts, something like, for instance, temperature, pressure, and so on, in a gas. It's my opinion that temporal and spatial concepts are meaningless when speaking of a *single* particle, and that the more particles there are, the more meaning these concepts acquire. I often try to push this further, but so far with no success."

Part of the difficulty is the crucial difference between classical and quantum ignorance, between the probability of thermodynamics and the probability of the quantum wave. In thermodynamics we assume that all the gas molecules, whose vibrations give rise to what we call temperature, whose collisions give rise to what we call pressure, really have exact trajectories; we are just ignorant of them. The Copenhagen interpretation of quantum theory dictates that a subatomic particle doesn't have a trajectory until it is measured.

And so we are confronted again with this seeming difference between quantum information—qubits—and classical information. We usually think of information as being *about* something. An ordinary object like a marble *has* color, shape, weight, size—attributes that cling to some kind of underlying stuff. But what about the underlying stuff itself? Once we get to that level, there seems to be nothing but attributes. A marble *has* mass, color, size. But an electron *is* mass, spin, and charge. A particle is completely defined by its quantum numbers. It is all information. Spin $\frac{1}{2}$ plus 1 unit of negative charge ($1.6021892 \times 10^{-19}$ coulombs) plus a mass of 9.1×10^{-28} grams *is* an electron. These are not just labels or qualities exhibited by something underneath. There is nothing underneath. A marble can shrink or grow and still be a marble. Change the attributes of an electron and it is not an electron anymore. Two marbles with the same attributes are still different things; they have nicks and scratches, individual identities. If your daughter's cat is killed or even her goldfish, you cannot replace it with an identical one. With electrons there are no such subtleties. Two electrons are completely interchangeable, like all 3s or all C's.

In both the classical and quantum realms, we can talk about information that is static and information that can change. Leaving aside relativistic considerations, the mass of an accelerating object is fixed while its position and momentum vary. But, again, there is an important difference between quantum particles and what our nervous systems have come to recognize as things. We think of a marble's dynamic attributes—this extra information about position and momentum—as something that travels along with it; the data inhere in the marble somehow. But with quantum objects it is not so simple. We are told by the mathematics that an electron doesn't have a position or a momentum unless it is measured. Until then it is represented by the various mathematical constructs—Heisenberg's matrices, Schrödinger's waves—that give the probability that, if measured, it will assume a certain position or a certain momentum. It seems as though the dynamic attributes are not just in the electron but in the measurement as well. With the marble, any ignorance we have about its position or momentum comes because the information has not been registered by our brains. An electron is all information, so perhaps it should not be so surprising that it remains incorporeal until it is measured. Information unregistered is nothing at all.

In trying to make sense of quantum theory, some have been led to believe that consciousness, or at least the act of measurement, is necessary to bring about what we consider the real world. But many scientists are suspicious of what sometimes seems like a self-centered attempt to elevate humanity and the classical world we experience to a special, almost God-given role. As Gell-Mann likes to say, "When it comes to quantum theory even the most intelligent people can start talking nonsense."

If we follow the approach of some of the people at Santa Fe and Los Alamos and admit information as another fundamental, along with mass and energy, then quantum theory can be viewed in a subtly different light. All that is required to break the symmetry of the wave function is information processing. Not only are conscious observers superfluous—the theory does not even require artificial observers like photographic emulsions or photoelectric cells. The universe itself might process information just as it processes matter and energy. Seen in this light, our role as informational spiders, stringing and restringing our conceptual webs, is as natural as anything in the cosmos. We try to set ourselves apart from the universe and pretend to see it whole. But we are inevitably a part of what we are observing, and our observations may be but a single circuit in a great web of flowing bits.

At Los Alamos, Zurek and some of his colleagues have been examining the difference between classical and quantum measurements in an attempt to better understand how we come to know the world. The trick, they say, is to follow the information. Where does it go when we make a quantum measurement? In addition to its static attributes—mass, spin, that which makes it an electron, a photon, or whatever—a quantum particle carries this huge complex of dynamic information: the wave function describing every possible state, and every possible combination of states, that it might assume. When it is measured and takes on one of these states, to the exclusion of all others, what happens to the extra information? Does it dissipate into the environment in an irreversible act of erasure?

One way to approach these questions is to reexamine Maxwell's demon from both the classical and quantum points of view. Recall Szilard's rendering of the thought experiment, in which a single gas molecule is trapped inside a cylinder. The demon inserts a movable barrier in the middle of the cylinder, then determines which side the molecule is on, A or B. Depending on the answer, the demon rigs a rope and pulley to one side of the barrier or the other, in just such a way that it acts as a piston: the molecule pushes against it and performs work. Szilard argued that the only way to ensure that the demon is not getting work for free is to propose that there is a minimum energy cost involved in processing information: recording which side of the barrier the molecule is on and (as is now believed) erasing the information so that another measurement can be made.

Suppose that, looking down upon the demon from a god's-eye view, we want to calculate how likely it is that the molecule is on *either* side A or side B of the partition. We would simply add probabilities: a 50 percent chance that it is on side A plus a 50 percent chance that it is on side B equals a 100 percent chance that it is on either side A or side B.

Szilard treated the single gas molecule as though it were a billiard ball. But the simplest gas molecule, hydrogen, is nothing more than a proton exchanging photons with an electron. It is better described by the rules of quantum theory. Until the demon makes the measurement, the molecule will not actually be on either side, A or B, but in a superposition of possible states. In the quantum version of the experiment, calculating the probability of where the molecule will be is no longer so straightforward. We must add A plus B plus a third term that expresses the fact that until it is measured the molecule is not on either side A or side B but on both sides at the same time. The two possibilities, side A and side B, stick together in superposition, interfering with each other. They "cohere." But once the demon makes a measurement, the possibilities come unstuck; they "decohere." The molecule assumes either position A or position B.

Suppose the demon and its apparatus are inside a black box and we are wondering what is going on inside. Before the measurement, our ignorance, like the demon's, is quantum and must be expressed by the three-term equation; A + B + the interference term. After the demon makes the measurement, we on the outside still don't know the answer. But simply knowing that a measurement has been made converts our ignorance from quantum to classical. We now can say that there is a 50 percent chance that the molecule is on side A or side B. Somehow in the act of measurement, the third term has disappeared. Now a particle can be said to have one position or another, not every possible position at once. The transition from quantum to classical seems to require this unsticking of possible outcomes, this phenomenon quantum theorists call decoherence.

In a more complex quantum system, we would have a longer equation with a term for every possible state of the particle; but always, tacked onto the end, will be the interference term, showing all the conceivable quantum juxtapositions that we never see in our world: particle in both positions A and B, in positions A, B, and C, in positions A and C, or B and C. Zurek wondered if the interference term could be thought of as information that is erased when we collapse the wave function. When we make a quantum measurement, these bits rapidly dissipate beyond our reach. Imagine the wave function as a sphere of glass. Hit it with a hammer and it shatters. The information in the interference term consists of the instructions for how to put the pieces back together again. Or, to stretch the metaphor, it is a kind of mathematical glue. In its absence, all the possibilities come unstuck and we are left with the familiar situation where something can be in only one state at a time.

Using this idea of decoherence—lost information—we can see the two-slit experiment, that paradigmatic example of quantum queerness, in a new light. First imagine the classical version, in which we are working with bullets. They shoot from a gun with a range of initial trajectories, some passing through slit A, some through slit B. The gun is not perfect, so if we open one hole, we get a probability distribution that might look like this:

BELL CURVE

What this bell curve tells us is that most of the bullets land in the center, many land near the center, but chance fluctuations cause a few to land farther away. Since we are dealing with classical particles, if we open both slits simultaneously the probability curve is simply A + B:

OVERLAPPING BELL CURVES

If we repeat the experiment with electrons, and open first one slit and then the other, we get the same two humps. But if we open both slits at once we get a quantum interference pattern:

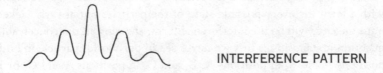

INTERFERENCE PATTERN

Mathematically, this can be represented by the three-term equation A + B + the interference term—the information that causes the possible outcomes to stick together, allowing the electron to act like a wave and flow through both slits.

When we put detectors at the slits so we can measure which way the electrons go, we get the two-humped classical distribution again. The detectors apparently cause decoherence. As they measure the particle, collapsing the wave, they siphon off the excess information, the interference term. The possibilities split apart and the particle goes through either slit A or slit B. Where does the information go? It is dissipated into the environment.

As a final variation on the experiment, imagine that instead of shooting the electrons through a vacuum, we shoot them through a gas of photons dense enough to ensure a reasonable chance that the quantum bullets will interact with the medium. The interference is gone and we get the classical distribution pattern again. Apparently we don't need a measurer or even an inanimate detector to cause decoherence. It seems that the environment itself can absorb the excess information and cause the possible outcomes to come unglued.

And there lies the beauty of this interpretation: there is no reason to give special status to an observer or to sanctify the measurement act.

Anything that can absorb information can be thought of as making a measurement. The collapse of the wave function can be shifted from the observer and placed on the environment itself.

"The essence is that the environment 'knows'—it has a record," Zurek said. "It takes the burden of having to collapse the wave function. We can blame the collapse on the watchful eye of the environment." Or as Zurek put it in his manifesto on information physics: "It is as if the 'watchful eye' of the environment 'monitoring' the state of the quantum system forced it to behave in an effectively classical manner."

Like the inevitable fall from order to entropy, this loss of quantum coherence is irreversible. According to the second law of thermodynamics, the entropy arrow points one way because, as the disorder of the universe increases, information is dissipated beyond retrieval. If we knew the speed and direction in which each shard of the shattered pot was flying, we could theoretically reverse their many courses and cause them to reembrace. This lost information is out there somewhere: in the vibrations of air molecules disturbed by the moving pieces. But these air molecules disturb other air molecules and the information scatters.

Likewise, the quantum-to-classical transition cannot be undone, because the extra information in the wave function is irretrievably scattered throughout the environment. The air molecules in the atmosphere, photons streaming from the stars—even in deep, dark space, the cosmic background radiation is there, interacting with quantum particles, causing decoherence, spiriting away the excess information at the speed of light. And, as Zurek says, you can't catch up with a photon. The environment is monitoring everything all the time, collapsing wave functions, bringing hard-edged classicality out of quantum mushiness.

Like a city saturated with sounds, the universe is saturated with the information from these constant measurements. We simply eavesdrop on the conversation. Like a mechanic measuring the pressure of a tire, or an electrician the voltage of a wire, we bleed off a tiny amount of information.

We are used to thinking of the difference between quantum and classical as the difference between the very small and the very large. Zurek sees it instead as the difference between a system that is closed and a system that is open. A completely closed system—one that is interacting with nothing—is represented by the pristine symmetry of the wave function, with every possible state and every possible combination of states in superposition, interfering with one another. In its unmeasured state, a quantum die would read six and one at the same time—and two and five, and three and four and five and six, and so on. But, in Zurek's view, these arbitrary superpositions are extremely unstable; they can survive only in a closed system. When we open up the system—forcing it to interact with the environment—they dissipate so rapidly that they cannot be retrieved. Only a small fraction of the possible outcomes that coexisted in the wave function are stable enough to endure on their own—the ones that behave classically, in which something can be in only one state at a time.

There are still many possible outcomes (six for the quantum die) left for us to contend with, but these are now probabilities that are knowable. So we go from quantum ignorance to classical ignorance, from inherent randomness to randomness rooted in what we don't yet know. The interaction with the environment acts like a sieve: it sifts the wave function, leaving a set of classical probabilities. Then we resolve our classical ignorance by measuring and gaining information.

The environment can be thought of as any system large enough to absorb the excess information and bring about decoherence. In this democracy of measurement, we cannot really say which is the observer and which is the observed. The photon leaves an imprint on our retina; our retina leaves an imprint on the photon. Measurement is simply the correlation of two systems, which go away from the encounter with a record of each other.

The result of all this is what we might call the classical illusion. Once decoherence occurs, all the pieces of the wave—even the weird juxtapositions—are echoing through the environment somewhere. But all we experience is the classical outcomes, where something is in one place at a time. For it is only these classical states that are stable, long-lasting, and predictable. We can measure them without fundamentally disturbing them, recording the information in our brains. There is no reason for our senses to perceive the superposed states, which dissipate so rapidly. They carry no meaning for us, they are invisible to our information processing. "Our senses did not evolve for the purpose of verifying quantum mechanics," Zurek has written. "Rather, they developed through a process in which survival of the fittest played a central role. And when nothing can be gained from prediction, there is no evolutionary reason for perception."

It is only when we can make stable records—memories—of something that it can be said to exist. Two things must leave imprints on each other to be mutually real—i.e., in the same universe. Decoherence makes possible information exchange, and it is only through information that we can know the world.

All of this goes a long way toward painting a picture in which classical probabilities arise from quantum uncertainty, without the need for observers. We are dealt this hand of premeasured, "decohered" states, but why do we experience just one of them? In the terminology of Everett and the many worlds interpretation, we make a measurement and the universe splits into branches: one in which the molecule is on side A, one in which it is on side B; six branches for each role of the die. But nothing in Everett explains why we are stuck on just one of the branches.

Zurek believes decoherence may hint at an answer. We have to remember, he says, that in this democracy of information exchange, our brains are also being measured by the environment. As we observe the die, neurons are momentarily placed in superposition between all the possible juxtapositions. But our brains are not closed systems. Blood brings in nutrients, carries off wastes. Networks of neurons send signals to one another. And so the superposed brain states decohere long before we are aware of them. The nonclassical states—the die simultaneously showing two, five, and six—are instantly spirited away. The brain, just like the die, will still be left with a variety of premeasured *classical* states, i.e., Everett branches. But here is the crux of the argument: there will be only one branch in which the die says six and our brain says six. That is the one that we perceive as real.

"Decoherence is preventing our brain from getting into these funny superpositions," Zurek said. "There is only one option in which I see the cup here and it *is* here. So we are in one state of mind at any one time and that state of mind is correlated with one state of the universe.

"It is sort of the reverse of the observer creating reality. The universe through your senses adjusts the record in your brain."

While Zurek is trying to use decoherence to understand the quantum measurement problem, some physicists are trying to apply the notion to the most fundamental wave function of them all—the one said to have emanated from the big bang. Zurek's colleague and intellectual sparring partner Murray Gell-Mann considers Zurek's interpretation a huge improvement over those that sanctify consciousness or the measurement act. But he feels that decoherence needs to be generalized for use in the exceedingly abstract realm of quantum cosmology. Working with James Hartle, a cosmologist associated with the University of California at Santa Barbara and the Santa Fe Institute, Gell-Mann has been attempting to understand that earliest instant when the universe was nothing but a primordial mass, no larger than a subatomic particle, emerging from the vacuum.

A world so tiny should by all rights obey the same laws as a photon or an electron. But how did the wave function of this ancestral "particle," with all its possible histories hovering in superposition, interfering with one another, give rise to our particular world? The old Copenhagen interpretation of quantum theory cannot begin to answer the question. "What was being measured when the universe was the size of an elementary particle?" Hartle asked. "You can't divide the system into observer and observed." It is fine to speak in terms of closed systems becoming open, interacting with their environment. But the universe, by definition, has no environment. There is nothing outside it. How do we think of decoherence in a system that remains forever closed?

Instead of talking about the environment, Gell-Mann prefers to cast quantum theory in terms of what he calls coarse-graining. One can always look at a system at different levels of abstraction. A very fine-grained description of a beach would include the position of every grain of sand. Viewed from a higher vantage point, the details become smeared together, the grains become a smooth expanse of brown. At this coarser level of description, different qualities emerge: the shape of the coastline, the height of the dunes. Could the qualities we call classical also emerge this way, a product of one's point of view?

A fine-grained description of the universe, the finest there could be, might include every possible position every particle could assume at every moment in time. (Or the description might include every particle's possible momentum—Heisenberg's uncertainty principle barring the possibility of knowing both.) In this most detailed of descriptions, all the information in the wave function would be accounted for, and all the possible histories would hover together in superposition. Like the possible histories in the two-slit experiment (particle through slit A, particle through slit B, particle through slits A and B), they interfere with one another, making it impossible to assign classical probabilities. In this state of pure potentiality, it is meaningless to say that a particle might be here or there. But if we look at the universe through a coarser mesh—following only certain things and ignoring everything else—the details of our description blur together like the grains of sand on the beach. If the description is coarse enough and done from just the right angle, the interference terms cancel one another out. The extra information is washed away. Now we can say that a particle is in one place or another and not in every possible place at once.

As Gell-Mann sees it, getting to the world of everyday existence requires still another level of coarse-graining. We live in a world where quantum effects are often negligible, where the motion of objects can be approximated to a remarkably accurate degree by classical Newtonian mechanics—a world he calls the *quasi*-classical realm, reminding us that nothing is purely Newtonian. The quality that allows for this stability is inertia: we must have objects that are large enough to resist tiny perturbations like those caused by photons bouncing against them. And so the description must be coarse enough that particles blur together into objects like billiard balls and planets.

Now this might still sound a little subjective. Who is devising these descriptions, these coarse-grainings? Physicists? God? But suppose, Gell-Mann asks, that there is a level of description coarse enough for all the possible histories to decohere, and coarse enough for inertia to emerge, but not the slightest bit coarser. This domain of "maximal" coarse-graining can be thought of as an objective feature of the universe, not dependent on the myopia of a particular observer. It is from this vantage point that physicists and astronomers behold creation.

Gell-Mann's interpretation is more general than Zurek's. The information that is ignored in coarse-graining can be thought of as the environment. Whichever way we think of decoherence, the essential point is that quantum information is lost somehow, banishing quantum interference and allowing a quasi-classical world to emerge.

But not every imaginable universe would decohere. In the primordial wave function emanating from the big bang, all possible histories—these potential universes—are suspended in superposition. Only some can be coarse-grained in such a way that the interference terms cancel out. The others remain stuck in quantum limbo. Whether a universal waveform includes histories, like ours, that decohere depends on the initial conditions of the big bang, the way the knobs were set at the moment expansion began.

One of the main goals of quantum cosmology is a theory of the universe's initial conditions, a compelling logic for why the knobs had to be set a certain way. Whether we are trying to describe the trajectory of an arrow arcing parabolically against the sky or the orbit of the earth around the sun, our laws of physics are cast in a certain form. On one end, which we can think of as the input, are slots for the initial conditions, the way something is at a certain time. Suppose that given the position at time t we want to know the position five seconds later. We plug the numbers into the equation and switch on the mathematical circuitry. Out from the other end the answer emerges. Unless we know the initial conditions, the law is useless.

But where do the initial conditions come from? We measure them, of course, but what if we want an explanation of how they arose? If we use the equation to extrapolate backward, we can calculate the conditions at an earlier time. Moving farther backward, we can show that these depend on conditions earlier still. But eventually our regression must bottom out: in the case of the arrow, the farthest back we can go is to the conditions with which it left the bow. And how do we explain the origin of these parameters? Only by considering the details of the bow and the archer. Unless we are willing to take these as given, we are led into the vast web of contingencies that conspired to put a bow of this type in the hands of an archer with these characteristics at this point in space and time.

In the case of the earth, we could trace the initial conditions back to the swirl of the cosmic dust cloud from which the solar system is said to have congealed. And if we wanted to know why the cloud was moving in just such a manner, we would have to extrapolate back further, eventually reaching the big bang. We always have to stop somewhere and take the conditions as given. Otherwise every calculation would have to take into account the history of the universe.

It is not just the past of a particular object that is abbreviated by the initial conditions. No classical system is an island, and so every measurement of the earth in its orbit or an arrow falling through the air reflects the gravitational perturbations of the rest of the universe, the distant tug of the stars. The effects of this vast web of influence, which we can lump together and call the environment, are implicit in the initial conditions we plug into the equations. The initial conditions act as a stand-in for everything we are leaving out.

But how are we to think of the initial conditions of the universe, of the big bang itself? This is where the buck stops, where all regressions must end. By definition, the universe is everything. Nothing is left out. The initial conditions, then, must be part of the fundamental laws.

Astrophysicists since Sir Arthur Eddington, in the years before World War II, have marveled at the ways in which the universe seems too good to be true. Anyone who follows popular science writing has heard the litany: if its expansion rate were a little slower, the universe would have collapsed in on itself; if it were slightly faster, there wouldn't have been the leisure for structures like galaxies to form. If something called the "fine structure constant," the square of the charge of the electron divided by the speed of light multiplied by Planck's constant, were about 1 percent different from what it is, then the universe would be unrecognizable, perhaps uninhabitable by anything remotely like us. These are the kinds of cosmological coincidences that so impressed John Updike's character Dale Kohler, a young computer scientist, in the novel *Roger's Version.*

What was it about the initial conditions that led to a universe like the one we see today? A universe that, viewed through the coarse mesh of our telescopes, seems fairly homogeneous—whichever way we look things appear pretty much the same. A universe that seems approximately "flat," with just the right density of mass to keep it delicately poised between open (expanding forever) and closed, doomed to eventual collapse. A universe in which the possibilities bound together in the wave function decohere, giving rise to a classical domain.

How did the knobs get set in just such a way? Why should we be so lucky? Why, against all odds, are we here, seemingly equipped with the neurological and mathematical tools to make sense of it all (or at least to give ourselves that illusion)? These seeming coincidences have been a comfort to those who seek scientific evidence for the existence of an Almighty God, who made the universe just so. Those more taken with Sartre than with the Holy Bible revel in another interpretation: we are flukes of a random, unfeeling universe. If the initial conditions had been set another way, there would be no stars, no planets, no astronomers.

In recent years, science has been reversing this perspective, looking through the other end of the telescope. Instead of attempting to explain how the universe gave rise to life, they start with life as a given and work the other way. Given that we are here, the initial conditions must have been a certain way. This approach is called the weak anthropic principle to distinguish it from the stronger version. In the weaker rendition, it is not that observers create reality, or are part of a symbiotic circuit in which the universe gives rise to observers which give rise to the universe by collapsing the wave function. The claim is simply that the existence of observers puts some constraints on the initial conditions, narrows the search.

Even this weaker version of the anthropic principle seems maddeningly tautological to some: the initial conditions had to be a certain way because otherwise we wouldn't be here to wonder about it. Some find the concept more satisfying if they are willing to buy the argument that there is not just one universe but many, maybe an infinite number, each arising from a different set of initial conditions. Only a tiny fraction of them give rise to life. Then the fact that life arose in this universe is no more remarkable than the fact that cities tend to arise at the juncture of rivers. There are still vast deserts—uninhabitable universes—where nothing blooms.

We could think of each of these other universes as possibilities rather than actualities, as paths through Everett space. Or we could take them quite literally. The version of the big bang theory called the inflationary scenario suggests a way multiple universes might be made: an instant after the big bang, a tiny patch of space-time could undergo hyperinflation, ballooning into an entire universe. Thus an arbitrary collection of random initial conditions would be amplified into what we marvel at as natural laws and constants. There is no reason why this couldn't happen many times, leading to a population of universes, each sealed off from the others by the upper limit of the speed of light and each operating according to a different physics. Or those who believe we live in an oscillating universe, with big bang followed by big crunch followed by big bang again, might suppose that each time the cycle begins anew the dials are randomly reset. In this view, the multiple universes are spread out in time rather than space. Naturally we find ourselves living in an incarnation where the laws favor life. But these laws, which we take as sacrosanct, would be accidents. The masses of the particles, the fact that there are three families of quarks, four forces, even the fact that there are three spatial dimensions (of the ten proposed by the superstring fans), might all be frozen accidents, outcomes of dice that could easily have landed another way.

For those who find the notion of multiple universes too spooky or the anthropic principle dissatisfying, but who are unwilling to see life as a miracle or a fluke, there is another recourse. They can adopt the approach favored by Hartle, Gell-Mann, Stephen Hawking, and others and look for some deep principle, an internal logic that dictated that the initial conditions had to be a certain way. When we look at the universe through our particular set of spectacles, we see laws and we see initial conditions. But why should this duality necessarily apply to the origin of the universe? If a way can be found to show that only certain settings of the knobs were possible, then the notion of laws and the notion of initial conditions would be one and the same. Cosmologists are still a long way from such a law, which would amount to no less than a precise description of the wave function describing the big bang. "The subject is in a rather primitive state," Hartle said. "It looks encouraging but there is a long way to go."

There is, for example, the problem of time. Wave functions evolve over time, but in the first instants of the universe the geometry of space-time is not fixed. It is subject to quantum fluctuations. "If you have two points you can't even tell if they are space-like separated or time-like separated," Hartle said. "So, in some sense, time as a fixed notion breaks down. That means you have to generalize the notion of quantum mechanics further so that it doesn't require a preferred time."

In collaboration with Stephen Hawking, Hartle has proposed a theory of the initial conditions—a wave function of the universe—in which, because of quantum effects, time and space are, in the beginning, indistinguishable. Most scenarios view the universe as a cone expanding from the singularity called the big bang. In this picture, time is represented by the vertical axis, the height of the cone, while the spatial dimensions of the universe are represented by the cone's circumference. Following along the cone's axis, the circle gets bigger and we have the notion of a universe expanding in time. But there is something arbitrary and inelegant about a universe that begins at a precise moment. Why then and not at some other time? And, as most cosmologists believe, if time, like space, began with the big bang, then what could it mean to say that the universe came into existence at a certain instant?

In the Hawking-Hartle picture, the base of the cone is not a singularity but a hemisphere. On this curved surface, time, the vertical direction, bends around, until it becomes indistinguishable from the horizontal, spatial dimensions. Time, like space, is simply a direction on the surface of a four-dimensional hemisphere. The universe could be said to unfold from any point on this surface. It makes no more sense to say that the universe began at a certain moment in time than it does to say that it began at a certain point in space. But begin it did; and as it unfolded, time gradually split off from space—the symmetry between them broke—leaving us with what Einstein called space-time. While the old-fashioned conical universe seems to require something outside of it to "decide" when to press the creation button, Hartle and Hawking contend that their universe is mathematically self-contained. And, as Gell-Mann has argued, the Hartle-Hawking wave function can be derived from superstring theory, suggesting that particle physics and cosmology might be tied together with a single bow.

There are still unanswered or unanswerable questions. Though the theory avoids the need for an outside Creator, it presupposes the prior existence of quantum theory and mathematics. Again the notion of a platonic netherworld beyond time and space sneaks in the back door. The best argument for the model so far seems to be its aesthetic appeal to those in search of a self-consistent, all-embracing system. If we want a universe that does not arbitrarily begin at a certain instant, we may have to restrict our search to those primordial wave functions that behave like Hartle and Hawking's round-bottomed cone. And if we want a creation story that can explain why the world we inhabit is so different from the world generated by the equations of quantum theory, then we must also restrict the search to wave functions capable of spawning universes that decohere.

But a wave function alone is not enough to explain how our own world, with its extravagance of detail, came to be. With his Eightfold Way and later developments, Gell-Mann showed how to reduce the hundreds of subatomic particles to a handful of quarks: at the roots of complexity was simplicity—if you are willing to allow that six flavors of quarks, each available in three colors and dancing with eight kinds of gluons, qualify as a simple system. In more recent years he has been gazing in the other direction and contemplating how simplicity gives rise to complexity. This is the question that drove him to help found the Santa Fe Institute. How, as he puts it, do you go from the quark to the jaguar? How did the universe become such an interesting place to live? In developing a theory of complexity, physicists are assuming that, in the beginning, there was very little information, just a few simple settings of the celestial knobs. How did we get to a universe teeming with structure, with galaxies and galaxies of galaxies? This is the question that has plagued so many cosmologists, compelling them to come up with more and more dark matter. In quantum cosmology the question arises in another, more fundamental form.

"According to the microwave background, which is as close as we can get to a picture of the universe in the early stage, there's practically no structure whatsoever," Hartle said. "What emerges is an early universe of remarkable simplicity. If there is a simple law of the initial conditions, it cannot contain much information, and therefore it certainly cannot describe the present complexity we see about us."

In fact, an explanation in which the information of the classical realm is implicit somehow in the wave function of the universe would constitute a hidden variable theory, which implies some kind of superluminal signaling. The implication is that the complexity of the universe was not encoded at the beginning but is injected along the way. "The present complexity we see about us arises mostly from the accidents of our particular history," Hartle said, "the particular rolls of the quantum-mechanical dice that have occurred since the beginning."

The rough features of our universe—cosmological flatness, the fact that it decoheres—may have been implicit in the initial conditions of the big bang. But the exact details—the order of the planets in the solar system, the shape of the Milky Way—are the accidents arising from a series of quantum splittings.

"In quantum cosmology we don't hope to predict from the wave function of the universe alone whether you sat in that chair or another chair or the number of planets in the solar system or the particular arrangement of galaxies in the sky," Hartle said. "Most things are not predictable from the initial conditions alone but only from the initial conditions *plus* the particular things that have happened in our particular history."

The history of trying to understand the world is replete with cherished fundamentals that were later abandoned as not so fundamental after all: the centrality of the earth was thrown out by Copernicus, the centrality of Euclidean geometry by Einstein's general theory of relativity, with its curved space-time; the wall between subject and object was breached by quantum theory; quantum cosmology may be on the verge of eliminating the distinction between initial conditions and laws. As Gell-Mann said in a speech at the Smithsonian Institution in 1987, important new ideas almost always are accompanied by "unnecessary intellectual baggage," prejudices that must be abandoned if we are to reach a deeper understanding.

We have seen how some cosmologists have concluded that we can't understand the universe unless we posit that as much as 99 percent of it is invisible. In a further attempt to squeeze creation into a simple set of equations, we are asked to see the classical world as an aberration, a special case of that which is held paramount: the equations of quantum theory. What we take for granted seems less and less significant all the time.

Still, there is something comforting about another implication of this scenario: the notion that information-gathering beasts may have been implicit in the big bang, written into the initial conditions. Only certain primordial settings would give rise to universes in which decoherence is possible and thus to a world with a classical domain, where it is possible to make partitions, to engage in coarse-graining, to divide subject from object, to process information—a world orderly enough for information-gathering-and-utilizing systems, "Iguses," as Hartle and Gell-Mann call them, to arise.

So perhaps, in a very general sense, we are not flukes of the universe. The particular form we have assumed may be the outcome of a series of accidents, a branching chain of contingencies. But the fact that we live in a universe that allows for information processing may be a necessary (or at least highly probable) outcome of the initial conditions. We don't need the anthropic principle, weak or strong, to feel at home. Given a universe that allows for classicality—where things can be measured without setting them hopelessly askew—it is not surprising that Iguses of some sort would arise to exploit the situation, beings that live off information.

If we rack up a set of billiard balls and send the cue ball racing toward them, balls bounce off balls and against the edges of the table. But the ricocheting soon stops. If we walked into the poolroom and saw the balls lying scattered across the felt, we couldn't work backward and calculate the initial condition: how and where they were arranged on the table. But though the balls stop, the bits keep flying. How fast the yellow ball was going when it struck the red one, where on the table the green ball bounced and at what angle—all this information is dissipated into the environment as heat. Though the signal gets weaker and weaker, its echoes remain; molecules in the environment have been rearranged, a recording has been made. We can imagine a beast whose perceptions are so acute that it could measure the changes caused by the thermal vibrations of the molecules, then use the data it collected to calculate backward and reconstruct the beginning of the game. Remember Maxwell's glass of water dumped into the sea. It is not really gone. We just lack the wherewithal to track all the molecules and retrieve them.

Nor does the information that dissipates during decoherence disappear. It is still in the universe somewhere, like the television shows that radiate forever into space. A being who could follow every bit of information exported into the environment when a wave function unraveled would be like the quantum version of the Laplacian demon: it wouldn't see a classical world with objects in single locations moving at certain speeds. It would just see all the possibilities.

In fact, it would be impossible for such a being to exist. For one thing, much of the decoherent information is carried away by photons; it is impossible to follow because it escapes from us at the speed of light. Einstein's special relativity helps bring about the classical domain by setting an upper limit to the power of perception. But even more basic is this paradox: the demon's acuity would ensure that the possibilities in its own wave function never decohered. It too would be stuck in the limbo of potentiality.

If the quantum theorists are right, we and the classical world we live in exist because it is impossible to be so acute as the demon. Information-gathering-and-utilizing systems can arise because it is possible to ignore a huge amount of detail, to engage in coarse-graining. We partition the universe into an area of interest and an environment to which we can banish excess information. And so we can make rough predictions. Iguses exist by virtue of this myopia, this inherent inability to keep track of every detail. If you know everything, you know nothing.

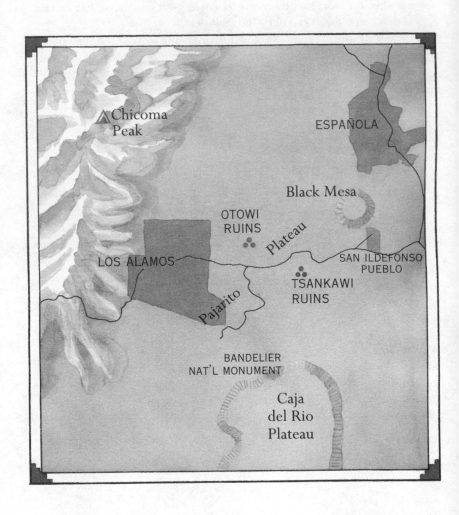

THE MYSTERY OF OTHER MINDS

On a cold January morning, a little more than a month after the winter solstice, the time when the sun ends its journey southward and begins its slow return toward the lake of emergence in the north, the people of San Ildefonso gather in the hills near their pueblo and wait for the animals to come home. San Ildefonso sits at the foot of the Pajarito Plateau, on which Los Alamos stands. As the first rays of sunlight come slanting over the Blood of Christ Mountains, lighting the mesas that step their way from the pueblo in the lowlands to the laboratory in the sky, the morning silence is broken by the steady low-frequency beating of drums. As a plume of smoke rises from a nearby hilltop, a chorus of men, dressed in headbands and shirts of many colors, begins singing an ancient song. Then the hunt chief, wearing buckskins and carrying a bow and a quiver of arrows, calls out sharp and high, and the dancers, their faces painted black and their heads bristling with horns and antlers, begin their descent.

Sprinkling cornmeal on the ground before him, cutting a sacred trail, the hunt chief and the buffalo   each with a gourd rattle in one hand and a small bow-and-arrow in the other—lead the way to the North Plaza, where a day of thanksgiving begins. The deer, hunched over and holding canes in their hands for forelegs, dance nimbly on all fours. Rising above them, the buffalo, tall and erect, meander in slow, deliberate circles, sweating beneath their heavy costumes. Darting in and out of this tight choreography are the antelopes and the bighorn sheep. Alternating between fast songs and slow songs, the animals complete the cycle of the dance, then retire to the kiva to rest and engage in rituals not open to the uninitiated.

Before one has long to contemplate these mysteries, the quiet is broken again as the South Plaza is taken over by the Comanche dancers, dressed in the brilliant feathered headdresses worn by their old enemies from the eastern plains. And when the Comanche dancers have returned to their own kiva, the animals emerge onto the North Plaza again. And so it goes for the rest of the day, the Comanche dance alternating with the buffalo dance, much as it has every January 23 for as long as anyone can remember.

Like most native rituals in the Southwest, the dances at San Ildefonso give a perfunctory nod to the religion of the Spanish conquerors. Early in the evening before the annual feast, when the dark, unpaved streets of the pueblo are lit with bonfires, the Roman Catholic priest assigned by the Archdiocese of Santa Fe celebrates mass at the adobe church built off to the side of the plaza. Then, to the sound of gunfire, an old Spanish custom meant to scare away evil spirits, he leads his own procession, carrying a statue of the pueblo's patron saint, Ildefonso of Spain, counterclockwise around the plaza and back to the chapel. Perhaps fifty people, a third of them tourists, follow behind him, singing hymns. Then, with the Catholic formalities dispensed with, the long wait for the dancers begins.

Before the Spanish brought the Gregorian calendar, laying its cycles over those of the moon, January was called the Ice Month. Once the sun has descended beneath the western lake, to spend the night in the underworld, temperatures drop quickly. Just when it seems too cold to stay a minute longer, more bonfires are lit, and the hunt chief calls out his distinctive cry. Illuminated by the firelight, the animals emerge from a kiva on the north side of the pueblo and parade slowly around the plaza, providing a preview of the drama to come the next morning.

Over the years, many of the men and a few of the women at San Ildefonso and the other Tewa pueblos of the northern Rio Grande have found work up the hill at Los Alamos National Laboratories. As electricians, carpenters, plumbers, secretaries, and laborers, they help provide the infrastructure that allows the scientists their long explorations and flights of speculation. A few of the scientists have befriended some of their Tewa neighbors, driving down the mesa to watch the dances. But on a deeper level there is little contact between these two adjacent worlds.

For the physicists who occasionally make the short journey to San Ildefonso, perhaps there is something comforting about descending from the stratospheres of abstraction to be among people commonly said to live closer to the earth. But the Tewa, too, have their system of abstractions, in some ways as lofty and intricate as those of the physicists. Over the centuries they have woven a tight web of concepts, invented imaginary spaces, laid a grid of geometry over the irregularities of the rough New Mexico terrain.

For those who wonder about the drive to find and impose pattern on the world, the Tewa might be seen, in retrospect, as participants in an unintentional experiment. We think of the science our civilization has developed as all but inevitable, an unearthing of preexisting truths. But suppose you take an equally curious society and isolate it from the Western philosophical tradition stretching back to ancient Greece. As they sift their environment for patterns, what kind of system would these people come up with to explain their world?

As outsiders we find it hard to believe that a dance, no matter how beautiful and intricate, can bring a more successful hunt, more bounteous crops, or better weather. How is it, then, that the Tewa system of beliefs has endured invasion after invasion, by Catholic missionaries, anthropologists, and the physicists of the Manhattan Project? While science is in constant flux, the finely etched universe of the Tewa has survived remarkably intact for centuries, its ideas so firmly lodged that no invader has been able to unseat them.

It is one of the curiosities of New Mexico that the faith of these descendants of the Anasazi has become joined, however tenuously, with the faith of the Church of Rome. In the Tewa world and throughout the Rio Grande, pueblos celebrate the day of their designated saint with dances. On Christmas they might dance the buffalo dance, the Comanche dance, or the turtle dance; on Easter, the corn dance, the basket dance, or the bow-and-arrow dance. Christmas itself was derived from celebrations of the winter solstice—the celebration of the sun became the celebration of the Son. And Easter was adapted from celebrations of the spring equinox, when the day finally becomes as long as the night. When, through a strange series of historical coincidences, the Franciscan priests came to this corner of the New World and found the pueblos dancing to the rhythm of the star whose energy feeds the earth, Christianity was rediscovering its pagan roots. The friars didn't see it that way, of course, and tried to realign the pueblos' faith with their own. In the end, it was the pueblos that prevailed, appeasing the foreigners by absorbing some of their rituals, then going on much as they had before.

Small details of San Ildefonso's buffalo dance may vary from year to year, but more remarkable is how much has remained the same. Toward the end of the twentieth century, the ritual seems very much like the one described in 1945 in the journals of Edith Warner, who came from the East to live in a small house on the San Ildefonso reservation, where the road up to the Pajarito Plateau crosses the Rio Grande on what once was called Otowi Bridge. What Warner saw was very much like what the anthropologist William Whitman described in 1937 and his predecessor Elsie Clews Parsons in 1926. And there is little reason to suppose that what Warner, Whitman, and Parsons saw has changed much in the hundreds of years the pueblo has sat at the spot where the Nambe River, flowing down from Lake Peak in the Sangre de Cristo Mountains, joins the Rio Grande. Long before 1617, when the Spanish missionaries decided that the pueblo, called "Where the Water Cuts Down Through," in Tewa, should be renamed for a former archbishop of Toledo, its inhabitants were gathering to celebrate a world in which animals are willing to lay down their lives so that people may live. Whatever the good and brave deeds Ildefonso was sainted for, on the day of his feast his subjects' thoughts are more likely to be focused on a time long before there was a Vatican or even a Christ—a time too long ago to remember, when people, animals, and spirits all lived together, in the underworld, and spoke the same language.

The exodus from this subterranean cosmos, where the sun, pale as the moon, shone all the time, began when beings called Blue Corn Woman and White Corn Maiden asked one of the men to explore the surface of the earth and see if it was fit for habitation. After refusing to go three times, upon the fourth request he emerged through a lake, somewhere around what we now know as the Colorado border, and walked north, west, south, and east, moving counterclockwise. Seeing only mist and haze in every direction, he returned to report that the world above was formless and chaotic—green, unripe, not ready for intelligent beings.

Not willing to give up so easily, the Corn Mothers sent their explorer back for another reconnaissance. This time he was surprised along the way by a pack of wild animals and insects—mountain lions, wolves, coyotes, bears, foxes, vultures, crows, dragonflies, and bees. Because he was afraid, they attacked him. Then, persuaded that he had learned his lesson, they became his friends and confidants. They gave him a bow-and-arrow and buckskins; they painted his face black and tied feathers in his hair. And he returned to his people as the hunt chief, the keeper of the magic bond between hunter and prey.

Summoning everyone in the underworld with a fox call, the hunt chief appointed two assistant chiefs, one to rule the people in the summer, one to rule in the winter. These chiefs then sent six pairs of brothers to explore the earth and map the terrain, to impose order on the world. The first pair, who were colored blue like the cold, headed north. They didn't make it very far because the ground was still so soft, but in the distance they saw a bluish shape which they named Hazy Mountain. And so it was that geography and cartography began. Having named the north, the exploration proceeded counterclockwise. The yellow brothers went west, the red ones south, and the white ones east. Each saw a mountain in the direction they were traveling and gave it a name. After that the only directions left to investigate were up and down. And so the brothers who were colored black were sent upward to the zenith to explore the darkness of space; instead of a mountain, they saw a large star in the eastern sky. The last pair of brothers, the multicolored ones, explored the nadir of this new universe and saw a rainbow.

Having prepared the upper world for habitation, the people were eager to live in this newly charted land. But first they had to organize themselves as neatly as they had organized the land between the mountains. It was not an easy task. They made four ill-fated attempts to colonize the land above the lake, returning each time to the underworld. And each time, to help with the next expedition, new categories of people were created: the summer and winter clowns (called the Kossa and Kwirana) to keep the people happy on their journey, the Hunt Society to keep the people nourished, the medicine societies to keep the people well. Finally they were ready to try the exodus for the final time. Dividing themselves into two groups, led by the summer chief and the winter chief, they left the lake. Upon reaching a big river, the Rio Grande, they divided themselves into tribes. The Navajos, Utes, Apaches, Kiowas, and Comanches went off to live as nomads, in houses of deer and buffalo hide; the others headed downstream to build villages of adobe. The summer people proceeded south along the west side of the Rio Grande, the winter people along the east side. After each made twelve stops, forming pueblos like Taos and Picuris in northern New Mexico, the ancestors of the Tewa rendezvoused to start a village, called Posi, on a mesa above the site the Spanish later named Ojo Caliente, or Hot Springs. After an epidemic forced them to abandon the village, they moved to the canyons and mesas of the Pajarito Plateau, at the edge of the Jemez Mountains, and to the flatlands to form the Tewa villages, including those that exist today: Nambe, Pojoaque, San Ildefonso, San Juan, Santa Clara, and Tesuque. According to their tradition, the San Ildefonso Tewa once lived in Otowi and Tsankawi, two fallen cities just up the road toward Los Alamos. The Santa Clara Tewa, who live just north of San Ildefonso, say their people came from the nearby Puye cliff dwellings. The San Juan Tewa consider Posi, by Ojo Caliente, their ancestral home.

To this day, the Tewa of northern New Mexico divide themselves between summer people and winter people, each with separate ceremonial responsibilities. Until recent years, each pueblo had a winter chief and a summer chief. The twins who explored the horizontal directions are still said to live atop the four sacred mountains. Hazy Mountain to the north is Canjilon Peak (though some say it is San Antonio Peak, northwest of Taos); the west is marked by Obsidian Covered Mountain (Chicoma Peak in the Jemez Mountains); south is Turtle Mountain (Sandia Crest); and east is, depending on whom you ask, either Blue Stone Mountain (Lake Peak) or Stone Man Mountain (Truchas Peak), both in the Sangre de Cristo range. Within the domain of the four mountains, each pueblo has its own four sacred mesas (created when the twins threw handfuls of mud in each direction), and within the ring of mesas, four sacred shrines—a field of overlapping, concentric tetrads each converging on plazas, which, to their people, have long been the center of the world.

Today, walking among the fallen walls of Tsankawi, the mesa-top ruins the San Ildefonso call home, one can't help but wonder: Why did the Anasazi ancestors so often build their cities on the hot, dry mesa tops instead of in the shade of the canyons, where water at least sometimes flowed? By day they would tend their crops down by the stream, then climb up the steep cliffsides to spend the night closer to the stars. Were they fortifying themselves against enemies, or just seeking a higher vantage point, a place to gaze out upon the world? Standing atop Tsankawi, the four sacred mountains rising majestically all around (Sandia is stretched out so that it indeed looks like a giant turtle), it is easy for one to pretend what it might have been like hundreds of years ago. The village filled with dancers imitating animals and the hunt. Artisans molding pots and painting them with the intricate black and white designs one now sees shattered in pieces all across the plateau. The priests sitting inside the kivas quietly recalling the story of the emergence from the underworld—how the gods ascended to the surface and found disorder all around, and laid out the four directions, a bubble of lawfulness amid the chaos, and gave the people the job of keeping it from collapsing, of imposing the geometry of the dance onto the messiness of the world. Surrounded by the rough, eroded landscape of mesas and arroyos, one can imagine the Tewa longing for the pristine harmony of their heaven underground.

In their most basic form, the stories of the people of Tsankawi—worn smooth with so many retellings in the never-ending effort to make sense of the world—are reminiscent in spirit of those told by the people who would later fortify themselves atop a neighboring mesa in a city called Los Alamos. Like the gods beneath the earth, the laws of science were said to exist in a hidden realm—the platonic world of pure mathematics—their perfect symmetries shattered, like those carefully crafted pots, in giving birth to the world. While the keepers of the laws of physics spent billions of dollars on giant machinery, spinning unseeable particles around and around the sprawling plazas of their giant accelerators, crashing them together at higher and higher energies, trying to re-create for an instant the unified perfection of an ancient time, the Tewa kept on dancing, as though the rhythm of their footsteps would awaken the hidden orders underground.

Like the keepers of the laws of physics, the Tewa's Anasazi ancestors were seeking their own compressions, a system to distill the essence of their variegated, capricious, often dangerous world. And so, in their attempt to explain the strange in terms of the familiar, they took what they knew and arranged it into a harmonious whole, an inner religious world that exists, though slightly faded, to this day. Having no familiarity with concepts like mass and energy, they began with their own fundamentals: the six directions, the colors, the seasons, the creatures that inhabited their land. North, where the ice lingered, was associated with the color blue and the mountain lion. West, where the sun set, was yellow, the domain of the bear. South, where the air was hotter, was the red direction, the land of the badger. East, where the sun rose, was the white direction and the lair of the wolf.

Each direction also had its own bird, snake, shell, and tree. Walking though this spiritual force field, generated by the poles of the four magic mountains, were the two kinds of people: summer and winter. And within each of these divisions were the clans. At San Ildefonso, most of the winter people were of the Turquoise Clan, while the summer people usually belonged to the Red Stone or Sun Clan.

And the system was still more complicated than that. Cutting vertically across the summer-winter duality, and across the concentric tetrads of mountains and mesas, the meridians with which the Tewa fixed themselves on the land, was a ladder of three levels used to assign each person a place in the universe. At the bottom level were the Dry Food People, the men and women not privy to the inner secrets of the religion. On the level above were the Towa é, the earthly representatives of the sacred twins who discovered the four magic mountains. The Towa é served in the various political roles necessary to keep the society functioning and acted as mediators between the Dry Food People and those on the top rung of the hierarchy, who were known as the Made People. Led by a high priest (usually known in recent times by the Spanish title "cacique") and by the summer and winter chiefs, the Made People were the keepers of the religion: the men who belonged to the Hunt Society, the Warrior Society, the Fire and Flint medicine societies, the Kossa and Kwirana clown societies, all meeting secretly in the kivas or communing with the spirits on private retreats. A few women sometimes belonged to the medicine and clown societies, and they had an organization of their own: the Scalp Society, which was charged with caring for the trophies taken in war.

There were various ways a Dry Food Person might become initiated into a society and become a Made Person. A sick child might be bequeathed to a medicine society in return for its being cured. Or someone might stumble upon a society on one of its secret retreats and be compelled to join to protect the knowledge. Curiosity was tempered by a deep fear of possessing unauthorized information: the societies guarded their secrets as jealously from one another as they did from the Franciscan spies. It was once said that at Acoma and Santo Domingo betrayal of religious secrets could mean death.

Even the clowns, or Delight Makers, as the anthropologist Adolf Bandelier called them in his novel about prehistoric life on the Pajarito Plateau, were believed to be possessors of powerful secrets. So important were they to the well-being of the pueblo that they were given a latitude of behavior no other Tewa would dare emulate. They would talk backward, saying the opposite of what they meant, or make fun of the shortcomings and transgressions of even the highest officials. In the middle of the most solemn rituals, they would stumble onto the plaza, mocking the dancers and the singers. They would grab women from the audience and make lewd gestures. They would eat (or at least pretend to eat) human waste. Their purpose was apparently to show people how *not* to behave, and to provide some comic relief from the responsibilities the whole pueblo shared in keeping their world in resonance with the gods.

The three-part structure of Dry Food People, Towa é, and Made People was mirrored in the subterranean spirit world. The commoners were associated with the spirits of their dead ancestors, the Dry Food Who Are No Longer. The Towa é were associated with the twins who lived on top of the four sacred mountains. The Made People were associated with the spirits who stayed behind when the people emerged from the lake: the Dry Food Who Never Did Become. Though these spirits never walked on earth, some of them journeyed across the sky, ascending and descending through distant lakes that led to the underworld. The sun, the moon, the stars, the constellations, fire and wind—all belonged to this highest form of being, as did the Oxua, or Cloud Beings, the Tewa counterpart of the Hopi's kachinas.

The star spirits congregated in constellations called Meal-Drying Bowl, Star House, Big Round Circle, Sandy Corner, Turkey Foot, the Hand. Ursa Major was called Seven Corners, Seven Tail, or sometimes just Dog Tail. But most of the heavens went unnamed. Rooted so closely to the earth and the underworld, the Tewa seemed to pay little attention to the arrangements of the stars. Far more important were the sun itself, whose excursions north and south determined the solstices, and the moon, whose phases marked the monthly cycles. Together they would tell the people when to plant, when to harvest, and when to dance. And the dances would ensure that the rhythms were maintained.

While much of the sky went uncharted, the Tewa categorized the world around them with a level of detail unmatched by the cartographers of Spain or the United States. When the anthropologist John Peabody Harrington explored the pueblos in 1910, he wrote of a land "thickly strewn" with names. Where a visitor might look out on a seemingly undifferentiated expanse of mesas, hills, and arroyos, the Tewa saw a map on which even the most unobtrusive features had been given a name—information recorded nowhere but in the heads of the people who lived there, fragmented among many minds. Any one person's knowledge extended only so far beyond the pueblo, then another's memory took hold. In trying to commit it all to paper, Harrington ended up with a report some six hundred pages long, filled with thousands of entries.

At San Ildefonso there were, to cite but a handful of examples, Deer Tail Mesa, Red White-Earth Arroyo, Mountain of the Great Canyon, Arroyo of the Fire Gully Gap, Cave Dwelling in Which the Meal Was Placed, Place of the Blue Water Man, Red Stone-Strewn Canyon, Place of the Two Arroyos, Down Where the Spider Was Picked Up, Arroyo with Chokecherry Growing at Its Little Bends, Mesa Where the Donkey Was Killed, Hill Where the Snakes Live, Gap of Sharp, Round Cactus. The Tewa's maps were drawn with the finest of grains. Next to a place called Where the Comanche Fell Down, one would be sure to find the Arroyo Where the Comanche Fell Down.

Like the antics of the Kossa and Kwirana clowns, some of the names reflected an attitude toward sexuality far more relaxed than that of the Europeans. The Franciscans were trying to impose their curious notions of original sin on a people who thought nothing of assigning names like Loathsome Penis Mountain (the words probably lose something in the translation), Little Corner of the Hard Penis, or Vagina Kiva Spring. In his report, Harrington proudly noted that he was the first non-Indian to be taken to a shrine called Stone on Which the Giant Rubbed His Penis, which was by the Arroyo of the Stone on Which the Giant Rubbed His Penis. The giant was said to live atop Black Mesa, the volcanic monolith that is revered as the northern sacred mesa of San Ildefonso. Parents had long warned their children that this beast was so mighty it would stride into the pueblo with four giant steps and grab disobedient children to roast for dinner.

Reading Harrington's report, one has to wonder whether the Tewa, confronted by a nosy Anglo anthropologist pointing to every hill and dale and asking what they were called, sometimes amused themselves by engaging in a bit of improvisation. Alfonso Ortiz, a Tewa of San Juan pueblo who became a professor of anthropology at the University of New Mexico, recalled that when he told his father and uncle some of the names in Harrington's report they howled with laughter. Much of Harrington's information came from two young Santa Clara men, who may have had a passing familiarity with the names around their own pueblo. Encountering the terra incognita of nearby San Juan, they apparently decided to improvise, knowing that otherwise Harrington would stop handing out quarters. As Harrington conceded in his report to the Bureau of Ethnology in Washington: "The difficulties encountered have been many. The Tewa are reticent and secretive with regard to religious matters, and their cosmographical ideas and much of their knowledge about place-names hard to obtain." Yet, though the details may be unreliable, Harrington's survey stands as evidence that the Tewa gathered information as assiduously as their later conquerors, mapping both the physical world around them and the imaginary mental spaces that they associated not with mathematics but with the gods, the same ones they revere today.

The details of the Tewa creation story may strike an outsider as whimsical and arbitrary. But it is not always clear that the anthropologists have done much better in explaining how the pueblo people arrayed themselves into the patchwork of cultures we find today. While the Navajos and other Apaches who live at the fringes of the Tewa world are Athabascans, relatives of the Inuit who migrated down from Canada hundreds of years ago, the many different groups of pueblo Indians are believed to have descended, one way or the other, from the Anasazi empires of Chaco Canyon, in western New Mexico, and Mesa Verde, in southern Colorado.

The Tewa are indeed believed to have come from somewhere in the north, as their version of Genesis has it, around 1300, after Mesa Verde and Chaco Canyon collapsed for reasons that are still obscure. The migration may not have been the continuous journey southward that is described in the myth. Similarities between the Tewa languages and Kiowa, spoken by a tribe on the western plains, suggests a bit of meandering. But the stories the Tewa tell survive a certain amount of scientific scrutiny. Before settling in pueblos along the Nambe River and the Rio Grande, the Anasazi indeed built stone-and-mud villages in the flatlands around Ojo Caliente and on the mesa tops of the Pajarito Plateau; they scooped dwellings from the soft volcanic tuft of the canyon walls.

Linguists tell us that the Tewa are part of the Tano language group. Northeast of the Tewa, the inhabitants of the pueblos of Taos and Picuris speak a closely related dialect called Tiwa. Tiwa is also spoken in two isolated southern pueblos, Sandia (just north of Albuquerque) and Isleta (just south of Albuquerque). The speakers of a third dialect, Towa, live in Jemez pueblo, southwest of Santa Fe, joined by the remnants of the other great Towa pueblo, Pecos, decimated by Comanches and Spanish smallpox and abandoned in the nineteenth century.

If all the pueblo peoples of the Southwest spoke dialects of the same language, the anthropologists' task would be much easier. But in between the northern and southern domains of people who communicate in Tano are speakers of a very different language called Keres. They live at Cochiti, Santo Domingo, and San Felipe pueblos, between present-day Albuquerque and Santa Fe, and to the west at Santa Ana, Zia, Laguna, and Acoma. Still farther west are two pueblos with languages completely unrelated to either Tano or Keres and completely unrelated to each other: Zuni and Hopi.

On the basis of language alone, it would be tempting to think of the pueblo world as consisting of several different cultures. But again, the situation is more complicated than that. Though their languages are different (even two groups speaking Keres can find it difficult to communicate across the dialectic divide), the pueblos share a structure of beliefs that is remarkably similar. Each pueblo has its own tetrads of sacred mountains, its own stories about the emergence from the underworld; each celebrates the exploits of the hero twins. To ensure good hunting, people throughout the pueblo world don animal costumes and dance the buffalo dance. To ensure a good harvest, they dance the corn dance—the men with bare chests painted black, wearing white kilts and moccasins, shaking gourd rattles; the women barefoot in their black dresses and red sashes, with a pine bough in each hand and a turquoise-colored crown called a *tablita* strapped atop the head. And pueblos all over the Southwest have their sacred clowns to make sure that no one takes the intricate system so seriously as to confuse it with life itself.

Despite all the similarities, there is nothing like a monolithic pueblo religion. While the Tewa are divided between summer and winter people, the Keres have squash and turquoise people. While the Tewa say north is blue and west is yellow, the Keres reverse these designations. While the Tewa speak of a single underworld which they left after four tries, other pueblo peoples talk of a four-level underworld. But the resemblances tantalize the anthropologists, with their incessant need to find patterns.

In a culture without a written language, ideas leave little trace. There are no texts that can be examined for clues to the origins of pueblo religion. It is tempting to take the similarities among the pueblos' beliefs as signs of an ancient Anasazi religion, whose splinters developed in different ways. But it is impossible to tell how much of the resemblance came from later borrowing, ideas bartered back and forth with the invisible hand of the marketplace spreading the wealth around.

Today, anthropologists find a gradient of beliefs that flows across the Southwest. The worship of masked gods—the kachinas, or Cloud Beings, as the Tewa call them—seems strongest in the western pueblos of the Hopi and Zuni, fading as one heads east. The dualistic divisions are strongest in the east with the Tewa's summer and winter people, a little weaker in the center with the Keresans' squash and turquoise people, and weakest of all among the Zuni and Hopi. The medicine societies seem strongest among the Keres, fading as they radiate outward. Like pieces of a puzzle, these scraps of data are fitted together by scientists searching for an all-embracing order. But in many ways the pueblo peoples and their complex world have resisted explanation in terms other than their own.

For the Tewa, mythology left off and history began when the Spanish arrived in the sixteenth century, marching up from Mexico, and began to write the Tewa into their own story. To the Spanish and their Franciscan priests, the spirits lived in the sky, not under the earth. Unaware of or uninterested in such designations as summer and winter people, these European visitors imposed a different duality on the natives of the northern fringes. The *indios de los pueblos* lived in villages like San Ildefonso and grew crops. Surrounding this civilized world were the *indios bárbaros,* the Comanches and Apaches, who moved from place to place and raided the pueblos for food. One group of Apaches, the Navajos, had come into the area only slightly before the Spanish.

It is not certain just when the people of San Ildefonso saw their first Spaniard. Coronado's troops rode into Tewa country in the early 1540s, but it was not until 1598 that Juan de Oñate, who led the first group of settlers to New Mexico, visited San Ildefonso. From then on, linear time proceeded with a vengeance. The Ice Month was renamed January, after the two-faced Roman god Janus; the Wind Month became February, after a Roman purification rite; the Month When the Leaves Break Forth became March, named for Mars, the god of war; the Month When the Leaves Open became April, after an Etruscan name for Aphrodite; Corn Planting Month became May, after another Roman goddess. Before long even the Tewa themselves were renamed. By the time the American anthropologist Elsie Parsons arrived in the early twentieth century, it was not unusual for a San Ildefonso man to be named both Star Mountain and Santiago Alarides; a man called Starbird was known outside the pueblo as Juan Jesús Piña.

The changes were not simply a matter of laying new names on top of old ones. Time became divided with a finer and finer grain. While the Tewa had been content to segment the year into months, the Spanish gave every day a number and a name, and then the days themselves were mechanically sliced into *horas, minutos,* and *segundos.*

Before long, the Franciscans were trying to change the very names of the Cloud Beings to correspond with the Catholic Church's holy saints. Throughout the pueblo world, kivas were burned or filled with sand; masks and other sacred relics were destroyed. In 1675, forty-seven pueblo religious leaders were accused of witchcraft and idolatry and brought to Santa Fe for public flogging. One of the leaders was Popé, the San Juan Indian who would lead the Pueblo Revolt five years later.

San Ildefonso was one of the strongest supporters of the revolution, killing two of the Franciscans who lived there. And when Diego de Vargas staged his Reconquest twelve years later, the pueblo was one of the last to succumb, the people fleeing to the top of Black Mesa, the lair of the legendary giant, to fight a worse monster. For two years after most other pueblos had surrendered, San Ildefonso staved off Vargas's men. But finally the Spanish prevailed.

It is often said that the Spanish had learned their lesson in the Pueblo Revolt, that after the Reconquest they were more tolerant of the Indians' religion. The Tewa didn't seem to see it that way. In a smaller uprising in 1696, four years after the Reconquest, San Ildefonso sealed the doors of the Catholic church that had replaced the one destroyed in the revolt and set it on fire, burning the missionaries trapped inside.

It is hard to imagine what it must have been like to craft so carefully an inner world, a mobile of colors, spirits, animals, and mountains, all in such delicate balance, only to have it capsized by interlopers from far beyond—a people who looked different from any ever seen, with far more riches, far more powerful weapons, and a religion that they insisted was the only true one. It must have seemed like an invasion from another planet.

Pueblo religion was not set in stone, or even written in a book. As they brushed up against other peoples—the Keres pueblos to the south, the Hopis and Zunis in the far west—the Tewa religious leaders seemed perfectly willing to learn about new gods and adopt them into their pantheon. And the borrowing went beyond the pueblos. Today Buffalo dances often begin with songs from Plains tribes. Even the Tewa's old enemies the Navajos speak of four sacred mountains and six colored directions. No wonder the pueblos were pushed to violence when the Franciscans, bringing these kachinas they called saints, not only refused to reciprocate by respecting the Cloud Beings but tried to stamp them out.

Catholicism tried to assimilate Tewa religion, but ultimately it was Tewa religion that assimilated Catholicism. At each pueblo, the church remains off to the side, away from the kivas and the plaza. At the corn dances, the pueblo's patron saint sits on an altar in a tent at the edge of the plaza, bedecked perhaps with turquoise necklaces and beads—a sideshow to the main attraction, the dancers who fill the pueblo with color and sound.

While Catholicism was more or less defanged, in some ways the gods of Western science were more insistent, and harder to fathom. The Catholics, like the Tewa, had their mysteries; knowledge was something sacred, a means of connecting with the spirit world. And the most powerful knowledge of all was to be carefully guarded by a priesthood lest it fall into the wrong hands. What then was one to make of these anthropologists with their curious doctrine of knowledge for knowledge's sake? While the Spanish had come seeking gold and converts, the anthropologists were plundering in their quiet but persistent way for information, not so they could join the Indians in harmony with the gods (or compel them to worship their own), but so they could explain them away, absorb their religion as a detail of their own growing belief system.

In 1879 the anthropologist Frank Hamilton Cushing came from the East to spend five years among the Zunis, whose villages the Spanish had mistaken for the golden cities of Cíbola. Cushing moved into the governor's house and broke down the Zunis' resistance by learning to speak and dress like them. Then, in an anthropological coup unmatched before or since, he was initiated into the secret Bow Society. One of the membership requirements was to obtain a scalp for the society's collection. To this day it is unclear where and how he got it. Some Zunis insist that he scalped a Navajo. One likes to think that he purchased it on the black market or, as he implied in a vaguely worded letter, obtained it somehow in an area recently raided by Apaches. In any case, to gain initiation as a bow priest he apparently had to give the Zunis the impression that he had come by it in the traditional manner. Cushing seemed to genuinely care for the welfare of his subjects. He helped champion the land rights of the pueblos against Anglo encroachers. But in the end, the Zunis felt betrayed when he published their secrets for all the world to see. Knowledge for the sake of knowledge.

Gradually the pueblos began to view the anthropologists as suspiciously as they had the Franciscans. The historian Marc Simmons tells how the prying eyes and camera of the anthropologist Adolf Bandelier made him persona non grata at Santo Domingo in 1881. And Charles Lummis, a colleague of Bandelier's who lived at Isleta pueblo in the 1890s, was ostracized after he published a book of pueblo stories. When Elsie Clews Parsons published her findings about Taos in the 1930s, some of her informants were punished for their loose tongues.

Parsons, who moved to the Spanish village of Alcalde in 1923, north of San Juan pueblo, to try to penetrate the Tewa mysteries, seemed especially exasperated with her San Ildefonso informants. "The women were particularly timid and not well informed; the man was a three-fold liar, lying from secretiveness, from his sense of burlesque, and from sheer laziness," she wrote. "Curiously enough, this man, whose social position is of the best, but whose veracity is of the worst, according to both white and Indian standards, has probably been hitherto one of our sources of authority on the Tewa."

Parsons simply couldn't understand that the people she so assiduously studied had a different attitude toward knowledge; they instinctively recognized that with dissemination came dilution. As they saw it, anthropology was a zero-sum game: knowledge gained by the scientists was knowledge lost by the Tewa.

Cushing's experience with the Zunis showed, moreover, that the anthropologists didn't always practice the scientific ideal of detached objectivity. As Maxwell's demon might have told them, there is no such thing as an immaculate perception: you cannot gather information without altering the object of your curiosity. Consider the story of María Martínez, the world-renowned potter of San Ildefonso, who got her start in 1908 when she camped in Frijoles Canyon for the summer with her husband, Julian, who was hired by archaeologists to help excavate the ruins there. One of the scientists brought María a potsherd decorated with a beautiful black pattern unlike any she had ever seen and asked her if she could duplicate it. Pottery-making had all but died out at San Ildefonso after factory-made cookware became available. The little that was still made was plain and utilitarian. But with help from Julian, who was something of an artist, Martínez made modern versions of the ancient pot. The archaeologists purchased them to show at the state museum in Santa Fe, and before long a worldwide market developed for pots with stylized versions of the old designs used by the people who once lived on the Pajarito Plateau. A link between the Anasazi and the Tewa, broken long ago, had been reestablished by these curious outsiders, whose temples were universities and museums. It was indeed a complicated new world.

Only a romantic would insist that if it hadn't been for outsiders all would have been harmonious in the Tewa world. The pueblos had their own internal fractures, though these were no doubt widened by the pressure of Spanish and American occupation. When Bandelier visited northern New Mexico in 1883, he found that not all of the pueblos' decline could be blamed on exotic European diseases. At Nambe he was told that the pueblo's dwindling population had been caused in part by the custom of executing "witches," who seem to have included many of the most intelligent citizens.

In the early twentieth century, San Ildefonso split in two when most of the pueblo refused to obey an order from the head religious leader, the cacique. Sometime after the Reconquest (some say the late nineteenth century, others say after the destruction wrought by the 1696 uprising), the pueblo moved slightly northward, closer to Black Mesa. Some of the people were concerned that they were moving against the grain, that migration should always be to the south, just as it had been ever since they first left the lake of emergence. After years of pestilence, worries grew that the pueblo was seriously out of alignment with the spiritual energies. Eventually (some say 1923, others say 1910, 1918, or 1921), the cacique persuaded six families to move slightly southward and reestablish their own village. For most of the people, however, practicality outweighed religious concerns; they stayed where they were, and so San Ildefonso became divided into a North Plaza and a South Plaza.

Most of the winter people, all but two families, died in the influenza epidemic of 1918. With the winter people all but gone, San Ildefonso divided into north and south, as though compelled to maintain the old duality. But this wasn't the harmonious polarity that had existed before. In 1930, some of the North Plaza men raided the South Plaza, retaking religious paraphernalia and beating the cacique. The South Plaza retaliated by burning the North Plaza's kiva. A split also occurred along religious and secular lines. The Spanish had imposed a new layer of hierarchy on the pueblos, making them choose a governor and other civil administrators to deal with the conquerors. The arrangement continued with the American overseers. One year, a governor whose term was ending refused to follow tradition and surrender to the cacique the ceremonial cane given to all pueblo leaders by Abraham Lincoln; instead, the outgoing leader passed the cane on himself. A new governor was elected by the North Plaza alone, and the South Plaza refused to recognize him. And so, for years, the religious leader, the cacique, lived on one side of the pueblo, while the political leader, the governor, lived on the other.

In 1935, when moviemakers paid the governor to shoot scenes of the pueblo, they parked their trucks in the South Plaza while they filmed in the North Plaza. When it came time to divide the fee, the governor gave it only to North Plaza families. Some of the disputes were even more petty. A South Plaza man built an outhouse blocking a corridor used by the North Plaza to get to a kiva that was shared by the two factions. After an argument, he moved the outhouse but left it facing the North Plaza, mocking the people watching the dances.

Though the move south was supposed to realign the pueblo with the spirits, it was the North Plaza that seemed to prosper. It was there that the most famous of all the Tewa lived, the potter María Martínez. She became so successful that she was able to hire local Spanish women as housekeepers. And when she became distressed that a local Hispanic man was giving homemade wine to her husband and other pueblo men, she bought his house with a roll of hundred-dollar bills simply so he would go away. Before long, pottery-making became such a big business that many women refused to speak to one another, afraid that someone would steal their designs.

The split is still obvious in the feast day ceremonies: the buffalo dance in the North Plaza alternating with the Comanche dance in the South Plaza. Ask a South Plaza man how to find the home of a North Plaza family and you may be confronted with a blank stare—this in a pueblo of 350 inhabitants.

As the twentieth century progressed, the closed system of the Tewa was becoming open to an environment voracious for information. As news about the pueblos spread, the road from Santa Fe to San Ildefonso became more widely traveled. One of the early visitors was the physicist Robert Oppenheimer, who as a young visitor from the East had taken a pack trip in 1922 from Frijoles Canyon to the Valle Grande, the vast, high meadow that was created a million years ago when the Jemez volcano blew its top. On another trip in 1937, he met Edith Warner, the woman who lived at Otowi Bridge, near San Ildefonso, and whose journals document the changes that came to the land. Oppenheimer remembered the beauty and isolation several years later when the federal government was seeking a site for the Manhattan Project. In 1942, Oppenheimer returned to introduce his new wife to Warner, who now ran a small tearoom at the bridge. Within a year a weapons laboratory was being built on a mesa top a few miles away, and Warner's journals tell of visits to her rustic salon by Oppenheimer, Enrico Fermi, Edward Teller, and her favorite, a kind, softspoken man named Niels Bohr.

After the explosion of the Hiroshima bomb developed by the scientists on the Pajarito Plateau, the laboratory of Los Alamos quickly grew into a city and became the single largest employer for the pueblos and villages between the eastern slope of the Jemez Mountains and the western slope of the Sangre de Cristos. Before long, physicists from Los Alamos were sitting at the table with families from San Ildefonso for their annual feast days. But all was not congenial between the pueblos and the laboratory. Land the Tewa considered sacred was now behind fences marked with No Trespassing signs. Decades later, when stories began to surface about radioactive waste left in the canyons, the governor's office at San Ildefonso calculated that since Los Alamos had moved in above them, 10 percent of the pueblo's people had become victims of cancer. With only 350 people living there, it was hard for statisticians to evaluate the significance of the claim. But to the people of San Ildefonso it was almost impossible not to suspect a pattern and consider nuclear research the worst kind of witchcraft.

And what, in turn, did the physicists think of the Tewa's magic? For many, the dances were little more than a local curiosity to share with visitors. Some were struck perhaps by the mystery. What was the meaning of this dance step, that song, that symbol? What was going on inside the kivas, or in the plaza on days when the whole pueblo was closed to outsiders? What was going on in the minds of the Tewa? How did the world look through their eyes?

In the true Baconian spirit, some of the scientists gathered scraps of knowledge here and there. East is white, north is blue. But without a field to charge them with meaning, the pieces themselves were of little significance. Anyone was free to learn group theory and penetrate the priesthood of particle physics, but with the Tewa one was always on the outside looking in.

No individual Tewa could be expected to hold the entire intricate system of colors, directions, stars, and animals in his head. Like the maps of the northern New Mexican terrain that Harrington uncovered, the knowledge was fragmented among many minds. Alfonso Ortiz, the Tewa anthropologist, put it like this: You cannot expect to stop a Tewa on the street and ask him to unfold his people's world view any more than you can expect most Europeans to be conversant in Western philosophy. The most basic assumptions run deeper than our awareness; they are the epistemological air we breathe.

The unspoken assumptions that sat at the foundation of the Tewa's grand architecture bore little resemblance to those that science had built upon. The separation of subject and object, the value-free nature of knowledge, the power of mathematics to reflect nature—all this was ground into the lenses through which the scientists focused the data streaming in from whatever it is that lies beyond the senses.

Confronted with an alien system, all we can do is lay our own conceptual grid over it and explain it in our own terms—build models of their models. But there will always be a gap between our theories and the phenomena we are trying to formalize. Even if so much of the Tewa religion were not secret, there would still be no way to think with the brain of a Tewa, to erase our own network of beliefs and immerse ourselves in another. In the end, two different world views can be as immiscible as oil and water: their very structure holds them apart.

Yet perhaps there is a level below culture, a commonality that extends back before the ancestors of the Anasazi and the other natives of the Americas left the Asian continent, migrating across the Bering Strait— and even earlier, before there were Caucasians and Africans and Orientals, when there was just *Homo sapiens*. At the deepest level we are all information gatherers—Iguses, in Gell-Mann's term. Dig deep enough through the layers of the mind and surely you will reach rock bottom, an impenetrable floor: the architecture of the brain as it was molded by evolution to find patterns, even if they are not always there. We all find ourselves in a world of randomness, where some seasons are wet while others are dry, where bad things happen to good people and enemies prosper. Surely the world isn't meant to be this way. We all share this belief in symmetries, and finding ourselves in a world where the symmetries have broken, we imagine a time before the fall from perfection, whether we call it Eden, the underworld, or the big bang.

And so we impose number on the world. The four fundamental forces. The fourteen stations of the cross. The concentric tetrads of mountains and mesas. The three levels of people mirrored by the three levels of the supernatural world. Even mythology is rendered with precision. The giant of Black Mesa takes four steps to get to San Ildefonso. The first people took four tries to leave the lake of emergence, and then twelve stops to reach the first settlement of Posi. Along with this drive to enumerate, we hold in common a compulsion to divide the world into dualities: positive and negative, matter and antimatter, good and evil, summer and winter, north and south.

Scientists insist that their own numbers and dualities are testable hypotheses, while others are articles of faith; that there are true and false compressions. As Murray Gell-Mann once put it, when you don't see compressions that are there, that is denial; when you see compressions that don't exist, that is superstition. Among his many other interests, Gell-Mann has developed a fascination with Tewa religion and language. But he has no doubt that the patterns he and his colleagues in theoretical physics are finding are the ones that are real.

For a scientist, knowledge is something to be discovered. It might remain secret until a paper is published, but after that it is free to anyone who can understand its hieroglyphs. The Tewa have not traditionally used their powers of induction to zero in on causal relationships, to unearth platonic truth. To them, the most powerful truths have been known since the beginning. The burden is to protect them and pass them undiluted down the line.

Undiluted, but also untested, a scientist might say. For some of the denizens of Los Alamos the overriding question was: How could their neighbors down the hill possibly believe these things? How did a system with so many seemingly arbitrary components remain so remarkably intact, throughout the domination by the Catholics, the scrutiny of the anthropologists, the invasion of the physicists? Part of the explanation is that the Tewa's world, like that of science, is not static. It absorbed new gods along the way. But have gods or dances ever been abandoned when they didn't prove effective?

Some parts of the system are tested every day. East is white each time the sun comes up; to the north is the bluish haze of the distant mountains. The pueblo dances in the spring, and before long it usually rains. In so precarious a world, people are not tempted to experiment, to forgo the dance one year and see if it rains anyway. After all, one could always blame a particularly dry summer on any number of things: maybe a dance was not done correctly, or the thoughts of the dancers were not pure. Failure of a ritual would make them dance all the harder. The system's very resilience makes it resistant to falsification. If challenged by the people, the priests can always hide behind the veil of secrecy. They have a vested interest in maintaining their power.

But is science always so different? Few people, even with college educations, are in a position to evaluate the intricacies of particle physics or cosmology. We accept them on faith, stories told by the high priests and priestesses. We are provided with the warranty that the speculations are grounded in observation and that even the most beautiful tower can be toppled by a single observation. But the relationship between the observable world and the theories we build is subtle, to say the least. In practice, we are more likely to add filigrees to our models than to see them overturned.

When confronted with an observation that stubbornly resists the reigning theory, a scientist is tempted to dismiss it as experimental error. Or the theory can be supplemented with what are politely called "auxiliary hypotheses." Beta decay violates the law of conservation of energy, so there must be invisible neutrinos. Galaxies spin in contradiction of Newton's laws, so there must be invisible dark matter. Science too builds self-sustaining systems, a tissue of concepts that, as the philosopher Willard Quine put it, "impinges on experience only along the edges."

How then can one avoid a retreat into relativism, a conclusion that one people's system is as good as another's? When it comes to predicting and controlling the world around us, physics has provided the more powerful set of tools. Where is the Tewa equivalent of television or the atom bomb? It is hard to imagine that given any amount of time the Tewa's way of explaining reality would have led to the digital computer or laser holography. And yet, if their way of carving up the world didn't provide levers powerful enough to move the earth, it gave them the inner strength to weather invasion after invasion—by Navajos, by Spanish soldiers and missionaries, by inquiring anthropologists, and finally by the physicists of Los Alamos. For the Tewa, the purpose of building mental orders seems less to control the environment than to control the world within. And until recent times they were largely successful, but the task is becoming more difficult year by year.

The system created by the Tewa provided enough comfort that, left on its own, it probably would have continued to thrive. But some reality-testing was inevitable. It became clear, case by case, that the doctors from the Indian Health Service were often more successful than the pueblo healers in curing sickness. And so, inevitably, the medicine societies began to die. Other sacred societies succumbed to the expediencies of modern times. In a world where one can take a paycheck from Los Alamos and cash it to buy groceries at a supermarket in Española or Santa Fe, there is little need for Hunt Societies, and so these too have faded. In the 1960s, Alfonso Ortiz interviewed the last hunt chief at San Juan pueblo. Feeling that his knowledge was no longer valued, the chief died without naming a successor. The chain was broken, except for the information Ortiz was able to record in his field notes.

This membrane the pueblos have erected around themselves keeps information from flowing out, but the protection is mostly one-way. Nothing can be done to block television waves. And the sheer force of the American marketplace has been all but impossible to resist. Absorbed into the economy that surrounds them, the pueblos have had to find ways to turn resources into cash. Even a religiously conservative pueblo like Tesuque finds itself opening a bingo hall; in 1994 it started constructing a full-blown casino. Santo Domingo is now the site of a factory outlet mall. Many of the billboards that obstruct the view from Albuquerque to Santa Fe and beyond stand on rented patches of pueblo land.

Some pueblos have succumbed more completely than others. At Pojoaque, which lies between Tesuque and San Ildefonso, many adobe houses have been replaced by trailers. In place of a central plaza is the asphalt parking lot of the Pueblo Plaza shopping center. Taking full advantage of its main resource—the Y-shaped intersection that connects the Santa Fe highway to Los Alamos and Taos—the pueblo has developed a strip of chain restaurants, gasoline stations, a gambling hall equipped with video slot machines, and a liquor store that includes the finest wine collection in northern New Mexico. At Po suwae geh Restaurant (from the original Tewa name for the pueblo) one can order sandwiches called Tesuque Tuna, Taos Turkey, San Ildefonso Ham and Swiss, and Picuris BLT.

In recent years, Pojoaque has tried to restore its lost traditions, relearning dances from other pueblos and opening a museum, which acts as both a cultural resource and a way to divert the flow of tourists traveling along Highway 285. Perhaps what is most surprising is that Pojoaque is an exception, that most of the Rio Grande pueblos look much as they have for as long as there are photographs to compare. A picture taken around the turn of the century of the corn dance at San Ildefonso might have been taken in 1995: the same costumes, the same low skyline of adobe houses, the same old cottonwood tree in the North Plaza.

Anthropologists distinguish between the exoteric part of a religion, that which is open to outsiders, and the esoteric, that which is secret and held within. In the Tewa world, the esoteric core is slowly withering, leaving the exoteric shell. The dances seem as vibrant as ever. To the older pueblo people, they are an invoking of magic, to the younger a way of preserving a tradition, keeping an identity intact. But even those who take the religion metaphorically—and the dances more as a cultural celebration than as an appeasing of the gods—they too are shaped by this world view, just as agnostics and atheists are shaped by the Judeo-Christian tradition.

Each year, the trappings of the world beyond the mountains become harder to resist. Even at some of the most traditional pueblos, like Santo Domingo, the feast day dances are accompanied by a traveling carnival with a midway and Ferris wheel. Recent Christmas festivities included a parody of the corn dance. Made up in blackface—not like the animals in the buffalo dance but like members of a minstrel troupe—the dancers, some of them wearing Afro wigs, some of them dressed in drag, performed what they billed as the Alabama Corn Dance. The racist undertones were only softened by the knowledge that the pueblo clowns make fun of everyone, including themselves. In a traditional corn dance, male and female dancers form facing parallel lines. In the Alabama Corn Dance, this configuration quickly degenerated into the childhood game called Red Rover.

Following the Alabama corn dancers was a march by a corps of pueblo firefighters, wearing yellow hard hats and singing an old cavalry song adapted from a John Wayne movie: "Around her neck she wore a yellow ribbon. She wore it for that firefighter far, far away." When it came time for the head firefighter to review the troops, he opened a briefcase and pulled out a giant rubber penis. Carrying it like a baton, he walked down the line shaking it at the troops as he asked each man his name.

Though the pueblos like to make fun of the puritanical ways of their Catholic and American overseers, their sense of humor has become more subdued. Throughout the pueblo world the clowns have cleaned up their act, though they still put on hilarious performances. At Picuris pueblo, up near Taos, they still storm the plaza during Fiesta, disrupting the chorus and mimicking the dancers. Dressed in colored jogging shorts, which clash loudly with their traditional black-and-white striped headpieces, the clowns perform a traditional pantomime in which they try to figure out how to get a watermelon and a bundle of food down from the top of a tall pole. They still grab women from the audience to hug and spin around, but the sexual undertones are muted.

The clowns seem to be having so much fun that it is easy to forget that the dances are not put on for the amusement of outsiders. When a Hispanic man recently attempted to photograph the clowns at Picuris as they tried to climb the pole, a young pueblo man who was in charge of security angrily seized the camera, returning it only after confiscating the film.

Many pueblos ban photography (or charge a camera fee). It is not that the people fear having their souls captured by the camera, as in that old white man's cliché. Some no doubt are photographers themselves. What they fear is assimilation. They know what happens when a closed system is opened to the world, how information dissipates across the barrier and order turns to entropy. None of the pueblos allow significant anthropological inquiries anymore. As Ortiz, the Tewa anthropologist, put it, "The attitude is, Why don't you go study yourselves?"

They are equally leery of the New Age seekers coming from Santa Fe in quest of spiritual knowledge. A recent governor of Tesuque pueblo, which is the closest to the city, was disturbed to pick up a weekly alternative newspaper and see an advertisement for Indian shaman lessons. On a visit to Trader Jack's Flea Market, a weekend bazaar on the highway near Tesuque and the Santa Fe Opera, he saw a crowd gathered around a man selling crystals that supposedly had healing powers. The governor picked up a rock from the ground and called out, "I'll sell you this one for half price." The magic isn't in the object, he said, but in what you believe. He seemed to find the whole New Age movement pathetic—these people without a religion trying to invent or steal one.

"Our beliefs are one of the few things that are still our own," he said. "There is always the chance that this knowledge will be what saves the world." He drew a circle representing the earth, then a line to represent its axis. Around the circumference he placed the various native peoples: North American Indians—the pueblos, Navajos, Apaches, Utes—as well as the tribes of Africa and the Australian aborigines. Their purpose, he said, was to keep the earth in balance. If their cultures were destroyed, the axis would tilt and the planet would fall into the sun. "We knew there was an axis before science did," he said. "We don't worry about what makes it go around. If we are meant to know that, it will be revealed."

One of the most radical proponents of the notion that different peoples carve up reality in very different ways was the American linguist Benjamin Lee Whorf. For Whorf even such seemingly ingrained notions as time and space were culturally determined. The Tewa have not let themselves or their language be studied closely enough for anyone but them to know how differently they might conceive of such basic concepts. Whorf studied the language of the Tewa's distant cousins, the Hopi, concluding that they have no words for time. For that matter, he contended, they do not even have verb tenses or any other constructions that would distinguish present, past, and future. And, Whorf insisted, they get along fine without them.

"Just as it is possible to have any number of geometries other than the Euclidean which give an equally perfect account of space configurations," he wrote, "so it is possible to have descriptions of the universe, all equally valid, that do not contain our familiar contrasts of time and space."

Though the mathematics of relativity treats time and space more fluidly, we instinctively think of ourselves existing in a rigid lattice of three dimensions, with time flowing through it—a steady, unidirectional wind from a dark, unknown future. Strained to roughly translate Hopi metaphysics into an approximation Westerners could understand, Whorf concluded that in place of time and space or past and future, the Hopi divided everything into what he called the Manifest and the Manifesting (or, alternately, the Unmanifest). For lack of more precise concepts we might call these the objective and the subjective, or adopt the names Descartes used when he split the universe into *res extensa,* things with extension, and *res cogitans,* things of the mind.

Everything that can be registered by the senses, whether in the present or the past, belongs to the Manifest realm, that which has already become. The future, along with everything else that is invisible to the senses—that which has not yet become—belongs to the realm of the Manifesting. The future is not out there flowing toward us, but hidden inside of things, striving to come forth. Included in this category of the Manifesting is everything we would call mental: thoughts, hopes, feelings, desires, intentions. All are striving to make themselves manifest. Likewise, the seed's striving to become a plant belongs to this subjective realm. Maybe this way of carving up the world gave rise to the Tewa's notion of Made People and Dry Food People Who Never Did Become.

The Unmanifest is "subjective from our viewpoint," Whorf wrote, "but intensely real and quivering with life, power, and potency to the Hopi." A better term, he proposed, might be the realm of hoping. The world is divided into the things that are and the things that are hoping to become real. The rituals in the kiva and the dances on the plaza are a means to make the hopes, the future become manifest.

As Whorf describes it, the implications of this world view seem almost Einsteinian. To the Hopi on First Mesa it would be meaningless to speak of something happening simultaneously on Second Mesa. One can only imagine. To find out what is going on, one would have to travel there or wait for a message to arrive. Only then does what happens elsewhere become real. Until then, it belongs to the realm of the Unmanifest, the subjective.

As an aid to thinking, like the mathematicians' imaginary spaces, the Hopi conceive of the universe as vertically structured. The world mankind inhabits stands halfway between the starry sky and the spiritual underworld. It is not that the gods literally live underground. They are just so removed from the realm of the objective, the things accessible to the senses, that one can only imagine what they are like. And the same is true of the distant stars.

It is hard to know how accurately Whorf carved up the inner world of the Hopi. His assumption that language determines and constrains how people perceive reality lapses in and out of favor, and his own mental filters may have ensured that he saw in the words of the Hopi only what he wanted to see. The very notion of abstract concepts—these mental constructs that exist somehow on their own—may be a concept (pardon the circularity) that other cultures do not necessarily share. To us, Tewa religion will probably always belong to the Unmanifest, standing as a reminder that even here on earth, among fellow humans, not everything is knowable. But there is one thing we share: we are all people struggling against the limits of our nervous systems, the computational power of the brain to compress and understand. We grasp at images and tell stories, trying to build ourselves a place in the world, but always with this uneasy feeling that we are skating on a thin crust of ice over a seemingly bottomless lake.

PART THREE

"A FEVER OF MATTER"

What then was life? It was warmth,
the warmth generated by a form-preserving insubstantiality,
a fever of matter, which accompanied the process
of ceaseless decay and repair of albumen molecules
that were too impossibly complicated,
too impossibly ingenious in structure.

—*Thomas Mann,* THE MAGIC MOUNTAIN

7

THE DAWN OF RECOGNITION

Like the Tewa Indians, biologists who ponder life's beginnings believe that we came from beneath the water, not a lake in northern New Mexico but an ancient, all-engulfing sea. And, like the pueblo people, these seekers of origins are shaped by a belief in the importance of dualities, not summer and winter or hot and cold, but the polarity of positive and negative electrical charge. For reasons perhaps too fundamental to fathom, the universe seems to come divided into two essences compelled by their very nature to obey a simple rule: like charges repel while opposite charges attract. When we run a comb through our hair and watch it pick up dust particles, we are witnessing a most ancient form of recognition.

Combine this fundamental affinity with the rules of quantum mechanics and we have our own creation story about the origin of atoms—how the carriers of the negative charge, the electrons, drawn to the carriers of the positive charge, the protons, could arrange themselves only in certain ways. First came hydrogen, a single proton with its positive charge counterbalanced precisely by its single negatively charged electron. Then came helium, a nucleus of two protons surrounded by two electrons. It is written that no two electrons with the same quantum numbers are allowed to occupy the same atomic orbital; so as not to run afoul of this dictum, Pauli's exclusion principle, each of these electrons "spins" a different way. When larger nuclei came along, forged inside the stars by the very power that lights the sky, they required more electrons to maintain their poise. With lithium (a nucleus of three protons) there was no more room in the bottommost orbital, so the third electron began a second shell. Then came beryllium with four protons, boron with five, carbon with six, nitrogen with seven, oxygen with eight, fluorine with nine, neon with ten. With neon the second shell, with its various suborbitals, was filled, and so with sodium a third shell was begun, a new row on the periodic table. And so it went, layers piled on top of layers, filling the rows and columns of Mendeleev's great chart, yielding the ninety-two naturally occurring elements, each with its peculiar properties, determined by the number of electrons in its outer shell.

Atoms whose outer shells were not completely full would carry an attraction, called a valence. And so, reacting according to their natures, atoms combined with atoms to form simple molecules. When two atoms together contained less energy than the two atoms apart, these new configurations were thermodynamically favorable. They occurred as spontaneously as water flowing downhill. It took energy to break them apart. Again we can think of this as a primitive kind of recognition. Oxygen, with two gaps in its outer shell (a valence of 2), would readily combine with two hydrogen atoms, each bearing a single electron, to form water. A silicon atom, in search of four electrons to complete its outer shell, would seek out two oxygen atoms to make silicate. These marriages were so convenient that on earth a silicate surface congealed and became covered with oceans, a medium in which other molecules could congregate, seeking their physically ordained mates.

Carbon atoms, because of their electrical nature, tended to hold hands with other carbon atoms, forming long chains, C-C-C-C-C-C-C. But carbon has a valence of 4, so many hands were left free—two for the carbon atoms within the chain, three for the ones on each end. And so the chains surrounded themselves with an electrostatic fur of hydrogen, oxygen, nitrogen, phosphorus, sulfur. Sometimes a carbon atom would attract another carbon atom, sprouting a side chain with a complexity of its own.

Bumping randomly against one another in the ancient waters, these writhing arrangements of positive and negative charge would occasionally find that they were complementary: the bumps and indentations along a segment of one molecule matched the indentations and bumps on another; electrically and geometrically, they formed a fit. Going beyond the mere recognition of positive and negative, the first pattern recognizers were born. A longer molecule might encounter two smaller molecules that fit different parts of its charged anatomy; and so it would clasp them together, in just such a way that they joined to make a new, more complex molecule. This new molecule might go on to act as a catalyst to marry other pairs of molecules. Or a small molecule might arise that clamped onto a large one and split it in two, creating the raw material other molecules might use to make new structures. The possibilities were endless. A small molecule created somewhere in this web might latch onto a longer one, causing it to twist in just such a way that it was now receptive to a third molecule that it wouldn't ordinarily recognize; or it might twist so that it now ignored molecules that it once had an affinity for.

Molecules creating molecules creating molecules, switching one another on and off. In time, chain reactions arose in which molecule A catalyzed the formation of molecule B, which catalyzed the formation of molecule C. Now if it so happened that molecule C catalyzed the formation of molecule A, a reaction loop would form. And then loops became interconnected with loops. Many of these webs were fragile, dissolving as quickly as they formed. But those that could sustain themselves over time became what we now call metabolisms. Taking in smaller molecules—food—from the world around them, these chemical networks constantly forged large molecules from small ones, broke them apart, built them up again. The oceans were abuzz with these ceaseless transactions, building order where there once had been randomness.

It takes energy, of course, to fight the tides of entropy. As Maxwell's demon taught us, a thermodynamic price must be paid for imposing order on the world. The oceans of emergence were bathed in the light of a nearby star, but the energy had to be captured, harnessed, stored. Powered by photons from the sun, one part of a metabolic web would build up energy-storing molecules, as though winding up a spring or pumping water to a reservoir uphill; another part of the web would break the bonds, releasing the energy flow to power more reactions.

Eventually, the story goes, some of these metabolisms became encased in sacs of molecules to form what we now call cells. The cells, being children of molecules, inherited their talent for recognition: a protein molecule embedded in the membrane of one cell would find its complementary mate in the protein of another cell. And so cells would link together in complex configurations to form organisms that, after billions of years of evolution, have become so adept at detecting patterns that they are driven to uncover the logic of their own cells, and to construct stories about how they came to be.

But the stories we tell leave the most troubling questions unanswered. What was the impetus behind this rise of order from randomness? Was it mostly a matter of luck, or an inevitable outcome of natural law? We can imagine a chemical logic according to which molecules fed by solar energy join with molecules and these molecules with other molecules, becoming complex, then alive—animated. But how likely was it that just the right pieces would come together in just the right manner to ignite the flame, this "fever of matter," as Thomas Mann called it?

From the time we first learn science, we are confronted with what seems the incredible unlikelihood of our own existence. Even if we can make peace with the chain of contingencies that resulted in stars giving birth to carbon, or the earth being poised at just the right distance from the sun to be a potential incubator of life, we are left to wonder at the additional coincidences that led molecules to congregate into these seemingly delicate arrangements called metabolisms. Even with the most modern laboratory imaginable and the symmetries of the periodic table at your service, assembling a single variety of biological molecule is a painstaking task. That it happened again and again, spawning proteins and nucleic acids, lipids and sugars—the staggering variety of molecules that interact so precisely to generate life—is, we have long been told, largely because of random jostling over eons of time, atoms bumping against atoms in the primal sea. Of course, once stable combinations of atoms came together, natural selection would ensure that the "fittest" molecules—those best adapted to the brackish environment—would flourish. And if, by chance, a molecule or assemblage of molecules with the power to copy itself—to replicate—came together, naturally it would be fruitful and multiply. But how likely is it that self-replicating molecules, the contenders in the Darwinian saga, would have arisen in the first place? Some scientists have estimated that the odds are so infinitesimally low that the flame of life might very easily not have been ignited at all.

So here we sit, at the end of a long chain of electrons recognizing protons, atoms recognizing atoms, molecules recognizing molecules, cells recognizing cells—this drive to find pattern rooted in our very core. And after divining the patterns around us, we are confronted with the possibility that we are flukes or miracles. But our brains, honed by evolution to forge long chains of causal explanation, cannot rest without a more satisfying story. Surely these beautiful patterns were somehow meant to be.

At the Santa Fe Institute one finds an unusually high density of people who dispute the notion that we are creatures of chance, who believe in what once was considered a heresy: that there is an inevitability to life. As we have seen, the new interpretations of quantum theory presented by Wojciech Zurek, James Hartle, and Murray Gell-Mann give reason to believe that Iguses, information-gathering-and-utilizing systems, are a natural part of the universe, this particular Schrödinger wave whose crests we are riding. Delving as deep as people have gone into quantum physics, they conclude that gathering bits—the thing people do best—is as fundamental as anything can be. For classical reality to arise from quantum possibility, information must be exchanged.

Nothing, though, in their theories requires that living creatures be the ones processing information. Bits flow just as easily in a completely inanimate world. To find a place for ourselves among the stars, we have to turn our sights in other directions. Following a very different route, some biologists at the Santa Fe Institute have come to believe that the emergence of life itself is written into the universal laws, that given conditions anything like those of the early earth—and perhaps even conditions wildly different—it is to be expected that creatures of some sort would emerge from the darkness and begin carving up the world into categories.

Of all the patterns that excite our neural webs, some of the most tantalizing are those we find inside our cells, and those of every creature on the planet: long helices of nucleotides, DNA and RNA, which carry the information for making proteins, the chains of amino acids that form the structure of our bodies and, as enzymes, channel the chemical confusion into the web of reactions we call life. Ever since James Watson and Francis Crick teased apart DNA's double helix, it has been hard not to think of cellular chemistry as a language, with the genetic instructions coded with four molecular symbols: A, T, G, and C, representing the nucleotides adenine, thymine, guanine, and cytosine. (In RNA, uracil is substituted for thymine, U for T.) A triplet of these nucleotides is said to "stand for" each of the twenty amino acids that participate in earthly life. Of all the ways that atoms can combine to form molecules, just four kinds of nucleotides and twenty kinds of amino acids have been sifted from the confusion to serve as the vocabulary. Hence the suspicion that all life, plant and animal, sprang from the same universal ancestor, which happened to carry this subset of molecules within the membrane of its single cell.

In the code that has been passed down the line, GCA stands for the amino acid alanine; CAC means histidine, AAA is lysine, UGG is tryptophan. And so a string composed from DNA's four-letter alphabet can encode a chain of amino acids, a protein. We marvel each time we hear how the molecular cipher, this sequence of A, C, T, and G, is so effortlessly copied every time a cell divides: how A fits snugly with T, G with C, so that sequences of nucleotides naturally form complementary pairs, the double helices. Peel apart the two strands of a DNA molecule and immerse them in a soup of single nucleotides and each half will recognize its missing pieces, seeking them out to re-create the whole.

So awed are we at the power of double helices that it is natural to suppose that the chemical dance that led to our emergence must have begun with replicating chains of nucleotides, coming together, splitting apart, copying themselves throughout the sea, eventually learning to make proteins and to encase themselves in cells. This, in fact, is the mainstream view among biologists, that life began in an RNA world (RNA being somewhat easier to make than DNA). The problem, as biologists are quick to point out, is that only with great difficulty and cleverness can the components of a single nucleotide be conjured from a flask of primordial soup: the sugar (ribose for RNA, deoxyribose for DNA), the ring-shaped purine or pyrimidine bases that give it its identity as A, T, G, C, or U. Given enough time, perhaps these unlikely reactions would occur without a chemist to help them along. But even assuming a supply of these components, it would be the most daunting of tasks to get them to assemble in just the right way to make the delicate structures we have come to call nucleotides. And even if nucleotides could be assembled, water would constantly break apart the bonds. How do cells do it? With protein enzymes, holding the pieces just so, catalyzing reactions that otherwise would have only the slightest chance of occurring.

Accept for the moment the unlikelihood that a small supply of nucleotides spontaneously came together in the waters of the early earth. Linked into chains, they would form nucleic acids that divide and replicate, a few seeds multiplying to populate the medium. But how did the first seed, the first chain, come about? We know how strings of nucleotides form pairs, G matching with C, A with T, dictated by nothing more ethereal than stereochemistry. But that doesn't explain where the strings themselves come from. Complementary pairing explains how GCATTA acts as a template to make CGTAAT. But how did nature get GCATTA to stick together in the first place, joining hands along the other, perpendicular dimension? To signify a wide range of proteins, A, T, G, and C must be able to string themselves together into any combination, to form any arbitrary message—that is the source of the power of the genetic code. Inside the cell, this is done with enzymes. In a pure RNA world, before proteins were born, all these reactions (A followed by A, T, G, or C; T followed by A, T, G, or C—sixteen in all) would have to occur spontaneously.

Suppose we accept this chain of flukes or miracles and, *mirabile dictu,* replicating chains of nucleic acids somehow assemble, copying themselves over and over again. The seas become filled with different species of nucleic acids, each with its unique sequence of nucleotide bases. An occasional mutation—a G where a C should be—would ensure an ever-changing pool of replicators. Now if some of these survivors could somehow develop the machinery to translate their sequences into chains of amino acids, proteins would be born. Those replicators that code for proteins that happened to enhance their ability to survive and reproduce would crowd out their weaker neighbors. Natural selection would enter the story, leading to fitter and fitter structures, culminating perhaps in the first cell.

But again, without protein catalysts, the likelihood of a chain of nucleotides (its very existence already a vast improbability) developing the ability to assemble a protein seems almost unimaginably small. So beautiful and elegant is DNA, its structure and function so perfectly entwined, that it is easy to fall prey to the myth of the genome as central controller, with the power to construct and operate a cell. In fact, the mechanism does not run so smoothly or independently as is sometimes supposed. It takes a very long chain of nucleotides to code for an enzyme, but a long chain cannot replicate itself without being overwhelmed by random replication errors—unless it is aided by enzymes, the so-called proofreaders, which ensure that the copy is as faithful as possible. Or consider the complex process by which the genetic cipher is translated into proteins: the two strands of the DNA double helix peel apart, re-

vealing a section of nucleotides that is copied by assembling a complementary strand of so-called messenger RNA. In the process, stretches of
apparently meaningless DNA sequences, called introns, are "edited" out.
Every step in this process requires enzymes. Carrying the genetic signal,
the abbreviated messenger RNA is then decoded by the ribosomes, structures made of nucleic acids and proteins, which match each nucleotide
triplet with the proper amino acid. How is this done? Floating in the cytoplasm of the cell is another kind of nucleic acid molecule, transfer
RNA, which acts as a kind of adaptor plug, matching up the two kinds of
molecules. One transfer RNA molecule, for example, consists of a
triplet of GCA bound to the amino acid alanine; in another, CAC is
bound to histidine. These assemblages of transfer RNA are matched up
with complementary sites on the long strand of messenger RNA, creating a sequence of amino acids, which are then joined to make a protein.

How does transfer RNA match up its nucleotide triplet with the
proper amino acid? It would be nice to think that the fit was as natural and
elegant as the match between complementary strands of DNA, that the
contours of each triplet in a transfer RNA molecule matched the contours of an amino acid, so that GCA, for example, naturally sticks to alanine. In fact, the fit is apparently not natural at all but mediated by still
more enzymes, which specialize in making these otherwise unlikely at-
tachments. One can easily imagine an entirely different set of enzymes
that would, for example, attach GCA to tryptophan instead of alanine,
CAC to asparagine instead of histidine. Given the vast space of possible
enzymes, it is likely that the genetic code is arbitrary, a frozen accident
arising from the enzymes that happened to be around when nucleic acids
began making proteins. In fact, hardly any of the reactions inside a cell
are "natural"; with different enzymes, the molecules would be steered
along entirely different pathways. It is the enzymes that provide the
buffer that allows the molecules to free themselves from the dictates of
chemistry and create complex, unlikely structures; enzymes are the
source of the wonderful arbitrariness of life.

Given its utter dependency on proteins, DNA begins to look less like
a master controller than a clerk, passively recording the information
needed by the real actors, the enzymes. Even so, only part of the information needed to make proteins is in the DNA. A nucleotide sequence is
not a simple algorithmic compression of a protein. The nucleic acid code
determines the sequence of amino acids, but a protein is not a linear chain
of molecules. Once it is assembled, it thermodynamically folds into a

lower energy state, resulting in the complex configuration that gives it its catalytic power. Trying to determine a protein's three-dimensional shape from its genetic cipher is one of the great unsolved problems of science, taxing the most powerful computers, programs, and human minds at Los Alamos and elsewhere. Oppositely charged regions are attracted to each other, similarly charged regions repel, the final configuration influenced to a large degree by the chemical environment of the cell. Without this context, the genetic code would be meaningless.

Given the inescapable necessity of protein enzymes, it is tempting to reverse figure and ground and consider the possibility that proteins came first, later giving rise to nucleic acids to help with replication. Amino acids are far easier to make than nucleic acids. In a landmark experiment in 1952, Stanley Miller, a chemistry graduate student at the University of Chicago, and his academic adviser, Harold Urey, tried to re-create the earth's primordial atmosphere in a jar. Filling their apparatus with water vapor, hydrogen, and methane and ammonia—two simple carbon compounds often thought to have been present at the creation—Miller and Urey exposed the mixture to electrical sparks (simulated lightning) for a week, and then examined the by-products. The first attempt failed to produce anything of interest except gooey tar. But ultimately, by tweaking some of the variables, Miller was able to produce glycine and alanine, the two simplest amino acids used in proteins. How long could it be before little creatures began crawling from the flask? First one would have to figure out how to form long chains of amino acids. Again you would need an incredible amount of luck, or enzymes. And where would *they* come from? From other enzymes? At some point we have to stop begging the question and propose some kind of construction mechanism for stringing together amino acids, and the only one we know of requires nucleic acid templates. But they are due to arrive on the scene only after we have enzymes to make them. And how, without nucleic acid templates, would a protein reproduce? Once a chain of amino acids is strung together and folds up spontaneously into a convoluted glob, how could it be copied? One would have to posit a mechanism to unravel and read it, amino acid by amino acid. There is no evidence that anything like that has ever existed.

Either way you look at it, life seems a miracle or a fluke. All creatures are built from two types of complex molecules, each of which seems to

require the other to exist: amino acids form what biologists call the phenotype (the body), nucleic acids form the genotype (the encoding). The
riddle of the chicken and the egg repeats itself on a molecular scale. If one
cannot exist without the other, how did this interdependency arise?

Those who insist that nucleic acids came first strive to show that,
under certain special conditions, RNA can exhibit weak catalytic powers.
In the meantime, the protein-first advocates strain to show how chains of
amino acids might conceivably have reproduced without nucleic acid
templates. But whether one believes the protein chicken or the nucleic
acid egg came first, when one multiplies all the slight probabilities that
seem necessary for chemicals to evolve into cells, it is hard not to conclude that life is indeed accidental—an irreproducible result. In fact, for
years, any suggestion that life was driven by anything that could be con-
strued as a purpose or an inevitability was dismissed as vitalism, the belief in a mystical, self-propelling life force. What had become the party
line was eloquently expressed by Jacques Monod in his book *Chance and
Necessity*. The title comes from Democritus: "Everything existing in the
Universe is the fruit of chance and of necessity." In modern computer science, this is called "random generate and test." Structures are thrown
together haphazardly, then filtered through the impersonal sieve of
evolution.

The cosmologist Fred Hoyle found this idea of life arising from randomness so incredible that he compared it to a tornado sweeping through
a junkyard and assembling a 747, a quote that has been kept alive by biblical creationists. Hoyle looked to outer space as the source of the earthly
infection, the creationists to a caring, cognizant God. And most scientists
dismissed Hoyle, the creationists, and the vitalists as people lacking in
imagination, blindered by an inability to conceive of a more powerful
god: vast geologic time. Anything, it would seem, can happen in the 4.5
billion years the earth is said to have existed.

But this too was begging the question; if you can't provide a plausible
explanation for how life arose, you can always appeal to the fathomless
eons. One might as well say life came from space or was created by an
omnipotent being. Not even geologic time is infinite. In fact, scientists
are continually finding evidence that the time in which life could have
arisen is not so long as they had once supposed. Radioactive dating puts
the age of the earth at 4.5 billion years. But in the scenarios the geologists
paint, the congealing planet was hot and bombarded by meteors capable
of boiling seas—so inhospitable, that the window of life probably

couldn't have opened before 3.9 billion years ago. To determine when the window closed, we need to find the oldest fossils of single cells. Today, off the coast of Australia, blue-green algae, believed to be one of the oldest organisms on earth, form colonies, accumulating in their lifetime a layer of mineral debris. Another colony of algae forms on top of the layer, leading tier by tier to domed structures called stromatolites. Formations believed to be ancient stromatolites have been dated to at least as far back as 3.56 billion years. And so we are led to believe that between the time when the conditions were ripe for life and the time when we find the first fossils, a few hundred million years passed—still a vast span of time but not quite so fathomless as before.

The width of this bracket should be viewed skeptically. It is dependent on the web of assumptions that go into radioactive dating and on the legitimacy of the fossil finds. Single cells leave no bones. What look like the impressions of primitive organisms to one group of scientists are sometimes dismissed by others as geological filigree. But the consensus of scientists who study these things is that the period in which life could have formed is shrinking. As the window narrows, the coincidences edge closer and closer to the realm of "too good to be true." Given a supply of self-replicating molecular systems, natural selection may be powerful enough to mold them into complex cells in hundreds of millions of years. After all, it took less than ten million years for apes to evolve into people. More worrisome is this chicken-and-egg problem of how the self-replicating molecular systems arose in the first place. As they compute the probabilities, some scientists have been led to conclude that there simply isn't world enough or time. Either we are children of chance, a once-in-a-universal-lifetime event, or we spring from some natural source of order.

Some have become so frustrated by their inability to tease apart the interdependency of genotype and phenotype that they look beyond organic chemistry for the source of order. The search for the origin of life can be thought of as the search for a Maxwellian demon, a tiny "creature" whose ability to recognize patterns allows it to build structures amid disorder, to temporarily overcome entropy—paying, of course, the thermodynamic cost by burning energy and exporting heat (random molecular vibrations) into the environment. Enzymes make wonderful demons, but they seem to need other demons before they can arise. And if these demons need demons, then one can quickly fall into an infinite regress. Some scientists, such as Francis Crick, have entertained the notion that the original demon was an extraterrestrial scientist who sent seeds to earth as part of an experiment. But that explanation is not a lot more satisfying than the one in Genesis.

The mystery of the origin of life, then, comes down to this: finding a demon that can arise without a demon. That would put an end to this dizzying hall of mirrors of phenotype and genotype and allow the potentially infinite procession of demons to bottom out.

But where do we dig for this ultimate wellspring of biological order, this bottommost demon? The most obvious place to look is within the laws of physics themselves. If life can be shown to be dictated somehow by the symmetries of chemistry, then the odds of our existence are not so terrifyingly slender. Pursuing this thread, some scientists have sought ways in which the geometric surfaces of minerals such as pyrites might have acted as the first catalysts, their naturally occurring molecular lattices acting like little chemistry sets, channeling otherwise unlikely reactions. Others have gone so far as to suppose that some kind of mineral life preceded organic life. Obeying nothing but the laws of chemistry, silicon and oxygen, the two most abundant atoms on earth, spontaneously form into crystalline surfaces and shapes. When they shatter, fragments can act as seeds to encourage the growth of more of these symmetric patterns. Or a crystal can act as a template: atoms will arrange themselves on its surface, forming layer on top of layer, like those in mica. Either way, we have replication of a sort.

But what about a genetic code? Perhaps tiny flaws in a crystal—a metal atom substituted for an oxygen atom or a layer slightly misaligned—would be propagated as a crystal grew and copied itself. Occasionally a flaw would happen to alter a crystal so that it was more efficient at crystallization or at spreading itself. Some shapes might be more likely to adhere to one another, damming a rivulet of water and forming a pool hospitable to crystallization. When dried by the sun, some shapes might be lighter so that they would be more easily borne on the wind like dandelion seeds. Eventually some of the crystals might "discover" organic chemistry. By chance, they would develop shapes that acted as primitive catalysts, encouraging the formation of carbon-chain molecules. At some point these carbon inventions would become more efficient than the crystalline replicators, crowding them out of the chemical marketplace. The mineral metabolism would usher in the carbon metabolism we find in our cells. With the new structure in place, the scaffolding would fall away. Graham Cairns-Smith, the British chemist who devised the most elaborate of these stories, involving replicating clays, calls this usurpation "genetic takeover."

The theory, of course, is pure speculation. Cells seem to contain no vestigial traces of an ancient mineral metabolism. Yet for those who seek a reason for our existence, this notion of rooting life in the orderliness of chemistry is hard to resist. Harold Morowitz, a biochemist from George Mason University and a member of the scientific board of the Santa Fe Institute, believes that the biological mainstream is underestimating the power of the periodic table to generate life. He is convinced that the metabolisms in our cells are shaped not by the particular history that happened to unfold on this planet, but by the laws of Mendeleev's's table.

Morowitz is skeptical of Cairns-Smith's attempts to use replicating clays as the Maxwellian demon that sorted the primal molecules. It is possible that all traces of the mineral metabolism were long ago swept away, the information dissipated into the infinite sink of the universe. But Morowitz believes that if we are to constrain the search for life's origins and avoid flights of fancy, we should look no further than the cellular chemistry bequeathed to all living things by the first single-celled organism, the universal ancestor. His hope, as he puts it, is that "metabolism recapitulates biogenesis," that the key to the origin of life is the chemistry in our own cells, a fossil more than 3.5 billion years old.

He hopes to show that if we are accidental, then at least the chance event that led to our existence occurred long ago, shortly after the big bang, when the laws of physics congealed. Once the universe was seeded with the elements of the periodic table, he would like to believe, metabolisms were a foregone conclusion, a natural consequence of chemistry. In this universe, anyway, life at the metabolic level is pretty much the way it had to be. "If we find life elsewhere in the universe," Morowitz once said, "I believe it will be carbon-based and that it will include the Krebs cycle," the familiar sequence of biochemical reactions studied in biology class.

The dense web of reactions that inhabits living cells can be represented by what Morowitz reverently calls the Universal Metabolic Chart, a tangle of lines and arrows that resembles the diagram of a computer chip, or a grand conspiracy theory—boxes connected to boxes connected to boxes. Included in this blueprint is, for example, photosynthesis, by which solar energy is harnessed to string the simple molecules of carbon dioxide and water into the molecular food called glucose. And there are glycolysis and the Krebs cycle, by which the glucose is broken down to release energy, yielding carbon dioxide and water as wastes. Every organism on earth is a subset of the chart. Photosynthesis occurs only in plants. Molds have the unusual ability to synthesize penicillin, thiobacteria to extract energy from sulfur—all of these chemical pathways are represented in the chart. Some pathways that are crucial to one creature have been closed off to another, presumably from a random mutation long, long ago. People lack the ability of other creatures to synthesize ascorbic acid and so we must ingest vitamin C.

More striking, though, are the similarities between metabolisms: every cell on the planet, whether it is a single-celled bacterium or part of a multicellular plant or animal, largely consists of the same fifty organic molecules, which we can think of as the core of the metabolic chart, the biological common denominator. Animals may get their energy from breaking down sugars, thiobacteria from sulfur, plants from the sun, but the most basic reactions are universal. Every cell employs just four kinds of nucleotides, twenty kinds of amino acids. Genetic information is invariably recorded in DNA or RNA. Energy is universally stored in phosphate bonds as part of that familiar schoolbook reaction by which adenosine *di*phosphate is turned into adenosine *tri*phosphate, ADP to ATP. Taken together, this web of reactions forms the infrastructure of life. As Morowitz puts it, "The chart is to the biochemist what the periodic table is to the chemist." But Morowitz sees more than a metaphoric link. He believes the Universal Metabolic Chart arose naturally from the periodic table, that this universal set of biochemical pathways did not assemble itself by chance but by necessity.

Finding the same chemical reactions in bacteria, trees, ants, and wolverines is a little like finding scattered tribes of Native Americans sharing the same fundamental system of beliefs. With pueblo religion, we are tempted to attribute some of the similarities to recent borrowings rather than a common ancestral origin. But it is difficult to imagine how a metabolic cog or pulley invented by one creature could be borrowed by another. Morowitz finds it far more persuasive to suppose that the most basic reactions—respiration, protein synthesis, genetic replication, and such—have remained largely unchanged for aeons. A mutation at this deepest level would probably cause the whole chemical house of cards to collapse. Naturally there have been alterations as environments change and organisms adapt. But it seems sensible to suppose that these changes have been filigrees added to life's underlying infrastructure, variations on enduring themes. When Morowitz stares at the core of the metabolic chart, he believes he is catching a glimpse of the universal ancestor.

The shape of this primal network could still be an accumulation of chance events. Of all the possible metabolisms that happened to arise in the primal sea, perhaps ours is simply the one that became locked in, only because it was lucky enough to survive. If we could start from scratch and, in a grand Miller-Urey apparatus, re create life's beginnings, chance fluctuations might lead to a very different network of reactions. In this view, the core of the metabolic chart is a frozen accident, like the fact that GCA means alanine and not histidine. Or we could propose, as Morowitz does, that these primary reactions are so basic that they were all but determined by the grammar of atoms and molecules. If so, then there is hope for what Morowitz calls a grand unified theory of biochemistry, deriving the core of the metabolic chart from the periodic table, showing that life is as natural an extension of chemistry as crystallization.

Morowitz's creation story turns those told by his colleagues inside out. Despite all the strategic difficulties in assembling nucleic acids in a test tube, most origin-of-life researchers believe that RNA indeed came first, followed by proteins, and finally by the encasing of the metabolism in a cell. An influential minority believe that proteins preceded nucleic acids. When asked which came first, the molecular chicken or the molecular egg, Morowitz answers neither. In the beginning, he believes, long before the emergence of enzymes or nucleic acids, were empty shells, lifeless vesicles made of fatty molecules called lipids that provided a haven in which metabolisms later evolved. Where did the vesicles come from? They are practically guaranteed by one of the most basic laws of chemistry, the fact that oil and water do not mix.

H_2O is what chemists call a polar molecule; the way its atoms are arranged, it is charged positively on one end and negatively on the other. If other polar molecules are added to water, they dissolve: their negative ends are attracted to the positive ends of the H_2O molecules, and vice versa. Molecules, like those of salt and sugar, that align themselves so readily with those of water are called hydrophilic—"water-loving"— molecules. With no charge that would compel them to stick to water molecules, nonpolar molecules, like the hydrocarbon chains of petroleum, are insoluble, hydrophobic. We see the results of this aversion in street puddles, where a thin film of gasoline or motor oil floats atop the water, fracturing the sunlight into fracturing rainbows.

A third kind of molecule is said to be amphiphilic—one end is polar, the other end is nonpolar. It swings both ways. Mixed with water, these molecules, a variety of lipid, naturally join, end to end, to form pairs. The hydrophobic ends of two molecules stick together, leaving their hydrophilic ends dangling into the water. These pairs then link side by side into sheets. And the sheets, seeking the configuration that requires the least energy to maintain, spontaneously fold into vesicles, trapping little bits of water inside. The result is a protected pocket shielded from the oceanic randomness, a vessel in which order might arise.

As chemists have discovered in their laboratories, the making of proteins and nucleic acids requires great precision. The making of these primitive membranes is child's play. "They come for free," Morowitz says, "a gift from nature," which perhaps is why virtually every cell on earth is contained within an enclosure of amphiphilic lipids and proteins. He takes this universality as reason to believe that it was the vesicles that came first. The result was a primitive kind of membrane, and, as we know from studying the membranes surrounding our own cells, these semipermeable barriers make very good demons. By allowing only certain compounds in and out, membranes select a subset of all the possible molecules, holding them inside. Once simple protein molecules become dissolved in a membrane, the selectivity becomes more precise: the proteins can act not only as gates for certain ions and molecules, but also as pumps, burning energy in order to transport recalcitrant molecules across the divide. Where does the energy come from? Floating in the surrounding sea, Morowitz supposes, were chromophores, simple molecules capable of a primitive kind of photosynthesis, directly absorbing solar photons, storing their energy by forging molecular bonds. If chromophores became absorbed in one of these primitive membranes, the vesicle would have an energy source.

Tapping into the energy of the sun and sorting molecules, each vesicle, with its various configurations of gateways and pumps, would give rise to different internal chemistries. Bumping into one another, some of these protocells would fuse like oil drops, combining their contents, making their insides more and more complex. And, long before the invention of either enzymes or nucleic acids, they would have the ability to replicate. Absorbing amphiphilic molecules from the surrounding waters, membranes would expand until they became thermodynamically unstable. Requiring too much energy to maintain their unwieldy size, they would divide in two, driven by simple physics. And then these progeny would divide. In time, there would be a population of species with different internal chemistries. Most of these vesicles would be bags of dead chemicals, nothing that could be dignified as a metabolism. But among the population would be vesicles that had stumbled upon the simple chemistry for making their own amphiphiles, chromophores, and other membrane channels; these would be especially fruitful, multiplying throughout the medium.

The earliest of these metabolisms would have to operate without the complexities of amino acid catalysts. The only components would be simple molecules made from carbon, hydrogen, and oxygen with a bit of phosphorus and sulfur thrown in. The only reactions that could occur would be those favored by the periodic table of the elements—those that are thermodynamically likely. Only later, when a vesicle came upon a simple recipe for using nitrogen, would it be able to increase its survivability by making amino acids and then primitive enzymes. In addition to the narrow range of natural reactions, constrained by the periodic table, nature would have at its disposal the limitless invention allowed by the arbitrariness of enzymes. More complex chemistries would be possible. And those reactions that originally ran without enzymes would be streamlined, made more efficient by the power of catalysis.

In the final act, vesicles would use the enzymes to make the most delicate of the organic molecules, nucleic acids, opening the way for the production of more complex, longer-chain enzymes—and for genetic memory. Now every cell would carry its own instruction book. With the metabolic recipe encoded within replicating helices of DNA, the cell would have a more efficient means of reproduction. Everything would be in place for the fine-tuning of Darwinian evolution. As Morowitz puts it, this is where physics ends and history begins. The winner of this early struggle is none other than the universal ancestor, with its chemical network preserved (like an ant in amber) in the Universal Metabolic Chart.

A test of Morowitz's controversial theory is to show by experiment that the fundamental reactions that power our cells, the core of the chart, are so natural that they might have once run without enzymes. So far, he has shown that the pathway by which all cells take in nitrogen, for use in making amino acids, can indeed run without the help of protein catalysts. If this can be demonstrated for the other reactions, there is reason to believe that the spontaneous eruption of life is not so mysterious, less of a miracle—that there are fundamental reasons for us, or at least our earliest unicellular ancestors, to be here.

Viewed panoramically, Morowitz's origin myth has a compelling logic to it. Life, in this view, arose through a series of levels, each more complex than the last. First were empty vesicles dividing and fusing like oil drops; then vesicles with simple chemistries inside. Among these were vesicles with the means for making their own components. When one of these cells "discovered" nitrogen, the next step was enzymes and the richer chemistries they entail. Finally came the enzymatic production of nucleic acids. With this development the cells had the ability to keep a separate record of their genetic information; they could mutate and evolve.

If Morowitz is right, the potentially unending regression of demons making demons making demons indeed bottoms out in the laws of chemistry, which arise, in turn, from quantum mechanics. In the end, it is simple physics that gives rise to these primitive Iguses, the vesicles. Providing a buffer against the randomness of the environment, they allow for the formation of the delicate chemical arrangements which otherwise would be unlikely to emerge at all.

Morowitz is an unabashed "carbaquist," a term coined by the biologist Robert Shapiro and the physicist Gerald Feinberg to describe one who believes that life is necessarily built from carbon atoms. To some of Morowitz's Santa Fe colleagues, this assumption seems overly restrictive. They seek the source of biological order elsewhere, in abstract "laws of complexity" that would drive simple components of many different kinds to form complex wholes, a kind of inverse of the second law of thermodynamics that would somehow make it natural for order to arise. Some of them believe these laws are so compelling that self-sustaining structures can arise inside a computer, what they tantalizingly call artificial life.

Fred Hoyle declared that the probability of molecules assembling into a cell was like that of a tornado in a junkyard assembling a 747. A creationist is essentially supposing an intelligent tornado, the Almighty, with the intention and wherewithal to put together a jet plane. But what if the "intelligence" was within these laws of complexity? There would be no need of a demonic membrane to sort molecules into life; for in this view, metabolisms are not so delicate after all. Given any sufficiently rich mixture of molecules, laws of complexity would all but guarantee that a metabolism would crystallize from the broth. It might not be anything like the metabolism we have in our cells, but a metabolism nevertheless—a self-sustaining chemical web that could generate all the catalysts it needed to keep its reactions going, to maintain itself against the harshness of its surroundings.

The notion of "autocatalytic sets" of molecules springing forth from the random ooze has been entertained over the years by a handful of scientists trying to find a source of order for life. As early as 1969, Melvin Calvin suggested the possibility in his book *Chemical Evolution*. Two years later, the notion was further developed in independent papers by the theoretical biologists Otto Rössler, Manfred Eigen, and Stuart Kauffman, who is now a member of the Santa Fe Institute. Eigen coined the term "hypercycle" to describe his version of the phenomenon. Suppose an RNA molecule, drawing on an enzyme available in the environment, acted as a template for the formation of a second RNA molecule, which in turn acted as a template for the formation of the first; Eigen called that a simple catalytic cycle. If, in addition, one of the RNA molecules happened to code for the very enzyme needed to complete the loop, it would be a hypercycle: a self-sustaining chemical circuit.

While Eigen and his colleague Peter Schuster, a frequent visitor to the Santa Fe Institute, have concentrated on how hypercycles of nucleic acids might have given rise to life, Kauffman and others have looked at chemistry more abstractly, ignoring the details of how many valence bonds carbon or oxygen has. They look at molecules as symbols and a reaction as a kind of algebraic relationship. In this idealization of chemistry, a reaction might look like this: $A + B \overset{D}{=} C$. Molecule A combines with molecule B in the presence of catalyst D to yield molecule C.

To see how a network might naturally form, start with two simple molecules, A and B—monomers, as they are called in the trade. There are no catalysts at this point to favor certain reactions, just a weak buzz of molecules randomly bumping against one another. Occasionally the monomers, the A's and B's, will join to form simple dimers: AA, BB, BA, and AB. Now there are six kinds of molecules in the soup. We can combine these ingredients, both monomers and dimers, to yield longer chains: AAA, AAB, ABB, AABB, BAAB, and so forth.

Once we achieve a rich mixture of different kinds of molecules, it is not too much of a stretch to suppose that some of them will be shaped in such a way as to act as catalysts, encouraging certain reactions. Suppose that BAAB catalyzes the pathway by which AA and B form AAB. The production of AAB will now outpace that of the weaker, uncatalyzed reactions. Now further suppose that AAB catalyzes the formation of something called ABA, and that ABA catalyzes the formation of BABA, which catalyzes the formation of AAAAA. A chemical chain reaction has been ignited. Now suppose that at some point in this progression a molecule is produced that catalyzes the formation of the first link in the chain, that AAAAA catalyzes BAAB. The chain would close in on itself to form a loop: BAAB ⇒ AAB ⇒ ABA ⇒ BABA ⇒ AAAAA ⇒ BAAB. Against the background noise of weak, random reactions, this loop would grow and grow in strength.

While some pathways would build up molecules, others would break them down into smaller pieces, recycling the A's and B's. AAAAA might be cleaved by another enzyme into AA and AAA, then the AA joined with a B to make AAB. Loop by loop, a network of self-reinforcing reactions would begin to form that looked like a corner of Harold Morowitz's Universal Metabolic Chart. Start the mechanism with a stock of simple food molecules, A's and B's, and it would sustain itself.

If putting together such a network were a delicate affair, requiring great precision, then autocatalytic sets would be subject to all of Hoyle's reservations about 747s and junkyards. Without a designer, it would be all but impossible to imagine them arising by chance. But if there were many, many ways that autocatalytic sets could spring together, then we would need neither a designer nor incredible luck. Networks would arise spontaneously whenever a mix of molecules was rich enough. As Stuart Kauffman sees it, there is something a little too good to be true about the neat fit between G and C and A and T in Watson-Crick pairing. What, he asks, if chemistry were a little different? What if nitrogen, for example, had an extra valence bond? Would life have never arisen? He would like to believe that our existence is not so dependent on such seemingly trivial details.

In the autocatalytic systems, no assumptions are made about which atoms or molecules come together to form networks. In its simplest form, Kauffman has demonstrated, the question of whether self-sustaining metabolisms are rarities or likelihoods can be approached as a problem in what mathematicians call combinatorics. As the length of the objects in the mix increases, the number of possible molecules rises exponentially. Six kinds of molecules can be built of two units or less: A, B, AA, BB, AB, and BA. If we allow a length of three, the number increases to fourteen: A, B, AA, BB, AB, BA, AAA, ABB, AAB, ABA, BAA, BBB, BAB, BBA. Raise the length to four and we can admit sixteen new members—AAAA, AABB, AAAB, AABA, ABAA, ABBB, ABAB,

ABBA, BAAA, BABB, BAAB, BABA, BBAA, BBBB, BBAB, BBBA—for a total of thirty. And as the number of kinds of molecules rises, the number of potential reactions between them rises even faster, increasing the possibility of forming loops and networks.

A set of lines connecting objects like those in a catalytic set is called a graph. Drawing on graph theory and the mathematics of combinatorics, Kauffman found that as the complexity of the mixture increases, a collection of molecules will almost inevitably undergo a kind of "phase transition": like ice crystallizing from water, a self-sustaining network will spontaneously arise. Forget for a moment the details of biochemistry. In any system in which objects combine to make other objects, connected sets will arise driven by nothing more than simple mathematics. "Order for free," Kauffman calls it. From randomness on one scale, order seems to emerge on a higher level of the conceptual hierarchy.

Is what is true for these alphabetic metabolisms true for the molecules of life? The graphs Kauffman was playing with when he made his discovery were static arrangements of lines connecting points. But a metabolism is anything but static. Reactions proceed at various rates, depending on the concentration of the chemicals in the mix. The laws of thermodynamics favor some reactions and discourage others. Even though an autocatalytic set might arise within a random collection of objects, it would have to rapidly bootstrap itself to a significantly high concentration in order to endure.

In the mid-1980s, Doyne Farmer, who had met Kauffman at a conference in Boston, invited him to Los Alamos to work with his Complex Systems Group. Autocatalytic networks were a natural extension of Farmer's interest in chaos. The loops in a metabolism give rise to nonlinear feedback: tiny changes in the concentration of a single chemical can become amplified, its influence propagated throughout the net. While the result of such hypersensitivity could easily be chemical chaos, a network that never settles down, perhaps under the right conditions the self-reinforcing loops would give rise to stable structures. Drawing on these ideas, Farmer, Kauffman, and Norman Packard (who was visiting from the Institute for Advanced Study in Princeton) built a stripped-down computer simulation that they called artificial chemistry. In the model, certain molecules were arbitrarily designated as catalysts for certain reactions and assigned specific strengths. They could either encourage condensation, combining two molecules to form a larger one, or cleavage, breaking a molecule apart. While Kauffman's original model was an all-or-nothing affair (a reaction either occurred or it didn't), with random chemistry a molecule could exist at various concentrations; the more there was, the stronger its effect.

At equilibrium, nothing interesting happens in a flask of chemicals: molecules are put together at the same rate at which they are broken apart, resulting in homogeneous, random noise. It is only when a system is driven away from equilibrium that there is the possibility of structures forming. In Farmer, Kauffman, and Packard's model, the chemical flask was driven from equilibrium either by supplying it with a constant flow of food molecules or by providing it with energy. As a kind of substitute for Stanley Miller's lightning bolts, the equations for condensation and ligation were adjusted to behave as though the system were energized by a primitive version of photosynthesis.

With this more dynamic chemistry, Farmer, Kauffman, and Packard found the kind of phenomena that Kauffman had predicted with his static sets: under a wide range of conditions, closed metabolisms formed, increasing their concentrations many times over that of the background noise. Later, a biologist named Richard Bagley significantly improved the model. Working with Farmer, he found that the networks that sprang forth were surprisingly robust. If he destroyed a loop, by erasing one of its molecules, the network would quickly regenerate the missing piece. And while some of the networks were dependent on having a certain kind of food molecule to draw on (Farmer and Bagley compared them to pandas, which only eat bamboo), others were like cockroaches: they could devour just about anything.

To believe that the ABCs of life began when carbon, hydrogen, nitrogen, oxygen, phosphorus, and other atoms joined into these symbiotic chemical alliances, one would have to find a way for the simple metabolisms to evolve and become more complex. One could hardly expect a primitive metabolism to spring from the soup with a full-blown genetic mechanism. But Farmer and Bagley found in the model what might be considered a crude genome. The metabolisms that crystallized from their simulated soup contained "seeding sets": a small piece of the network had the power to regenerate the whole. Place one of these subsets of molecules into the reaction flask and it would blossom into a whole autocatalytic net. In fact, autocatalysis was such a robust phenomenon that each metabolism had a number of seeding sets. Taken together, they could be thought of as a kind of genotype, a list of the information necessary to generate the "organism." Random changes in a seeding set would lead to mutated networks. A spontaneous, uncatalyzed reaction might add a new metabolic loop or short-circuit one that was already there. While most of these changes would be neutral or detrimental—they might cause the house of cards to collapse—some would occasionally result in a more resilient metabolism. And so the changes would propagate themselves; the network would evolve.

This is not Darwinian evolution: there are no creatures competing against one another for resources, just a single structure becoming better and better at what it does. Evolution in this sense is indistinguishable from development, in which an embryo unfolds into an adult. At this stage of protolife there would be no such thing as an individual. In artificial chemistry, a metabolism thrives by making more of itself, but there is no sense in which it reproduces, forming two metabolisms where there once was one. If a drop of liquid from a pool occupied by one of these metabolisms sloshed over into another pool, the network would replicate itself, as long as all the elements of a seeding set were present. Then we might sensibly speak of two individuals, but they would exist in different worlds. If we took the contents of each pool and recombined them, the individuals would merge back into one. In a pool with several different metabolisms, each might be thought of as an individual, though it is likely that they would share many chemical reactions; their "bodies" would overlap. So do we think of them as individuals or as parts of a single organism?

The result, it seems, is a sophisticated version of a flame: a reaction that feeds on molecules in its environment and spreads. Unlike fire, it can improve itself, using random variations to explore new ways to burn brighter. Individuality would come when a metabolism chanced upon a way to produce its own enclosure or to inhabit a preexisting one—perhaps Morowitz's vesicles. The difference here is that one need not assume a carbon-based chemistry. These self-sustaining structures—whatever they were made from—would compete for resources. Darwinian evolution would take over, distinguishing among variations, honing them to fit their niches. Arising from randomness, these structures would survive by learning to seek regularities in the chaos around them, to make maps—genetic and later neural—of the world. Or so goes the autocatalytic version of the origin myth.

The drive to seek and impose order in the world has given birth to the sciences of biology, geology, particle physics, astronomy, cosmology; it has generated grand cathedrals of abstraction like quantum theory and

Tewa religion. Our abhorrence for randomness has driven us to seek, beneath the rough surface of the universe, hidden symmetries and to describe precisely how they became shattered. From the time positive first recognized negative, or oxygen recognized two hydrogens, we have been devourers of information, seekers after patterns. And so we are compelled to tell these stories of how and why we sprang from the primal waters—and of what happened after the grand emergence.

8

THE ARRIVAL OF THE FITTEST

U p to the point where simple, single-celled creatures flourished in the waters, the notion of autocatalytic sets finding refuge in Morowitz's spontaneously made shells provides a plausible explanation for how matter might have become animated on the early earth. One can see why self-maintaining chemistries encased in membranes would come to dominate the waters.

But the scenario stops with the development of these simple creatures, which biologists call prokaryotes. A prokaryote is essentially a bag of molecules with the ability to sustain itself and divide. The next step in evolution is the development of the so-called eukaryotes. These larger, more complex cells have internal structure: the genetic material is not simply floating in the cytoplasm but encased in a nucleus; the machinery for burning sugars to yield energy is encased in organelles called mitochondria; the photosynthetic apparatus is in chloroplasts. Algae cells are eukaryotes, as is each and every cell that forms the body of all the plants and animals of the earth.

Moving beyond the question of how the first cell—the universal ancestor—sprang together, more mysteries arise when we seek a scientific rationale for the emergence of life. How did the simple prokaryotic cells develop into the eukaryotes, and then how did these complex cells come together to form multicellular organisms, which are more complex still? Again one confronts the question of whether the rise of pattern—and patterns of patterns—was largely accidental or driven by some kind of natural law.

The biologist Lynn Margulis of the University of Massachusetts has proposed that the mitochondria and other organelles inside eukaryotes were once freestanding, unicellular creatures themselves—prokaryotes. Under various evolutionary pressures, they formed symbiotic alliances with one another which were codified for all time when they somehow merged into single eukaryotic cells. We can extend this story and suppose that eventually symbiotic alliances of eukaryotes joined into the first multicellular organisms. Why this happened is, again, something of a mystery. But if we forget for a moment all the difficult details, we can see a compelling pattern to the development of life: communities of molecules uniting to form a prokaryotic cell, communities of prokaryotes uniting to form a eukaryotic cell, communities of eukaryotes uniting to form an organism.

But what caused these great leaps to occur? If we can believe the fossil record, the simple prokaryotes dominated life on earth for more than two billion years. Why, then, didn't life remain stuck in this limbo? What spurred the prokaryotes to band together into eukaryotes? And why, for that matter, did the eukaryotes later band together to form the multicellular organisms that exist today?

In trying to explain this trajectory, it is easy to get caught in the vortex of teleological reasoning, assuming that there is a purpose to evolution, that the end somehow causes the means. Multicellular organisms arose, we say, because they provide division of labor, as though the idea Division of Labor were hovering out there somewhere in Plato's universe, somehow compelling the cells to pay it heed. As Walter Fontana, a young Italian-born chemist at the Santa Fe Institute, once said, "This is backward. You can't have division of labor *until* you have multicellularity." His point is subtle. Once multicellularity developed, we can invoke natural selection to explain why organisms that efficiently divided their tasks among many cells would flourish. More difficult to explain is why multicellularity arose in the first place. Was it simply dumb luck—one advantageous accident after another—or something that was compelled to happen?

Ever since Darwin, there has been a tension between two schools of thought about how to carve up the biological world. The strict adaptationists believe that natural selection alone is powerful enough to explain most of the order, while a school called the structuralists insists that something extra—laws of self-organization—is required. Darwinism, they point out, is a theory of how traits were selected, not how they were generated. Where, the structuralists ask, does evolution get the grist for its mill? As Stuart Kauffman once put it, Darwin tells us how frogs change, but not how we get frogs in the first place. And the same argument goes for other structures, other patterns—eyes, brains, kidneys, vertebrae. What we need, the structuralists tell us, are laws of complexity that would explain how biological structures effortlessly emerge, only then offering themselves to evolution for fine-tuning.

Most biologists would grant that the patterns of life arose because of a mixture of these two tendencies, adaptation and self-organization. The question is where one draws the line. The biological mainstream is generally content to see the rise of multicellularity primarily as one more example of evolution in action—filtering randomly mutated structures through the sieve of natural selection. Fontana and some of his colleagues have come to believe, however, that the odds of multicellularity arising this way are extremely slim. They believe that, like the origin of life, the origin of the whole biological hierarchy—prokaryotes, eukaryotes, multicellular organisms, the towering architecture of life—rests on odds that are too good to be true. They believe some deeper law must be involved.

Those who believe in a direction to evolution are often accused of engaging in teleological reasoning. While Aristotle rejected the notion of pure ideas existing independently in another realm, he wrote of equally mysterious phenomena called final causes: the acorn grows the way it does because it contains within itself the potential of becoming a tree. In constructing a Darwinian explanation, one starts with a tale that sounds, at first, very teleological—evolution is driven to develop Division of Labor. But then Teleos is banished by declaring that the end "caused" the means not by ineffable forces but simply because it offers a survival advantage.

The adaptationists' explanations can sound most persuasive. Because of some random fluctuation in the environment, a tiny prokaryote that burned various food molecules to release energy was forced to find refuge inside a larger cell, living on its internal wastes. At first this tiny invader may have sickened its host, but evolution punishes parasites that are too deadly—they perish along with their victims. The survivors would be those that could exploit their host's chemical resources without endangering it, those that became domesticated. In time, perhaps, the host would even benefit from the arrangement. The extra energy produced by the invader would give the larger cell a survival advantage. What began as an infection would become a symbiosis. And this, perhaps, is why the eukaryotes that make up our bodies harbor organelles called mitochondria that specialize in respiration. We can come up with a similar tale for how plant cells got their chloroplasts, the organelles that convert sunlight into edible energy. In *Microcosmos,* the beautiful paean to microbial life she wrote with her son, Dorion Sagan, Margulis goes so far as to suggest that the flagella or fringe of cilia used by simple eukaryotes to propel themselves through the water were once the rapidly swimming bacteria called spirochetes. They, too, formed a symbiotic relationship with other prokaryotes, offering locomotion in exchange for other goods and services.

In the next chapter of the adaptationists' story, the fruit of these chance alliances, the internally elaborate eukaryotic cells, diversified through random variation and selection, developing various specialties. Some were adept at using their cilia for locomotion, others for sensing the existence of harmful chemicals, others for responding to light. And then these eukaryotes formed alliances of their own. A light-seeking eukaryote that happened to stick to a eukaryote with a swiftly lashing tail would beat its competitors in the race to find the brightest sunlight, the nectar for its chloroplasts. And so is born a primitive organism. With more feats of imagination one can come up with a story of how more complex animals with kidney cells, liver cells, and brain cells came to be. The stories are driven by a compelling logic. But at every step of the process, a great deal of luck is involved.

To the believers in laws of complexity, these rather ad hoc explanations begin to sound like Rudyard Kipling's *Just So Stories.* How the Rhinoceros Got His Skin, How the Tiger Got His Stripes. How the Algae Got Its Chloroplasts, How the Sperm Cell Got Its Tail. When they hear their colleagues straining perhaps a little too mightily to squeeze everything into the Darwinian framework, some biologists call the result an evolutionary Just So story: a compelling tale based on scant evidence, which sometimes has the facile ring of reasoning after the fact. Could the complexity we see around us come from nothing more than one chance event after another, selected by the filter of evolution? This strikes some skeptics as an awfully long string of "too good to be true's." Again, it seems, life as we know it would be largely a fluke, more a matter of history than science.

There is something about hierarchical organization that seems almost inevitable to us, a natural part of the universe. Our whole scheme of carving up the world is rooted in this image of simple units joining to form complex units which join to form units that are more complex still. We are made of organs which are made of eukaryotic cells which are made from prokaryotes which are made from molecules made from atoms made from leptons and quarks. The hierarchies also extend in the other direction: animals band together in flocks, herds, societies. Throughout northern New Mexico, the honeycombed ruins of the Anasazi are a reminder of how ancient is this drive to join together into larger, more complex wholes. Perhaps this order we find is simply a matter of projection: we are built hierarchically, so we can't help but see the universe that way. In their search for laws of complexity, some of the scientists at the Santa Fe Institute are seeking evidence that hierarchies are indeed natural, a pattern ground not simply into our eyeglasses but into the world, an antidote to randomness and disorder.

A favorite game of biologists is to imagine what the living world would be like if the slate were wiped clean and the evolutionary drama repeated from scratch. As the Harvard biologist and paleontologist Stephen Jay Gould likes to put it, What would happen if we rewound the tape and played it again? Would the life forms that resulted from the chance variations of evolution, the throws of the Darwinian dice, look anything like what we now see in our world? If the order of the living world is indeed produced largely by random variation and selection, then the patterns we see are mostly contingent, not necessary; historical, not universal. Even the hierarchies we marvel at might be miracles of chance, no more required than having ten fingers instead of twelve. Gould himself leans strongly in the direction of contingency. If the tape were played over, he believes, the flora and fauna of the earth would be nothing like what they are today; and he doubts that there would be anything like biologists with the compulsion to make sense of it all. Complexity, he believes, and probably consciousness itself, are accidental. After all, prokaryotes thrived for half the lifetime of the earth without showing any drive to become more complex. And today most of the mass of the biosphere consists of bacteria.

Yet when we ponder the ease with which life seems to organize itself into these wedding cakes of ever more elaborate layers, it is tempting to believe that it is not all accidental, that there is something fundamental about the rise of complexity. In the early 1990s, Fontana, the Santa Fe Institute chemist, and his colleague, the Yale biologist Leo Buss, began collaborating on a theory of how life arose and then arrayed itself in a grand architecture of tiers piled upon tiers. In their earliest conversations, Buss and Fontana found that they shared a deep suspicion that hierarchies were more than a fluke of evolution. If the tape were run again, they believed, the creatures that resulted would indeed be very different from what we see around us. But they would inevitably arrange themselves in ladders of increasing complexity.

Buss had made a name for himself in 1987 with the publication of a book called *The Evolution of Individuality*, a lucid exploration of this problem of how and why hierarchies arise. How is it that actors pursuing their local, selfish interests can give rise to organizations in which the good of the individual is subordinated to the good of the whole? Molecules pursuing the laws of chemistry, that complex web of compromises arising from the simple, deeply rooted duality between positive and negative charge, seem to combine naturally into metabolisms and cells—and cells into organisms, and organisms into societies. But to do so, each must cooperate, giving up the blind pursuit of personal advantage. For each of these hierarchical jumps to occur, individuals must subordinate their selfish goals to the good of the organization, creating, in effect, a new higher-level individual. The overarching question, then, is how does this individuality arise?

Some of the answers to these questions have become clearer because of the theory of kin selection. People who study animal behavior were long puzzled by the problem of altruism: bees laying down their lives in the defense of the hive, birds exposing themselves to predators by sounding a warning cry. Why would natural selection reward such self-destructive traits? Then came the sociobiologists, who argued that if we look at evolution from the point of view of the gene instead of the creature, altruism is not so mysterious. Many different creatures harbor identical genes; if the purpose of the game is for the genes to propagate as many copies of themselves as possible, then it might pay for one creature to perish so that a dozen might live. Creatures that are closely related will subordinate themselves to the greater good if the result is the dissemination of more of the genes they hold in common. According to kin selection, "superorganisms" like anthills and beehives arise because the insects have three-fourths of their genes in common. It makes sense to cooperate. The biologist Richard Dawkins found this perspective so compelling that in his book *The Selfish Gene* he proposed that we can think of ourselves as "survival machines," invented through evolutionary tinkering by ancient replicators, snippets of DNA, to propagate themselves more efficiently. "Now they swarm in huge colonies," Dawkins wrote, "safe inside gigantic lumbering robots, sealed off from the outside world, communicating with it by tortuous indirect routes, manipulating it by remote control. They are in you and in me; they created us, body and mind; and their preservation is the ultimate rationale for our existence."

Altruism isn't limited to animals; the puzzle extends all the way down the line. Why do self-replicating molecules subordinate their selfish interests—making as many copies of themselves as possible—to form metabolisms, cooperative chemical societies? How do cells that have joined together to form an organism work together in such precise harmony? What keeps cells from engaging in an orgy of replication, unleashing the uncontrolled growth called metastasis, or cancer?

Kin selection, again, clears up some of the mystery. If ants, with three-quarters of their genes in common, find it advantageous to live together in anthills, then perhaps it is no surprise that cells—with identical genomes—would be compelled by Darwinian logic to subordinate their interests to those of the organism. But again, the question is how these structures come together in the first place. Once these alliances are struck, kin selection helps us understand why they endure. But what compelled the organizations to arise? How much was due to random variation and natural selection? How much was due to self-organization? Or, to put it another way, how much was contingent, accidental, unlikely to happen again? And how much was to be utterly expected?

In puzzling through the confusion, Buss suggested, in *The Evolution of Individuality,* that we look at the organism in a subtly different way. In kin selection, the genes, rather than the creatures, compete for survival. Going a step further, Buss proposed that within a single creature, evolution goes on at many different levels at once—not only at the level of the individual and the level of the gene, but at the level of the cell and even the molecule. At the same time that an animal is competing and cooperating with other species, playing its part in the ecosystem, inside it different "species" of cells are woven into their own web of alliances. And within the cells are ecosystems of molecules—metabolisms. Communities are nested within communities like Chinese dolls. At every level, the best strategy may be to subordinate one's selfish interests for the good of the organization, for if the organization dies it takes its parts along with it.

From this perspective, one could tell a creation story that goes something like this: In the beginning, perhaps, were self-replicating molecules that were dependent on the presence of an enzyme, a replicase, that happened to be in the waters around them. Then environmental disaster struck. There was a replicase famine, caused perhaps by overpopulation, too many replicators competing for the enzyme's use. In this new environment, there would be a huge survival advantage for replicators that could learn to make their own replicase. This would probably require the help of others, so suppose several replicators chanced to come together in an autocatalytic web that produced as one of its products the replicase enzyme. As the naturally occurring replicase in the environment continued to fade, the synthetic replicase became all the more valuable. And so each molecule in the network was increasingly dependent on the survival of the whole. Suppose again that a mutation made one of the members of the metabolism more efficient at copying itself, at the expense of the other molecules. The metabolic society would collapse. On the other hand, chemical mechanisms that somehow discouraged individual molecules from frenzies of self-reproduction would have great selective advantage. A metabolism that chanced upon a mechanism for controlling its own members, regulating their ability to reproduce, would thrive.

Eventually, though, the price of this stability would be stagnation. While a single molecule is free to explore a huge space of possibilities, mutating into every conceivable form, the molecules in a network are much more constrained. Only certain arrangements are stable enough to endure. At some point, every evolutionary niche available to the metabolism would have been explored, the waters filled with every viable kind of network, competing for molecular food. If evolution was to continue, the only direction to proceed would be up the hierarchy. All the niches available to single metabolisms might be occupied, but the niches available to combinations of metabolisms would remain unexploited. Metabolisms that chanced to come together into alliances could colonize this terra incognita. Just as molecules increased their survivability by banding together into metabolisms, so might several metabolisms join to produce a kind of meta-metabolism that synthesized many more of the nutrients necessary for survival. Now, in time of chemical famine, these more complex networks would have an edge. They might even diffuse to other parts of the pond once inhospitable because of a lack of nutrients. Metabolisms with mechanisms for neutralizing poisons would have still other spaces to explore. Their world would become larger.

Venturing into strange territories is safest if one has a vehicle. Those supermetabolisms that stumbled upon a way to encase themselves in vesicles could colonize a whole new universe—all the niches in which different variations of cells can thrive. And so, the story goes, the prokaryotes were born. But again there is a price for this new power. Once one metabolism joins with another, it is no longer as free as it was before. Its fate is tied inextricably to the fate of the cell. Again, stagnation eventually sets in. All the new niches are occupied and the only direction to evolve is upward. Self-replicating prokaryotes are driven to form communities: the eukaryotic cells.

Though they are now parts of a larger whole, these former prokaryotes have not completely given up their old ways. In this new arrangement of cells within cells, selection proceeds at more than one level. Variations of the eukaryote itself are selected by the environment, but within the cellular environment, the cytoplasm, the prokaryotic organelles also continue to evolve. What is good for the organelle is sometimes good for the cell, but more often there will be conflicts: a mutation that favors a mitochondrion might be detrimental to other organelles whose functions are critical to the operation of the whole assemblage. Compromises must be struck. From the point of view of the cell, there will be a selective advantage to limiting variation at the lower level. Organelles lose their own replicative machinery while a single organelle, the nucleus, takes over control of their reproduction along with that of the entire cell. Today we see what may be fossils of an earlier time: the mitochondria and chloroplasts, with their own separate stock of DNA.

As before, the price of stability is eventual stagnation, and if evolution is to proceed, it will have to climb another step up the hierarchy, creating a more complex individual, a new vehicle with which to explore. The opportunities available to combinations of different kinds of cells are much greater than to those of one cell alone. Thus there will be a huge selective advantage to mutations that allow cells to differentiate and form an organism. The environment is once again expanded, opening up a new space of possibilities.

Within these new creations—multicellular creatures—evolution again proceeds on several levels. The creature as a whole is selected by its environment, with all these new niches, but the cells continue to compete within the internal environment of the body. Again some mutations may favor a lineage of cells at the expense of the organism. But what is good from the point of view of the cell, encouraging it to make as many copies as it can, is, from the point of view of the body, a malignancy. Compromises must be made so that the many varieties of cells can live together and, at the same time, preserve the whole.

Buss hopes this vision of multiple levels of selection may provide a window into the mystery of development: how an embryo unfolds into a body. We think of development as an exquisitely orchestrated unfolding, but perhaps it began as a wild and combative process, as different varieties of cells jockeyed for advantage. Eventually a web of mutually beneficial relationships evolved, a system of checks and balances that would preserve the whole, without which all would die. Think again of an ecosystem: Darwinian competition leads to a network of compromises, overlapping treaties that allow each creature to make the best living it can without destroying its delicate system of support.

Whether we are talking about molecules, cells, or organisms, alliances of simpler entities can explore niches not open to a single entity alone. But the price is always a subordination of freedom. Sacrifices must be made for the commonweal. At each step of the hierarchy, new possibilities arise, but equilibrium inevitably sets in and the only way to evolve is upward, to create a new vehicle to explore new regions of evolutionary space. As Buss wrote in *The Evolution of Individuality,* "Life, once established, will come to array itself in an ever-increasing wardrobe of 'vehicles.' "

Where do the hierarchies stop? Might an entire ecosystem develop into an individual, suppressing the freedom of each member until they became like organs in a body? Could this tightly linked system develop the power to replicate itself, spreading its spores to other planets? "The notion sets the mind to fancy," Buss wrote. "If earth—or some earthlike planet—did not occur so close to an errant asteroid zone, then these arguments would suggest that life would continue to develop hierarchical entities unabated."

For all the enlightenment it provided, Buss began to find his vision unsatisfying. He too was spinning Just So stories. His theory helped explain the evolutionary rationale for the various steps of the hierarchy. Each one made Darwinian sense. It was understandable that once formed, the alliances would endure. But after taking adaptationism as far as he could, he still wasn't able to say with any conviction why each level sprang into existence in the first place—why, instead of stagnation (surely the path of least resistance), the transitions occurred. What would happen if the tape were run again? Was the climb up the hierarchy inevitable or simply a string of good luck—one advantageous event after another? He was still stuck with the question of why the two-billion-year reign of the prokaryotes came to an end.

Discouraged by his conclusions, Buss began to toy with the idea that in generating the order, more than the random variation and selection of Darwinian evolution was involved. He was not seeking the kind of holistic explanations one can find among the healers of Santa Fe—the belief that life cannot be explained by chemistry, that we are animated by mysterious crystalline energies, the Life Force, an ineffable *élan vital*. As Stuart Kauffman and his collaborators showed with their autocatalytic sets, the spontaneous formation of order is not as unlikely as we are accustomed to thinking: when a chemical soup reaches a threshold of complexity, metabolic tangles are mathematically bound to arise. Perhaps it was possible to take the argument a step further, to show that once a metabolism had sprung up, laws of complexity made it *natural* for it to join with other metabolisms into an organization, and for the organizations to join with organizations, and so forth up the scale. And that is where Fontana comes into the story, for he offered a way to test this assumption, with a computer simulation. By stripping chemistry to what he considered its bare essentials, he would try to re-create something like genesis and see for himself whether hierarchies seemed inevitable or accidental. In this limited way, he would take Gould up on his challenge and rerun the tape of life.

Fontana had begun working on artificial chemistry in the late 1980s as a member of Doyne Farmer's Complex Systems Group at Los Alamos. Many biologists thought that in trying to simulate prebiotic chemistry, the Los Alamos group had gone too far into the cave of abstraction, throwing out so much detail that their model was unlikely to say much very useful about life on earth. Their "molecules" of A's and B's were much too short to be capable, in the real world, of catalytic powers. But as abstract as it was, with its alphabetic atoms, the model made some attempt at earthly realism, so much so that Fontana felt it was not abstract enough, that it adhered too closely to the chemistry that happened to have developed on this planet. By pruning away everything he considered distracting detail, he hoped to show that living organizations are not simply a function of the kind of molecules that accumulated in this backwater of the Milky Way, but a general property of the universe.

The mark of a good simulation is that it separates the essential from the incidental, cutting through what is deemed irrelevant detail to get at the heart of a problem. This involves making instinctual judgments about which details are crucial and which can be ignored. Depending on the tastes of the theorist, one can imagine a gradation of possible approaches to simulating prebiotic chemistry: at one end of the scale, one could try to duplicate earthly chemistry in all its rich detail. Simulated carbon atoms with four valence bonds would congregate with simulated hydrogen, oxygen, nitrogen, and other atoms. In reality, no programmer is clever enough nor any computer powerful enough to pull off either task. And even if scientists could develop a fine-grained simulation of the ways atoms interact, what would it show? Chemistry would have been transplanted from the physical world to the simulated world inside a computer, along with all its mysteries. Simulated molecules might come together to form simulated metabolisms and even simulated cells. But then the simulators would be left with the same puzzles to explain. Like a theory, a simulation must be more general and abstract than what it is intended to explain. It must serve as a compression.

If a hypothetical model containing carbon atoms, hydrogen atoms, and the other components of organic molecules anchors the left-hand, most concrete end of the scale of simulations, then the Los Alamos version of artificial chemistry might hover somewhere toward the middle. At the far end of the spectrum would be Fontana's austere abstraction, which he called algorithmic chemistry—"Alchemy," for short. Unlike his colleagues' simulation, Alchemy contained no information about chemical concentrations or reaction rates; there was no simulated solar energy source to forge chemical bonds. There were simply strings of symbols (triangles, circles, squares) that randomly collided with other strings of symbols. From these collisions, new strings were born. Using this as a framework, Fontana was able to show, like his colleagues, that under the right conditions, self-sustaining networks—metabolisms—emerged, providing more hope for those who believe that life is not so delicate or unlikely an occurrence after all. But Fontana went a step further. When two of his spontaneously formed metabolisms were put together, they joined to produce a symbiotic organism, a kind of meta-metabolism. The next step, it seemed, was meta-meta-metabolisms, and so forth, step by step up the ladder of organization.

The talk in which Fontana and Buss described these developments was, for many, the high point of the third Artificial Life conference, which was held in the summer of 1992 in downtown Santa Fe. The presentation began in the guise of a lecture by Buss, then erupted into what seemed like a spontaneous discussion between Leo and Walter over the very meaning of biology. Wearing suspenders and wire-framed glasses, Buss, a young, heavyset man with a beard and shaggy brown hair covering the tops of his ears, strolled around the stage in what Stuart Kauffman calls his "country lawyer mode." Occasionally scratching his head and pushing a shock of hair out of his eyes, he made clear his conviction that when it came to explaining the origin of structure in the biological world, Darwinian evolution was hardly more convincing than Just So stories. Once molecules have come together to form prokaryotic cells, we can invoke Darwinism to explain how prokaryotes evolve, he said. Once prokaryotes unite to form eukaryotes, we can use Darwinism to explain how eukaryotes evolve. And once eukaryotic cells come together to form organisms, we can use Darwinism to explain how organisms evolve. But what impelled these transitions to occur in the first place?

"It's really a very curious thing," Buss said. "We all seem content with evolutionary theory in its current form. We have this form that is virtually universally accepted. We teach out of the same textbooks. But I would argue that the form that we have begs the fundamental question." Then, interrupting his own reverie, Buss looked at Fontana, a thin, neatly groomed man sitting in the front row of the auditorium. "So, Walter, is that a fair characterization?"

"Well," Fontana replied, in his deep, precisely formed syllables, "I would say that what you want is a theory of organization . . ."

"Exactly," Leo replied. "The problem is one that periodically recurs. Hugo De Vries, the turn-of-the-century macromutationist, ended his 1904 book with a wonderful line that really summarizes this—that he's not interested in the *survival* of the fittest, he's interested in the *arrival* of the fittest."

"You're not just interested in the origin of life, I suppose," Fontana interjected.

"No," Buss replied. "The origin of life is really just the first instantiation of this problem. We're interested in the origin of all these organizational classes. How you get genes, how you get cells, how you get multicellular individuals . . ." Then, as if suddenly overwhelmed by the magnitude of the task, he looked straight at Fontana and said, "Walter, why don't you just come up here."

By now it was clear that what had seemed a spontaneous encounter had been carefully scripted. Fontana walked up to the podium, watching patiently as Leo laid transparencies on an overhead projector, explaining the standard view of the history of life: how self-replicating molecules (Level 0, he called this) gave rise to prokaryotes (Level 1), which banded together to form eukaryotes (Level 2), which joined to form organisms (Level 3).

"Wait a minute," Walter interrupted, as though he had heard this story too many times before. "We're not just interested in understanding the historical progression. We're interested in understanding what is necessary and what is contingent." We need to know, in other words, not just how life unfolded but *why* it unfolded the way it did. If we could take two planets and, in a controlled experiment, seed them with the same molecules, would they produce similar biospheres?

"That's right," Leo said. "It's the old 'control earth' problem. If you ran the experiment again, I want to know what features would necessarily arise and I want to know which organizational classes would necessarily arise."

But what did Leo mean by an organization? What, for that matter, was a molecule? Walter began prodding his colleague to closely examine the terms he and other biologists so cavalierly threw around.

"Well, a molecule is a molecule and . . . an organization is an organization." As Leo quietly struggled for answers, Walter put a transparency of his own on the overhead projector: "Warning, Big Abstraction Ahead." And suddenly the audience was confronted with bewildering diagrams of circles, squares, and triangles—abstract molecules combining into increasingly complex structures.

As Fontana outlined the details, Buss looked on with the mix of skepticism and fascination he had probably exhibited when Walter first laid out the plan several years before. In Fontana's Alchemy, the strings of triangles, squares, and circles stand for enzymes, and like enzymes they serve a double role: as both objects and processes. A protein is, in the most familiar sense, a thing—a string of amino acids all tangled up—but it also carries out a function, catalyzing a particular chemical reaction. So it is with a computer algorithm: it is an object (a string of 1s and 0s represented perhaps as magnetized spots on a disk drive or charges in a RAM chip) that embodies a process, the carrying out of a specific task. As Fontana once put it, for centuries mathematicians studied functions that were not objects: pure abstractions said to exist in the platonic heavens. And for centuries, physicists studied objects that were not functions: everything from colliding billiard balls to quarks. With computer science, people found a way to talk about objects that are also functions. Mathematics became embodied as computer science, a perfect vehicle for talking about how structures interacting with structures generate complexity.

Specifically, Fontana took his inspiration from the self-referential way in which certain computer languages, like Lisp, operate. Though all algorithms are, in a sense, both objects and functions, in most computer code there is a sharp distinction between algorithms and the data they manipulate. To add 2 and 2, the data, 2, 2, are fed to the addition algorithm, +, yielding 4—another piece of data, which might be fed to another algorithm that finds its square root. In Lisp (the name stands for "list processing"), the line between data and algorithms disappears. One moment, a string of Lisp might be called upon to act as an algorithm, performing the function coded in its string of symbols; the next moment it might be treated as data, fed to another algorithm and modified somehow. Take one string of Lisp and feed it to another and you get a third string, which might act as an entirely new function. Thus a program written in Lisp can change itself as it runs, making up new rules as it goes along.

As a result of its recursive nature, Lisp is a natural language for the development of artificial intelligence programs: generating a new string of code can be seen as a form of learning, the acquisition of a new rule. Fontana saw that Lisp, or something very much like it, would also be a good tool for modeling chemistry. An enzyme can act as a function, altering other molecules. Or it can act as an object, something to be altered by other enzymes. Another way to think of an enzyme is as a little computer. In computer science, the paradigmatic example of a digital computer is the Turing machine. An input tape, encoded with a string of symbols, is fed to a black box, which reads the symbols and alters them according to a few simple rules. Give a properly programmed Turing machine a tape that says 2 + 2 and it will grind away until it replaces the expression with the symbol 4. If we think of an enzyme as a Turing machine, the input tape is the molecules it acts upon and the output is the new molecules that are produced.

Lisp was invented by John McCarthy, the Stanford University mathematician who is often said to have coined the term "artificial intelligence." He developed the language from a system of symbolic logic called lambda calculus, which also exploits this powerful ambiguity between object and function. As Fontana explained to Leo Buss, lambda calculus (which he mercifully declined to describe in detail) contains the bare bones required to allow objects to collide with one another and make new objects. In Fontana's abstraction, chemistry is based on just two ideas: construction and sameness. A chemical system must be able to build things, taking simple components and combining them into more complex wholes, and it must be able to recognize when two objects are the same. There might be many pathways for making a molecule—alanine, for example—but it must be recognized by the system as the same object no matter how it is made; regardless of its roots, it must function in the same manner. If A makes B and B makes C and C makes A, then the system must know that the A's at each end of the sequence are identical. Thus the chain can be closed to form a loop, a circuit in a metabolic web. As Fontana puts it, construction leads to diversity and the recognition of sameness leads to network formation.

In the physical world, atoms and molecules cannot combine arbitrarily; there is a syntax that arises from the laws of physics: the shape of the enzyme, the distribution of charges. Unwieldy combinations of atoms will fall apart or rearrange themselves into thermodynamically stable forms. In Fontana's Alchemy, the rules of lambda calculus stand in for the laws of physics, dictating the ways in which his atoms and molecules are allowed to combine. The result is a chemistry far more general than that which happens to prevail on earth, describing how strings of symbols, whether they consist of amino acids, nucleotides, or expressions in Lisp, react with one another. Fontana's hope was that under a wide range of conditions, symbol strings of all varieties would organize themselves into greater wholes.

The first step, he told Leo Buss, was to randomly generate two thousand or three thousand objects—strings of circles, squares, and triangles—then see what happens when they mingle in what we might think of as a simulated flask. Two strings are chosen, also at random, and allowed to interact according to their shape and the physics embedded in the rules of lambda calculus: i (the catalyst) collides with j (the substrate) and produces k (the product of the reaction). Then k is available for collisions with other strings. A new object has been added to the pool, and since objects are also functions, the system has made up a new rule. Reactions are possible that weren't possible before. To avoid overcrowding, once a new string is created, another is randomly chosen and killed.

This image of strings haphazardly colliding with strings suggests the metaphor of a gas. Instead of a homogeneous mix of simple molecules bouncing around in a container, this gas consists of enzymes, strings that can act as little Turing machines, taking one another as input and producing new strings. And so Fontana's system is sometimes referred to as a Turing gas, a random mix of all these little programs colliding and altering one another.

When Fontana first ran the simulation, it became quickly apparent that something was wrong. Before anything like an autocatalytic web crystallized from the confusion, the system became overrun with parasites: self-copying strings so prolific that they took over the flask, crowding out everything else. In some cases, simple little autocatalytic loops would form. But if the system was perturbed slightly, removing or adding a molecule, it would almost invariably collapse to a population of these self-copiers, so driven by the imperative to procreate that they had no incentive to participate in a metabolism.

To avoid this problem in future runs, Fontana installed a filter that prevented self-copying. If i collided with j and produced i again, the filter would void the transaction. The second i would be removed and the initial reactants restored. In the language of Turing gases, self-copying collisions are elastic. While this might strike some as a rather ad hoc solution, Fontana offered a biological rationale. Imagine that a primordial pool of molecules was taken over by a prolific self-copier, barring the possibility of anything more complex arising in the soup. Then suppose a fluctuation in the environment, perhaps the loss of a needed replicase enzyme, suddenly made self-copying difficult. This might smooth the way for a metabolism to form.

The effect of the change was dramatic. As Fontana explained to Leo, when self-copying was prevented, the system began producing self-maintaining networks of reactions, autocatalytic sets. His Alchemy, this means of transmuting elements, turned lead into gold, taking lifeless strings of symbols and generating a metabolism.

"That's great," Leo said, pointing at a diagram showing the pool of simple self-copiers and the more complex autocatalytic sets. "Let's call this Level 0 and this Level 1."

As Fontana went on to explain, the strings were interacting according to the rules of lambda calculus, the physics of the system. But the networks that emerged behaved according to their own, higher-level rules. Three circles might invariably be followed by a square but never a triangle. A grammar had emerged. Physics gave rise to chemistry.

"That's great," Leo said again.

Like the autocatalytic systems of Kauffman, Farmer, and their colleagues, Fontana's Alchemical structures were quite robust: if a handful of molecules was removed from the network, it would repair itself, restoring the lost circuitry. When Fontana explained that each of the metabolisms could also be reduced to a seeding set, a small subset of strings that would interact to regenerate the whole, Leo seemed barely able to contain his enthusiasm. "You know, biologists have a word for objects that are capable of coding all the information necessary to elaborate the organization," he said. "They're called genes."

"No," Walter said, content to stay within the abstraction. "We will call them seeding sets."

The next step, he explained, was to generate "an entire zoo of such organizations." By introducing more filters, like the one that prevented self-copying, one might prohibit certain arrangements of circles and squares. As a result, a completely different metabolism would emerge, with different characteristics, different rules. Three circles might be invariably followed by triangles alternating with squares. Each of these structures could be thought of as a different kind of prokaryote.

"Outrageous," Leo said. "Let's go to the next level."

Level 0 was self-copying molecules. Level 1 was self-sustaining metabolisms. What if the system was used to construct two different Level 1 organizations, which were then allowed to mix their components, to have sex? The result was no less than a self-maintaining system containing the two original systems joined by a newly emergent set of reactions that Fontana calls "metabolic glue."

"Okay. This is great," Leo said. "This looks a whole lot like a eukaryote."

The development of this artificial eukaryote was not contingent on the accidents of the system's history. Using his syntactical filters, Fontana could generate a variety of metabolisms with different grammars. But when he combined two of them, they would almost invariably fuse into a cooperative whole. In the world of Alchemy, anyway, hierarchy seemed to spring up naturally.

"I would maintain," Walter said, "that these organizational features would appear again with necessity if the tape were run twice, not only in this world but in any life." Restrained from simple self-copying, strings would give rise to metabolisms which would couple to form organizations.

Leo pretended to have one reservation. Where was natural selection involved? In Darwinian evolution, reproducing entities of some sort are randomly altered and selected according to some criterion of fitness. But there were no self-reproducing systems in Walter's Alchemical flask, no Darwinian competition between metabolisms, nothing representing survival of the fittest. "So where, when we were generating these organizations, did we have a self-reproducing entity?" Leo asked. "And when did we select out that entity? We never did."

"That is true," Walter replied.

"So what you're saying is that with the principal organizational features of evolution, the major features, the fundamental shifts, there is no necessary role for Darwinian selection."

"That is true."

"Well, the biologists are not going to like that at all."

Walter nodded, and the audience, top-heavy with iconoclasts, applauded. The presentation was unlikely to persuade a diehard adaptationist that a eukaryote or a multicellular creature could not be thrown together incrementally by random variation and natural selection. But for those inclined to give a more central role to self-organization, Alchemy was a hit. He and Buss were not suggesting that Darwinian evolution had nothing to do with the development of biological order. Once an organization was formed, whether it was a prokaryote, a eukaryote, or a multicellular organism, it would begin to reproduce and be shaped by Darwin's machinery. Survival of the fittest would come into play. But in the arrival of the fittest, the organizations themselves sprang forth, the levels were created, for reasons of their own. The hierarchies were not imposed from the outside by the trial and error of evolution. The order arose from within.

Given the opportunity, Plato might have said that life arranges itself in hierarchies because the ideal form, Hierarchy, exists in a separate realm, the realm of ideas. All hierarchies in the material world are rough approximations of this idea in the mind of the gods. Aristotle might have said that the notion of hierarchy is a "final cause," immanent within all living things, compelling them to arrange themselves in these ladders of increasing complexity. In Fontana's Alchemy, the hierarchies are more Aristotelian than Platonic; they are not imposed from the top down, they bubble up from below. The world might be a sea of random elements, but with the ability to construct, to take two elements and combine them to make a third, and the ability to recognize when two of the constructions are the same, a hierarchy of organizations can bootstrap itself into existence.

Two of Fontana's heroes are his fellow Italians Galileo and Umberto Eco. Before Galileo, people looked at the skies and saw this unique white orb they called Luna, the Moon; they thought of her as a goddess. Then Galileo learned of the magic of lenses and put together a telescope so powerful it revealed that Jupiter also had moons. And so "moon" became a category, a lowercase noun, and we understood something more general, more universal, about the space around us. Fontana says he would like to do the same with chemistry. The chemistry he learned in college, how the atoms of the periodic table interact, is Chemistry with a capital C. But he believes there is a more general, lowercase chemistry which includes not only Chemistry itself but all systems that exhibit the two characteristics he calls construction and sameness: molecules interacting with molecules to form cells, cells interacting with cells to form organisms, organisms interacting with organisms to form ecosystems and perhaps even societies. His Alchemy is an attempt to strip away the conceptual clutter and produce a minimal system that captures what he believes is a tendency of all these phenomena to form organizations that can combine into hierarchies. In his loftier moments he hopes that Alchemy may point beyond chemistry and biology to a general theory of organization, which would explain how pattern arises from randomness, and how we get meaning in the world.

That is where Umberto Eco comes into the story. Eco is a professor of semiotics, the science of signs. We have come to think of signs as lifeless abstractions, labels assigned arbitrarily to things in the world. But in medieval times, as Eco shows in his novel *The Name of the Rose,* signs were believed to resonate with magic. With the proper incantation one could invoke the powers of the universe. As a semiotician, Eco tries to restore some of this magic to twentieth-century linguistics, showing that signs are not empty labels—mere reflections of what we think of as hard-core reality—that they form a world unto themselves, a kind of cyberspace in which they take on a life of their own. When we buy a pair of Guess jeans or a Gap T-shirt, we are not simply buying cloth cut and sewn with thread; we are buying a symbol that stands for a whole world of messages we are trying to convey. The chief natural resource of New Mexico might well be something called Santa Fe style, a look one can purchase in the form of silver and turquoise jewelry or homes with Saltillo tile floors, ceilings made with aspen vigas, and R. C. Gorman or Georgia O'Keeffe posters on the walls.

Or think about the signs we string together to make a conversation. They too seem to have a dynamics of their own. At a party we might find ourselves trading stock phrases with little consideration of their underlying meaning: "Beethoven is too bombastic, Mozart too measured and ethereal." The signs emitted by one speaker elicit signs from another speaker, a chain reaction that continues until the drinks run out or fatigue sets in. At the end of his book *The Selfish Gene,* Richard Dawkins gives a name to these coins of the cultural realm: memes. Our bodies are vehicles for propagating genes; our brains are vehicles for propagating memes. In both cases, signs interact with signs according to a system of rules. We can almost get a feel for what an avant-garde literary critic means when he insists that it is not we who speak language but language that speaks through us.

Just as matter and energy affect each other through the laws of physics, signs affect signs—perhaps through laws of semiotics. To a semiotician, signs, like matter and energy, are not human artifices but an integral part of the world. Perhaps this is just another way of saying that information is physical, a necessary ingredient for carving up the universe.

In the language of life, molecular words interact with one another, and together they produce sentences, paragraphs, texts—the hierarchy of life. It is easy to think of the genome, written in its code of A, T, G, and C, as a text. Its expression—its meaning—is the phenotype it gives rise to. But what do we mean by "meaning"? It is at this point that biologists and semioticians alike are confronted with the mystery of interpretation. In breaking the genetic code, scientists developed a set of rules, a kind of grammar or syntax, for specifying how the symbols A, T, G, and C spell out the twenty amino acids. But nothing in the genetic code tells how to translate from genotype to phenotype—how to take a string of genetic symbols and predict the nature of the organism that emerges into the world. One cannot even predict the shape of a single protein from its genetic code, only the sequence of its amino acids.

In linguistics this would be called a problem in semantics. How do you get from syntax—the shuffling of meaningless symbols—to a statement that is meaningful? The relationship between syntax and semantics—what is written on the page and what becomes written in the brain—is maddeningly subtle and nonlinear. Tiny changes can lead to unpredictable, cataclysmic results, or they can have no effect at all. In a sentence one can randomly change a single letter, making a typographical error, and the meaning usually remains intact. But sometimes the whole meaning of the sentence can hinge on a single letter. "Your laughter is beautiful" becomes "Your daughter is beautiful." In one of Fontana's favorite examples, "The kids are flying planes" becomes "The lids are flying planes." Changing a single letter alters the way every other word is used.

In a string of nucleotides, a typographical error, a mutation, will often have no discernible effect, but it might cause a hideous mutation, or lead to the development of a whole new architecture. A theory of organization like Fontana has in mind—a supersemiotic Alchemy—would seek to make sense of this puzzling relationship between syntax and semantics, telling how meaning—both literary and biological—arises in the world. Strings of molecules interact to produce organisms that interact to produce societies by generating strings of words. Could language simply be another form of Alchemy's metabolic glue?

If life can be thought of as an intricate system of symbols, then it is tempting to turn the tables and ask whether the structures that arise in Fontana's simulation are in some sense alive, examples of what has come to be called artificial life. Though Christopher Langton, a Santa Fe Institute scientist who organizes the biennial Artificial Life conferences, is often credited with coining the term, some trace the idea to John Von Neumann. After he came to New Mexico to help solve the equations for the hydrogen bomb, Von Neumann and his Los Alamos colleague Stanislaw Ulam developed a kind of mathematical kaleidoscope called a cellular automaton. As many who follow the culture of popular science know, a cellular automaton consists of a grid of cells that turn on and off or change colors according to a few simple rules. In the simplest case, a cell can be either white or black, on or off, and a typical rule might call for a white cell to turn black if more than half of its four adjacent neighbors are black, or for a black cell to turn white if all four of its diagonal neighbors are white. There are any number of variations on the nature of the rules and the number of colors, or states, each cell can have. Displayed on a computer screen at lightning speed, cellular automata can generate astonishingly complex patterns, some capable of navigating around their checkerboard universe and even cloning themselves. The lesson, which has been taken to heart by the A-life movement, is that rich, often unpredictable behavior can emerge from simple, local interactions. The situation is reminiscent of what is going on in a metabolism: molecules blindly interacting with their neighbors give rise to a complex system.

Von Neumann himself proved that it was possible for a pattern in a cellular automaton to reproduce itself. To do so it would have to include a blueprint of itself, a genome coded within the colored squares of the cellular array. Referring to these instructions, the pattern would grow arms into an unoccupied region of its environment and reconstruct itself piece by piece, including its genetic code. Von Neumann didn't actually make one of these abstract beasts—according to his proof it would have required a cellular automaton consisting of some 200,000 cells that could each be in one of twenty-nine states. It was left for later mathematicians and computer programmers to design simpler self-reproducing cellular automata. One of Langton's claims to fame is a small loop of cells, shaped like the letter Q, which can extend its tail into unoccupied territory and duplicate itself.

Embracing this belief that life is a process that can be skimmed from its carboniferous substrate and transplanted into a world of pure abstraction, some of the visitors to the A-life conference in Santa Fe—the one where Fontana and Buss gave their presentation—demonstrated artificial ecologies in which digital creatures compete for "resources": memory space and processing time. In a simulation called Tierra, developed by an ecologist named Tom Ray, self-replicating digital organisms hone themselves through random mutation and selection into more efficient forms. An original Ancestor, consisting of eighty lines of computer code, is supplanted by simpler self-replicators of seventy-nine lines, then seventy-eight and seventy-seven. They flourish because they can live on less "energy": they can duplicate themselves using fewer cycles of the computer's central processor. These programs then give way to even smaller versions, but at some point a profound change occurs. The organisms are preyed on by parasites, compact little programs that, like viruses, have developed the ability to copy themselves using their hosts' replicating machinery. But the hosts then develop defenses against the parasites, and the parasites develop defenses against the defenses. An evolutionary arms race ensues.

By releasing his self-replicating programs into a more complex, unpredictable environment, consisting of a web of computers linked by the worldwide system called the Internet, Ray hopes to create a digital wildlife preserve. Driven by the need for nourishment—free cycles of processing time—the creatures would migrate in search of idle computers, always staying on the dark side of the earth. What he hopes will evolve is a menagerie of creatures with different specialties, which might unite to form a multicellular organism.

Can Ray's creations be considered alive? In the construction of systems like these, there is always the danger of confusing the map with the territory, of mistaking the network of concepts we lay over the world for the world itself. If life is simply a process—an orchestration of bits—then it is hard to see why a self-sustaining, self-reproducing structure inside a computer would not be alive. Or is this a hideous case of reification, in which we have become so enamored of a concept—information—that we have elevated it to the status of a real thing? At the A-life conference in Santa Fe, the psychologist Steven Harnad argued that it is ridiculous to confuse artificial creatures with biological ones. To be real, a simulated creature would have to interact with the environment, he said, and simulated ones don't count. What is being billed as artificial life may mimic some biological processes, but unless it has contact with reality it is nothing but a simulation. A robot with artificial intelligence programmed into its silicon brain and the ability to sustain itself by seeking out energy and to reproduce itself with random variations, allowing for the gradual improvement of the species—now that might qualify as life, Harnad argued, for it would be living in what is still widely considered the real world.

Very few A-life enthusiasts find this argument convincing. So sure are they that information is fundamental, not just an abstraction, that they have little trouble believing that the environment inside a computer is every bit as real to its information-based creatures as ours is to us. Harnad bases many of his arguments against artificial life on the ones the Berkeley philosopher John Searle has aimed against artificial intelligence. Searle is well known among both philosophers and computer scientists for provocative remarks like this: A simulation of a rainstorm won't get anybody wet; why in the world would anyone think that a simulation of intelligence would really think? Now the A-lifers believe they have an answer: simulated creatures would experience the conditions that, in their world, qualify as wetness. How very parochial of us to believe that ours is the only universe that counts.

It was the mathematician Claude Shannon, among others, who first embraced the concept of information to help us better understand how to send signals through the random noise of a telephone line. Information theory proved to be a powerful tool. Thinking in terms of bits has allowed us to develop the field of computer science, in which we learn how to represent the world with patterns of information. So successful are our endeavors that some physicists and computer scientists believe that perhaps information is not a human invention but something as real, as physical, as matter and energy. And now a handful of researchers have come to believe that information may be the most real of all. Simulated creatures would have no way of knowing they are simulations, the argument goes. And, for that matter, how do we know that we are not simulations ourselves, running on a computer in some other universe?

Nature, it seems, has honed us into informavores so voracious that some can persuade themselves that there is nothing but information. Samuel Johnson rejected Bishop Berkeley's solipsistic views of reality by kicking a rock. Little did Johnson know that he might have been pure information himself, "kicking" a data structure called rock, "feeling" processes referred to as hardness and pain.

9

IN SEARCH OF COMPLEXITY

From many parts of Santa Fe, the most prominent feature on the eastern horizon is a piñon-and-juniper-dotted foothill called Talaya. With its steep sloping sides and its narrow rounded top, Talaya Hill looks so symmetrical and precisely carved that it sometimes seems less a happenstance of geological forces than the deliberate work of an intelligent hand. Whenever pattern, deliberate or accidental, leaps forth from an otherwise random field, the mind is jolted with a sense of recognition and a compulsion to explain. An economist who regularly visits the Santa Fe Institute says that the first time he saw Talaya he was struck by a feeling that it was a holy place—a reminder that there are forces and meanings beyond human comprehension. He had recently begun exploring Eastern religions and had come to believe in powers even more mysterious than Adam Smith's "invisible hand," with its ability to apportion goods and services according to the laws of supply and demand. Another visitor to Talaya, who perhaps had similar ideas, planted on its summit a stick wrapped with colored string, a spiritual antenna for some unknown religion, perhaps one he invented himself.

It is unclear from the history books how Talaya got its name. To the south of it, toward St. John's College, is the higher but far less striking visage of Atalaya Peak, named after the Spanish word for watchtower. The mutation that changed "Atalaya" to "Talaya" was probably accidental, though Seth Lloyd, the Santa Fe Institute physicist, once proposed that the progression be continued, so that the smaller rise in front of Talaya, which the locals call Apodaca Hill, would be known instead as Alaya, and the one in front of that Laya, and so on until the supply of letters was exhausted.

While almost everyone who visits Santa Fe sees Talaya from the distance, punctuating the skyline, very few make it to the top. There is no well-established trail, so hikers have to improvise. Drive east on Canyon Road, which leads from the town plaza toward the mountains, following it to a dead end at the Randall Davies Audubon Center, a small wildlife preserve. There you can park and hike up one of the side canyons or ridges that lead in the direction of Talaya's top. The problem is that you cannot always see Talaya from this far below—there are too many intervening hills—and so it is easy to go astray, following a canyon for a mile or so to the top of a ridge only to find that you have missed the goal entirely, that Talaya rises tantalizing and unreachable behind you. Or you might slowly scale the side of what must surely be Talaya only to reach the crest and find that it is merely an unnamed foothill, with a deep canyon separating it from another hill beyond. Is it worth descending into the canyon and then climbing the next hill, hoping that it will put you in sight of Talaya? Or is it best to turn around and start over again? Who knows? The next hill might be even farther from the goal.

The attempt to reach the highest point of a landscape without a map and compass is similar to what mathematicians call hill-climbing, a method they use to explore imaginary mathematical terrains. An equation with two variables, x and y, can be graphed as a curvy line, with two-dimensional peaks and valleys. For each point on the line, the coordinates represent the values of the variables. For $y = x$, we get a straight diagonal line. For $y = x^2$, a parabola. For $y = \sin(x)$, an undulating wave. A three-variable equation yields a surface—the three-dimensional extrapolation of a curve—which might look very much like a relief map of the foothills of Santa Fe. Each point on the landscape can be associated with different values of three variables, x, y, and z, giving longitude, latitude, and altitude. Equations with more variables cannot be directly visualized, but mathematicians think of them as describing multidimensional landscapes, with peaks and valleys in four, five, six, a hundred—as many dimensions as you like. In a hill-climbing problem, one wants to find the combination of numbers that, when plugged into the equation, will generate the highest output, that will take you to the highest point (or points) on the landscape. The constant problem is fooling yourself into thinking you've reached the summit when, alas, you are stuck on a lower hill—what is known in the trade as a local maximum.

Or, to change metaphors, think of an equation's variables as knobs on a black box that is hooked to a meter. Finding the maximum solution is equivalent to fiddling with the dials until you discover the combination of settings that causes the needle on the meter to rise to the highest point. With even a few knobs, the number of possible settings becomes astronomical and you might be left feeling as though you were seeking to open a safe by trying every possible combination. Suppose you are attempting to find the maximum solution for a six-dimensional equation. As you hold five of the variables, u, v, w, x, and y, constant, you slowly increase z and watch the needle simultaneously climb higher. At some point you increase z one more increment and the needle begins to fall back the other way. You have reached some kind of maximum, but is it a foothill or a summit? If you keep incrementing z will the solution eventually go higher again, crossing a valley and scaling another peak? And if it does, will you be on your way to the summit or wasting your time scaling another foothill? Perhaps you should be incrementing y instead of z, following a ridgeline along a different dimension, or readjusting u and v and starting the climb from an entirely different point on the landscape. It is as though you were blind and must feel your way to Talaya, sensing with your fingers if the slope is up or down. The simplest strategy in such a case is always to take a path that brings you higher than you were before. But more likely than not you will get stuck on a local maximum, separated by a chasm from a goal you cannot see.

Mathematicians have discovered that one way to avoid getting stuck on a foothill is to inject noise into a system. Every so often in your explorations, twiddle a few knobs at random; throw the dice or yarrow sticks and, depending on the way they fall, strike off in a new direction, even if it seems to lead downhill. While there is a danger that these random changes could steer you off course when you are inches from your destination, they can also knock you off local maxima, nudging you toward an indirect route that may bring you to the summit.

This, in fact, is the "strategy" that most biologists would say is used to explore the risks and opportunities opened by Darwinian evolution. Here the coordinates of each point on the imaginary landscape can be thought of as a different combination of genetic traits. Suppose we have an imaginary creature with a genome consisting of a mere one hundred genes, each encoding a different characteristic. It will evolve on a fitness landscape of one hundred dimensions. Each gene can be thought of as a knob that can be adjusted to various settings. One knob might adjust the length of the creature's legs, another the number of vertebrae in its spinal column. Or we can imagine more subtle parameters: processing speed and memory capacity of the brain. Depending on the genetic settings it was born with, the creature will occupy a specific point on the landscape. Because of random variation, a twiddling of the genetic dials, its offspring will have slightly different genetic coordinates, occupying points nearby. Favorable mutations cause an offspring to sit a little higher on the landscape. Unfavorable mutations place it slightly downhill. Reaching the top of Talaya is equivalent to achieving the state of maximum fitness, the best fit with the environment, the highest point on the terrain.

If creatures are represented by points on the landscape, then a species—a population of creatures with similar genomes—would be a tightly packed cloud. Generation by generation, the cloud of creatures changes shape, flowing across the hills and valleys of the landscape, driven by the engine of random mutation. Of course, there is no way to rise above the landscape and see it whole. So the species explores the invisible terrain by twiddling dials at random—sending out tendrils to feel for higher ground. Since favorable mutations increase the chance that a creature will survive, attract a mate, and reproduce, they tend to spread through the population, pulling the cloud a little higher. Unfavorable mutations tend to be weeded out by the filter of natural selection. And so, for the cloud as a whole there is an average increase in fitness, a slow, inexorable pull upward, toward the tops of the hills and peaks.

In this simplified rendition of the classical view of Darwinian evolution, order emerges from disorder through the filter of natural selection. Yet the process of evolution itself is directionless and blind. In the long run, a species will tend to increase in fitness, becoming better adapted to its environment. But with so much noise in the system, there is no way to predict which part of the terrain it will explore.

The problem of prediction is complicated by the fact that a species' environment includes not only its physical surroundings but the other creatures it must live among. When one species develops wings, another might suddenly become its prey. Its landscape is deformed as though hit by an earthquake; its fitness peak crumbles and it falls to a lower part of the terrain. On the other hand, a series of random mutations that cause the species to be camouflaged against its surroundings would increase its own fitness while eroding the fitness peak occupied by the winged predator. But then this species might "respond" (through the blindness of undirected mutation) by developing sharper eyes.

In an ecosystem, thousands of these many-dimensioned landscapes, one for each species, are coupled to one another as though by a web of invisible threads. A tiny mutation causing a change in one species resonates throughout the system, causing a cascade of unpredictable effects. Each of those effects unleashes more cascades, which unleash still more. Peaks are flattened, driving their denizens to change radically, or become extinct. New niches are created, allowing for the emergence of new species. One is reminded of chaotic equations, in which tiny changes in the input lead to wildly different behavior. Rewind the tape, as Stephen Jay Gould suggests in his book *Wonderful Life,* and there seems to be little reason to believe that anything like us or the creatures we see around us would emerge a second time.

If evolution could be wound up and started over again, we might predict that life would diversify and natural selection would, on average, drive species up higher fitness peaks. But if dumb luck plays so powerful a role, then many of the details would elude our power of reckoning. The earth might be populated by creatures beyond our imagination. Perhaps the ability to gather information confers such a powerful survival advantage that even the strangest of creatures would be endowed with some kind of intelligence. Or perhaps having a brain, or some equivalent, is as accidental and contingent as having a thumb. Given another run, there might not be anything on the planet with our talent and compulsion for dividing the world into categories, for seeking patterns amid the swirl.

This view of Darwinian evolution is very much like the classic random walk theory of the stock market, in which an investor following the day-to-day fluctuations of a stock is essentially receiving a random string of numbers. Put your money all in one stock and you might as well be playing bingo at Tesuque. But in an expanding economy, the market *on average* will tend to rise, so by diversifying investments, one can have a reasonable chance of making money. Similarly, each mutation to a single species is random and unpredictable, but, on average, the creatures of the biosphere will grow more and more fit. Put all your bets on one species and you will probably lose your shirt. Spread your bets across the biota and you will likely prosper.

For most biologists, Darwin's magnificent theory is enough to explain the stunning order and complexity of the biological world. But even a few of Darwin's admirers feel a tug of dissatisfaction at the possibility that the life we know lies at the end of such a long chain of contingencies—that, anywhere along the way, a different roll of the Darwinian dice might have meant no Darwin, no theory of natural selection, no science at all.

It is somewhat comforting to think that some kind of life is likely to have arisen, that laws of combinatorics might have caused autocatalytic systems of molecules to spring forth effortlessly from the randomness of the primal seas. And if Fontana and Buss are right, perhaps it is just as natural that these metabolic webs would stack themselves into the hierarchies of life. But could the order run even deeper? A few biologists, who have made the Santa Fe Institute one of their unofficial headquarters, entertain the possibility that evolution's explorations are not so blind, that they are driven by laws of complexity. In the debate over whether natural selection or self-organization is largely responsible for sculpting the biosphere, they, too, come down strongly on the side of self-organization. We, the pattern finders, arose not by accident, they say, but because we are somehow written into the laws of nature.

Just as some scientists, like those at the Prediction Company, believe that some of the randomness of the stock market might consist of deterministic chaos, a feeble signal hidden amid the noise, these biologists believe they can find subtle patterns within the seeming randomness of evolution—attractors steering it in certain directions, encouraging it to follow some paths while avoiding others. If we become lost in the mountains at night, we might begin with a random hill-climbing procedure, striking out this way and that, groping in the dark. But eventually we spot lights in the distance, or find a stream and follow it to civilization. Suppose that somehow the hills and peaks of the fitness landscapes were marked with beacons, like that colored antenna atop Talaya, emitting strange signals, drawing creatures toward them. Once a species was in the vicinity of the peak that represented the development of wings or eyes or spinal columns, it would be pulled inexorably up it, like a hiker spying Talaya across a ravine. Rewind the tape as many times as you like and the world would be different in detail, but eerily familiar, with the same general life forms popping up again and again. And there to behold it all might be vertebrates with pairs of arms and legs and brains complex enough to wonder and try to make sense of it all. The universe would be compelled to generate not only life, but life as we know it.

This, in fact, is what most of us seem to believe instinctively. Our contemporary folklore is filled with the idea that evolution is somehow pulled in the direction of creating creatures very much like ourselves. For all their grotesque embellishments, the menagerie of beings sitting around the bar in George Lucas's movie *Star Wars* tend to be bilaterally symmetrical, with two arms, two legs, a head equipped with a brain and two eyes. In the various incarnations of the television series *Star Trek,* some sense of cosmic affirmative action ensures an occasional part for creatures who do not fit the mold—one encounters intelligent mists, and the like. But usually a journey to the most remote quadrants of the universe reveals creatures that are but minor variations of the humanoid body plan. A late episode of the series tried to explain away this anthropomorphic prejudice by positing an ancient civilization that seeded planets throughout the universe with similar DNA. But it is hard to let go of the idea that there is something inevitable about the way we are.

Walking into the halls of a good natural history museum, we see, trapped in glass cases, the skeletons of birds, reptiles, amphibians, mammals—all so similar to one another, and to the skeleton of *Tyrannosaurus rex* towering above the atrium floor. All so obviously are variations on a theme. Would all this really be utterly different if the tape were run again? Would there be entirely different similarities to wonder about, on the off chance that there would be such a thing as wonder?

Before Darwin, a school of biologists we call the rational morphologists looked for mathematical laws—a kind of periodic table of the flora and fauna—that would explain why nature seemed to generate the same structures over and over again. Species were considered to be natural kinds, ordained by hidden laws of nature. In 1817, the French taxonomist Georges Cuvier divided life into four classes, each based on what he considered a fundamental body plan—the vertebrates (including mammals, birds, reptiles, and fish), the mollusks (including octopuses, snails, and oysters), the articulata (including jointed animals like lobsters, spiders, and insects), and the radiata (including animals like the starfish, which have radial instead of bilateral symmetry). A wing of a bird was similar to the leg of a lizard not, as Darwin would later say, because both could be traced to a common ancestry, but because both were independently generated according to the same principles; they were reflections of symmetries carved deep into the substrate of natural law.

Then Darwin came along and twisted the biological kaleidoscope. The harmonics were all quite accidental, not the reflection of laws of form but of an earlier body plan distorted over and over again by random variation, filtered by natural selection. The notion that organisms had to be the way they are, as though compelled to fill the squares of the taxonomic charts, was soon considered as myopic and misguided as geocentrism. Why should we think evolution was striving to produce us and the creatures around us? To believe that evolution was directed by anything other than natural selection was to be guilty of the teleological fallacy, the idea that an acorn strives to produce the oak tree, that the end somehow brings about the means.

Darwinism is so powerful in its ability to array the isolated facts of biology into a coherent whole that few scientists have found reason to resist it. But one doesn't have to be a fundamentalist Christian to feel a sense of disbelief at the idea that something as complex as a human is the result of a chain of accidents stretching billions of years into the smoky past, that if a cosmic ray hadn't caused a point mutation in one of our evolutionary ancestors, we and all our inventions—our glorious architectures and imaginary spaces—might not be here at all.

The rational morphologists never really died; they went underground, to emerge in a more sophisticated form in the twentieth century. Drawing on earlier work by theoretical biologists like C. H. Waddington and D'Arcy Thompson, these so-called structuralists argue that evolution is shaped not by God or by mysterious teleological strivings, but by the mathematics of complex systems. If one spends enough time around the Santa Fe Institute, it is not uncommon to find biologists who consider strict Darwinism in much the same light as many psychologists consider strict Freudianism: less a science than an ideology. At a conference on laws of complexity sponsored by the institute in the summer of 1992, a British biologist named Brian Goodwin flatly declared that natural selection is not primarily responsible for the order we see in the biological world.

Like theories of cosmology, the theory of evolution cannot be tested with a controlled experiment. There is no way, except in our imaginations and our computers, to rewind the tape and play it again. We are confronted with a world, a *fait accompli,* and we must work backward, piecing together a story of how it came to be. Marbled with this instinct to connect cause and effect—lightning is followed by thunder—our brains forge long chains of explanation. The stories we spin may sound eminently plausible, but how can we know whether they are true?

In his book *Beyond Natural Selection,* Robert Wesson has collected a number of these tales. In the alpine tundra above the timberline, in the Sangre de Cristo Mountains and throughout the West, marmot families cling together in groups because, we are told, food is very scarce; evolution has favored closeness and cooperation. But, as Wesson points out, if instead it happened that marmots lived the solitary lives of hermits, we could just as logically argue that evolution favored animals that spread themselves far and wide to forage for nourishment. In one species of bird, large-throated males are said to have evolved because the females prefer the mates best equipped for gathering food for the young. Yet in another species of bird, the female is said to prefer mates that ignore their young, because they are less likely to attract predators to the nest.

Like the cosmologists, the adaptationists look for predictions that can be used to test their scenarios. One might, for example, study different subspecies of marmots and see whether there is a correlation between scarcity of food and togetherness. But for the structuralists, who instinctively seek wellsprings of hidden order, the adaptationists' tales are about as convincing as Genesis, the Tewa creation myth, Kipling's *Just So Stories*—or the morning-after explanations for why the Dow Jones Industrial Average rose or fell.

We read on the front page that the dollar has reached a new low against the Japanese yen; turning to the business section, we see that the stock market has gone down. The explanation we are offered might go like this: a more expensive yen will make Japanese imports costlier, fueling domestic inflation and tempting the Federal Reserve Board to raise interest rates, stifling the American economy. But one can search the back issues of the paper and find cases in which a dollar falling to a historic low against the yen has been accompanied by a rise in the market. The rationale? More expensive Japanese imports give American companies a competitive edge, and a cheaper dollar draws tourists from the Orient. With the slightest effort we can construct a logical story. In 1993 when, after reaching twenty-year lows, home mortgage rates first began creeping up, the change was blamed on a number of factors, including rising world gold prices (a signal of inflation fears) and, again, the dollar falling precipitously against the yen. Several weeks later, interest rates temporarily turned around again and went even lower than before, but the dollar kept on falling and gold kept on rising. When IBM announces layoffs and its stock goes down, the rationale is obvious. But just as often, such events will cause a stock to rise. Why? Because, we are told, the market had already "discounted" the bad news and investors were encouraged that the company had become serious about cost cutting. There are so many variables to tweak, so much slack in the network of concepts, that you can easily find a neat explanation for anything that happens.

There have always been those who argue that Darwinists are also guilty of arguing in circles, that the theory of evolution is based on a tautology: survival of the fittest, with the fittest defined as those which survive. Gould suggests that when scientists find themselves spinning Darwinian tales, they should pause and look for solid evidence that the variants they are calling survivors are truly fitter than those that perished. Often they may find that which creatures perished and which survived had less to do with fitness than with the random swing of the grim reaper.

The structuralists go much further than Gould, arguing that the circular nature of so much Darwinian reasoning is a symptom of the weakness of the theory's explanatory power. They believe there are ways to explain the rise of biological order in which natural selection is supplemented or, a few would argue, supplanted by self-organization. The autocatalytic sets studied by Kauffman, Farmer, Packard, and Bagley show how molecules might spontaneously organize themselves into metabolisms, without the help of random variation and selection. And Buss and Fontana's Alchemical experiments suggest how self-maintaining structures of molecules might organize themselves into hierarchies, again without the guiding hand of Darwin. In both cases, order seems to arise from within because of the nature of complex systems. Perhaps some of the most familiar patterns we find in the biosphere also arise this way.

A favorite example of those trying to find evidence of self-organization is the human eye. So exquisitely designed, with its adjustable lens and iris, with its retina capable of rendering images better than any camera—the eye surely could not have developed from the blind meanderings of evolution. Or so it seems to Darwin's critics. The eighteenth-century theologian William Paley considered the eye and other precisely engineered organs as proof of an intelligent creator. But, again, one doesn't have to be a creationist to have difficulty accepting that eyes arose purely from random variation and selection. As Brian Goodwin recalled at the conference in Santa Fe, even Darwin said that every time he looked at the vertebrate eye his blood ran cold. Imagining the millions of tiny experiments that led to the honing of animal vision taxed his credulity.

The Darwinists respond with the familiar reminder that our brains are simply not wired to conceive of vast, geological time. Accept that our own lives, and the brief lifespan of modern science, are but twinkles, barely discernible against the backdrop of the eons, and one can put together a plausible scenario. We find in the world today single-celled creatures with a light-sensitive patch that acts as a primitive eye. Shielded on one side by an opaque pigment, it allows the creature to orient itself toward light. Imagine that clusters of these light-sensitive cells joined to form a retina in a primitive organism. With this crude photodetector in place, evolution would hone it, increment by increment, into a fully functioning eye. Suppose that a random variation caused the light-sensitive cells to become slightly recessed; this might provide a limited amount of protection for the eye and allow for better directionality. Creatures with this mutation could tell not only whether a light was somewhere in front of them, but roughly where it was. Because of this slight survival advantage, they would spread themselves more widely than the others until most of the population had slightly recessed eyes. Now among some of these creatures, random variation might make the recession a little deeper, and so we pull this more visually acute subset from the pool and let it multiply. Again a random variation might cause an even deeper recession among these members, and so on, until we have creatures with light-sensitive cells at the bottom of a deep cup. Now imagine a variation, or series of variations, that causes the cup to narrow at the top, the smaller and smaller opening providing sharper and sharper focus, greater visual acuity. At this point, a random variation that led to a transparent covering over the pinhole would put the species on the road to making a lens. Still other variations would lead to the musculature allowing the creature to flex and change the focus of the lens for different distances; others would lead to the honing of the iris, allowing creatures to operate in different levels of light.

There is, of course, little real evidence for this explanation. It is an evolutionary Just So story. How persuasive it is depends on the taste of the listener and the rhetorical skills of the storyteller. As presented by one of Darwinism's most eloquent explicators, Richard Dawkins, in his book *The Blind Watchmaker,* the story of the eye sounds utterly compelling. Evolution is pulled toward making eyes through millions of incremental steps each offering a slight advantage.

Without a way to test these hypotheses, however, they strike some biologists, like Brian Goodwin, as not very scientific. Even if you invoke vast geologic time, the series of fortuitous mutations leading to an eye, a kidney, or a brain seem too good to be true. In the case of the eye the dubiousness might escalate beyond belief when one learns that eyes have apparently developed not once but many times in the course of evolution. Each version is so different in structure that biologists conclude they are not variations on a single, ancestral eye but unrelated evolutionary experiments. "The eye developed independently in more than forty lineages during evolution," Goodwin said. "Now that suggests that there is something pretty damned robust about this process of generating eyes. It's not something that happens once. It looks as if they pop up naturally over and over again."

A strict adaptationist might say, "Of course, that is exactly what happened." Detecting light confers such a powerful advantage that any mutations leaning in that direction would be encouraged and amplified. But Goodwin is one of a small group of biologists who believe that there are organizing principles in biology that all but guarantee that certain structures will form. Eyes are not random accumulations of accidents, but patterns that arise "as waves and spirals arise naturally in water," as Goodwin once put it. Or, to use a more technical term, eyes and other organs are attractors in a dynamical system.

And this brings us back to the image of a fitness landscape in which the hills are topped with beacons sending out signals drawing evolution in certain directions. The metaphor seems less mystical if we invert the landscape so that the hills become basins. And instead of hikers think of marbles poised on the ridges ready to roll down the nearest incline. A marble that finds itself on a slope leading into a depression is said to be within a basin of attraction. Give it the slightest nudge and it will roll to the bottom, the point associated with the combination of genes that produces eyes or some other structure. Other basins might represent features like bilateral symmetry or a spinal column leading to a bulb of neurons with two photoreceptors protruding out. On this landscape of hills and valleys, a marble cannot stay stationary on just any point or move randomly in any direction. Its choices are limited. Once it is within the basin of an attractor it will be pulled toward the bottom. Rerun evolution as many times as you wish and it is almost inevitable that eyes and other structures will appear again and again. They are implicit in the dynamics of the organism.

Of course, as we have seen, the landscape for a single species will be constantly deformed by other creatures seeking their niches. Each species is on a roller-coaster ride on tracks that are constantly changing. One species is randomly twiddling its genomic dials, sending out feelers to explore the abstract terrain. But other species are twiddling their dials, constantly changing the landscape. It seems that all would be chaos, yet somehow order shines through. Perhaps this too can be explained with the image of attractors.

One of the phenomena Darwinism finds difficult to explain is speciation. We can see how a series of selected mutations would gradually change the morphology and behavior of a species, but how does it make those great leaps—from, say, lizard to bird? If the eye and other organs are attractors steering the dynamics of organisms, then perhaps different species are attractors in the dynamics of an entire ecosystem. Move up a level of hierarchy and imagine a landscape in which the depressions represent bears, dogs, cats, sea urchins, and *Homo sapiens*. Each would be implicit in the system—"ghost species," Goodwin calls them.

Sixty-five million years ago, when continental drift separated the crustal plates that would become North America and Australia, a group of warm-blooded, furry creatures was separated into two populations; evolving independently, one group became the marsupials and the other group became the placental mammals. The two groups are different in many ways, but marsupials include versions of wolves, cats, squirrels, groundhogs, anteaters, and mice. One can surely offer purely Darwinist explanations for this parallel development—similar genomes in similar environments yield similar creatures. To the structuralists, however, the phenomenon can best be understood if we think of the two isolated systems as being pulled by the same attractors in evolutionary space. Perhaps dynamical systems naturally fall into certain stable states for reasons that have little to do with natural selection. Rerun the tape, the structuralists argue, and the world would be far more similar than Gould believes. The details might be different, but the same general forms would endure. And we, the pattern finders, would be here to try and make sense of it all.

To understand how attractors might steer the course of evolution, favoring the rise of biological order, Stuart Kauffman uses computer models to study the dynamics of networks. As a medical student at the University of California, San Francisco, in the early 1960s, long before he came to Santa Fe, Kauffman found that under certain conditions, networks would behave in a surprisingly orderly manner, even if they were strung together at random. One could think of these webs as networks of molecules interacting to form a metabolism or networks of genes interacting within a cell or networks of cells interacting to form an organism. As long as the networks exhibited certain characteristics, Kauffman found, they would organize themselves spontaneously.

He began by designing simple models of the genome as part of an attempt to understand the mystery of cellular differentiation. Whether we are speaking of plants, insects, or human beings, we know that the cells of an individual generally contain the same sequence of nucleotides, the same genetic encoding. In people this consists of some 100,000 genes, segments of DNA each of which stands for a different enzyme or other protein. But if they have identical genomes, how can it be that some cells function as kidney cells, some as brain cells, some as immune cells, and so forth? In each type of cell only certain genes are active. We are asked to imagine a genetic piano with 100,000 keys. Play certain chords and you get the proteins that characterize a kidney cell, play others and you get a neuron. As a fertilized egg divides, and its daughter and granddaughter and great-granddaughter cells divide, each somehow bequeaths to its progeny the proper score to play.

Where are the fingers that play the keys? Beginning in 1961 the French molecular biologists François Jacob and Jacques Monod showed that, among the many proteins that the genes encode, are ones that turn other genes on and off. The function of some keys is to play other keys. But what is the source of this intricate order? The image of genes switching one another on and off also suggests the analogy of electrical circuitry. Kauffman, the medical student, wondered what would happen if you took one hundred lightbulbs, representing a string of genes (a thousandth the length of the human genome), and wired them together so that each bulb could turn on or turn off several other bulbs. The result would be a crude artificial genome. There are any number of ways such a network could be wired. How then did Nature settle on an arrangement that gave rise to an orderly system like a cell? If you believed in God the Great Designer, then you would have to suppose that he in his infinite wisdom figured out the exact wiring pattern for each cell in the biological symphony. If you believed that all the order came from natural selection, you would have to accept what Kauffman considered an equally unbelievable notion: that so sophisticated a device was thrown together, connection by connection, entirely through the trial and error of natural selection. Kauffman hoped to show that between these extreme views there was a middle ground. Perhaps the order of the genome was not so unlikely and fragile, imposed solely from the outside, whether by a creator or by the haphazard pressures of evolution. Perhaps the order was spontaneous, radiating from within. Just as metabolic order might arise from a random mix of molecules through the mathematical magic of autocatalysis, perhaps it is all but inevitable that a collection of genes will organize itself into the complex biological machinery needed to run a cell.

Using a computer to simulate genetic circuitry, Kauffman designed his network so that each gene received signals from two other genes. To ensure that any order that might emerge was not imposed by the designer but generated internally, he made the connections at random. And he randomly assigned to each bulb a simple rule for when it should be on or off. The rules were very much like those of a cellular automaton: "If input gene A is on and B is off, then turn yourself on," for example, or "Turn yourself on only if both input genes are on." In computer science these are known as Boolean functions, after George Boole, the nineteenth-century mathematician who invented the logic now used by digital computers. In Boolean terms, a gene that was activated only if two inputs, A and B, were both active was called an And gate; in an Or gate, the gene would fire if either A or B was active. At each tick of the clock, every bulb would examine its inputs and determine which state it should be in, on or off. Run the network at high speed and a complex sequence of patterns would unfold.

Now one might think that in a network wired at random, the bulbs would twinkle haphazardly, with no discernible order. If each bulb can be either on or off, there are 2^{100} (approximately 10^{30}) different states—patterns of illumination—that a network of one hundred bulbs could be in. If it simply cycled through each and every possibility, one after the other, it would not complete its trajectory before the projected death of the sun—even if the clock ticked billions of times a second. Kauffman found, however, that the network's behavior was far more constrained. After cycling through several patterns, it would quickly begin to repeat itself, settling down into a stable sequence of patterns, what is called a state cycle. It was pulled by an attractor.

Now it was possible that this was just a matter of luck, that Kauffman had happened to launch the network with an initial configuration of 1s and 0s that led to this settled state. Perhaps with different initial conditions, it would follow an entirely different course, never settling down. But Kauffman found that whatever input pattern he chose, the network would tend to fall into one of seven repeating patterns. Faced with an overwhelming array of possible trajectories, it favored these few. Or as a mathematician would say, there were seven attractors that pulled the network into seven stable patterns of behavior.

But not all networks operated in such an interesting and well-behaved manner. If the network was wired so that each bulb received an input from only one other bulb, it would quickly freeze up, displaying a static pattern of bulbs turned on and off. It was stuck in a state cycle that consisted of only one state. Or if one increased the number of connections, so that each bulb received inputs from many bulbs, the network might never settle down; it would essentially cycle forever, the state cycles so huge that it might take many lifetimes of the universe to traverse them. Through our finite eyes, the behavior would be fathomless.

These densely connected systems are, Kauffman says, essentially chaotic. Once it falls into an attractor, a sparsely connected net will tend to stay there. Nudge it by changing a 1 to a 0 in the input string and it will probably fall back into the same basin of attraction. In a densely connected network, however, the slightest perturbation will bump it from one attractor into another, sending it veering in a completely different direction. It was only those networks that received a few inputs from other cells—two seemed to be ideal—that settled down into a small number of stable cycles. As long as this criterion was obeyed, one didn't have to carefully handcraft the network, stringing it together just so. Slap it together at random and the orderly behavior would emerge.

Even for a sparsely connected network with 100,000 genes—the size of the human genome—there would be but a handful of stable cycles. Kauffman calculated that out of the astronomical number of states such a system could assume, it would naturally pick about 317 trajectories. Kauffman likes to believe that it is not a coincidence that the number of stable states in his rough model of the genome is of the same order of magnitude as the number of different cell types that have been found in the human body, about 250. Perhaps each of the stable cycles, each attractor, represents a different kind of cell, churning out a different sequence of proteins. In fact, after examining the data for a number of species, he believes he has divined a rule, a natural law arising from the dynamics of the genomic network: the number of cell types is roughly the square root of the number of genes.

Even after the insights of Jacob and Monod, one of the most perplexing mysteries of biology is how a developing embryo, with its expanding fronts of cells, "knows" when to stop making liver cells, for example, and to start making kidney cells. Chemical signals are believed to trigger the changes, switching certain combinations of genes on and off at the right moments. But how does this result in an orderly unfolding? In Kauffman's model, a skin cell would be a very stable affair, sitting at the bottom of a basin of attraction. Perturb it slightly with some kind of molecular messenger and it would be pushed up the hill a short distance, then naturally settle back down to the bottom. This is the essence of what we mean by homeostasis. But if it is perturbed enough it might be bumped over a ridge and into a neighboring basin, where it becomes a neural cell. During various stages of development, chemical signals would push cells into different basins of attraction. But the network dynamics would ensure that not all basins are near to one another. Bump a neuron as hard as you like and it won't become a blood cell. Neurons and blood cells are represented by widely separated valleys in genomic space.

Kauffman's work supports a growing realization among biologists of how dynamic the genome really is. It does not merely reflect, indirectly, the pressures of the environment; it is itself a generator of pattern and change. The genome is often described as a text. A better, more dynamic metaphor might be a computer program, but one that is very different from those the software engineers regularly unleash on the world. Start with a word processor consisting of a string of, say, one million bits, and flip just one of them from 1 to 0, or vice versa. It very probably will not be a word processor anymore. If this happens while it is running on a computer, the system will likely crash. The word processor lives on a landscape with a single, needle-thin peak. Each point of the terrain represents a different string, one million bits long, so the space consists of one million dimensions. The word processor is perched at the very top of the precipice; the tiniest nudge and it falls to the bottom. If we flipped enough bits we might eventually turn the word processor into an accounting spreadsheet program or a video game. We can imagine these as more needle-thin peaks in the far distance. But stretching in between is a vast, barren desert of all the combinations of one million bits that perform no function at all. Suppose that we were trying to evolve a word processor—or any program—on a landscape like this. Starting with an arbitrary string of 1s and 0s, we would randomly flip a bit, then test to see if we had achieved our goal. But since the vast majority of mutations would be disastrous or meaningless, it is hard to see how improvements could be accumulated. On this featureless terrain, we would receive no clues about the best direction in which to explore.

Kauffman's genomic computer is of a very different breed. As he likes to describe it, it lives on a landscape more like the Alps, with high peaks nestled next to one another and an abundance of clues about which ways lead up to Mont Blanc. The result is a network that is robust, not brittle, responding gracefully to mutations—one that is capable of evolving. Unlike the word processing program, the genomic program is unlikely to respond to a point mutation by plunging to the bottom of a precipice. Change it slightly by modifying a connection or replacing an And gate with an Or gate and it will move a little higher or a little lower on the terrain. Driven by these mutations, it treads carefully across the evolutionary landscape, inching its way toward the goal.

Seen this way, the genome is not a random-number generator, capable of producing any arbitrary output string, any phenotype. It unspools a tape of lawful patterns, not a smattering of forms. Niels Bohr used quantum mechanics to suggest why electrons surround nuclei only in certain lawful ways. You cannot have an atom halfway between hydrogen and helium. There is not an infinitely fine continuum; instead there are discrete quanta. Perhaps the dynamics of the genome is steered by these quanta called attractors. The lesson Kauffman takes from this story is that natural selection is not enough on its own to make complex affairs like organisms. It can work only on systems that already generate their own internal orders for selection to choose among. A system, like the word processor, that lacks this power of self-organization is incapable of evolving. The natural order of the genome provides the grist for selection's mill.

Generalizing this notion, Kauffman argues that spontaneous order was present at the very beginning of life. Think of the lightbulbs in the network as molecules instead of genes. A molecule that catalyzes the formation of another molecule can be thought of as switching it on. A molecule that impedes another's production can be said to switch it off. Sparsely connected autocatalytic networks should exhibit internal order. They would naturally fall into certain configurations, and selection would choose among these. Eventually—and this is a huge gap in the story—natural selection and self-organization would conspire to make cells with sparsely interconnected genomes. And they would begin to generate their own internal order. With only certain kinds of cells to play with, natural selection would assemble organisms—networks of cells ruled by attractors that ensure that eyes and other organs arise again and again. In the debate between the disciples of adaptationism and self-organization, Kauffman leans toward the latter extreme. Natural selection is important, but on every level, it is the internal order of the networks that is primarily responsible for the patterns we see.

So here is a new creation myth. In the beginning, the chemical mix in the primal sea reached a critical level of complexity. With the number of potential reactions increasing faster than the number of types of molecules, the waters were pregnant with life. Self-sustaining metabolisms spontaneously arose. Those with too few connections froze up and fell by the wayside. They weren't fluid enough to survive in a world where environments sometimes change drastically and you need to make huge leaps to keep from going extinct. Those with too many connections were too responsive. The slightest nudge sent them spinning off into chaotic tantrums. They too were abandoned. But sparsely connected networks responded to perturbations gracefully. Most of the time they would react conservatively to a mutation, but occasionally they would make the risky leaps. And so they evolved, but largely on their own terms. Because of their internal dynamics, their attractors, they could only assume certain forms. These are the orders that selection sifted among. Among the structures that naturally arose were membranes and nucleic acids and genes—the components of cells. Once a genome had sprung into existence, its genes organized themselves into the little computers that now inhabit all our cells, offering up a palette of cell types that organized themselves into the creatures of the earth.

If we think of an organism as another kind of network, whose nodes are cells linked by chemical signals, we can see why a structuralist would believe that there might be laws that give rise to certain body plans. The various phyla—arthropods, mollusks, annelids, echinoderms, chordates—can be seen as attractors in evolutionary space.

And if a cell is a network of genes and a body is a network of cells, then an ecosystem can be seen as a network of species. If the arrangement is too orderly, with one species affecting and affected by only a few others, then the ecosystem will be stagnant, unable to weather sudden environmental change. If an ecosystem is too closely interconnected, then it will become chaotic; the tiniest change in the environment will thunder through the network, causing it to collapse. But ecosystems that are sparsely connected will be able to organize themselves into stable structures that are capable of evolving.

If we ask why we see the same patterns over and over again, Kauffman's answer would be much closer to that of the rational morphologists than to Darwin's: because there are natural forms, laws of the organism. Hugo De Vries believed that the reigning question in biology was not the survival of the fittest but the arrival of the fittest. How does nature make each leap up the hierarchy of complexity? Like Fontana and Buss, Kauffman believes that self-organization plays the primary role.

If this spontaneous order is real, does it guarantee that life on other planets would be recognizable after all? Is it plausible that the Earthlings, Vulcans, Klingons, Romulans, Bajorans, Cardassians, and Ferengi of *Star Trek* would look so much alike? In Kauffman's origin myth, autocatalytic sets spring up so readily from any complex mix of molecules that there could be many different kinds of ancestral metabolisms. But he believes that all of these networks would exhibit the same general dynamics. Thus, each time the tape was played, evolution would hone metabolisms into cells with similar genetic networks. So perhaps we can expect to see structures like arms and legs and spinal columns arise over and over again. Or perhaps the bodily features we take for granted fall into the category of details that are not conserved. Even so, we might at least expect the repeated emergence of sparsely connected networks, stable enough to preserve themselves while unpredictable enough to generate the continuing variety of structures and strategies that help them survive. Gazing deep within abstract realms, Kauffman and the structuralists find hints of a reason for us to be here—not as creatures of randomness but as creatures of natural law. Humanity finds its place in the universe as an attractor in the space of possibilities.

Once a filter becomes installed in the brain, it bends everything we see. Gazing out on the jungle, a Darwinist sees the beauty of natural selection, an invisible Maxwellian demon sifting order from randomness in a Sisyphean effort that ultimately cannot succeed. A structuralist imagines instead a multidimensional fitness landscape, the vortices of its basins ensuring an orderly world. Like all of us, both are faced with never knowing the extent to which the patterns they see are out in the world or imposed by the prisms of our nervous systems.

10

IN THE EYE OF THE BEHOLDER

Sitting in the audience of Santa Fe's open-air opera house, one can look miles beyond the music—north, over the heads of the singers, to the Jemez Mountains, where sometimes real lightning bolts flash to the booming voices of the gods onstage. With nothing but sky overhead, you can lean back in your seat and, lulled by the harmonies and the melodies (notes arranged in space and time) try to count the random spray of stars. The arias booming from the stage—these elaborate arrangements of just twelve notes, constrained even further by the contours of the key—radiate outward for a few hundred yards. Then their structure dissipates in the random jostlings of the molecules of air.

Complexity—this delicate tension between order and surprise—is a very fragile thing. And yet, in a universe seemingly ruled by the second law of thermodynamics, we get structures and structures of structures. As energy flows through a system, entropy is expelled into the environment and pattern can temporarily arise. Are these pockets of complexity mere aberrations wrested from a world dominated by randomness, or do they arise effortlessly from laws of self-organization like those being explored at the Santa Fe Institute?

One of the most controversial proposals to come out of the institute is that there is a continuum between order and chaos, and that complexity lies in the middle of these extremes. Between the wild, unpredictable behavior of unruly systems like thunderstorms and the rigid, uninteresting behavior of highly structured systems like crystals is this phenomenon we call complexity—rich, unpredictable, bordering on chaotic, but always with just enough order to delight the mind. Christopher Langton, Stuart Kauffman, and some of their colleagues have come to believe that evolution naturally takes systems to this transition. They have become so enamored of the idea that they have turned it into a slogan that many science writers have found irresistible: "Life evolves to the edge of chaos."

This image of a continuum from order to complexity to randomness seems so compelling that some people at the Santa Fe Institute speak excitedly of a new fundamental law, one that would do for complexity what the second law did for entropy—show that it is a natural tendency of the universe. To make this notion concrete they sometimes use the metaphor of the phase transition in which a solid, its molecules rigid and crystalline, melts into a flowing liquid, and finally vaporizes into the random movements of a gas. Complexity is like the liquid phase—orderly, to be sure, but capable of behavior so rich and unpredictable that it constantly flirts with becoming unfathomably intricate, or chaotic.

It is this delicate balance, they reason, that allows complex systems to process information. The argument goes like this: Information processing requires storage and transmission. The solid regime is stable enough to store information; it is a land of enduring structures and patterns. But memory is not enough; it is necessary to move bits around. In the chaotic regime, information spreads too quickly; there is no memory at all. But the liquid regime is just orderly enough for memory, just fluid enough for transmission. Some even speculate that it is systems on the edge of chaos that are capable of universal computation. Like a digital computer, they can theoretically simulate any other system, perform compressions, make maps of the world. They can act as Iguses, information-gathering-and-utilizing systems.

Zurek, Gell-Mann, and Hartle followed an austere chain of abstractions to conclude that Iguses are rooted in quantum physics—that they are a natural part of this universe. Langton, Kauffman, and their colleagues seek to show—through an entirely different but equally abstract trajectory—that the emergence of pattern finders is to be expected. At the "edge of chaos" complex systems arise, and they are capable of information processing. So perhaps there is some ground for hope that this power to find order and build systems is not accidental but a natural consequence of physical laws. There is something curiously circular about all this: science, the art of compressing data, turns its gaze back on itself and finds, surprise, that the very ability to gather and compress data is fundamental. We, the pattern makers, reign supreme. And no wonder. Driven to spin our gossamer webs, we can't help but put ourselves, the spiders, at the very center.

For those seeking an all-embracing map of the reality our senses register, it is tempting to believe that complex systems may all be governed by the same invisible hand. Could not the same rules that steer a network of genes shape the behavior of the circuits of neurons that fill up our heads? With their computers, scientists can model the brain as a random network. Here the "lightbulbs" are neurons instead of genes and they are connected into an intricate web. Just as some genes activate and suppress other genes, the connections between neurons can be excitatory or inhibitory—a signal will encourage a neuron either to fire or to maintain a state of quiescence. In the simulations, the neurons, like Kauffman's simulated genes, turn one another on and off according to the simplest of rules. Again, complex order sometimes seems to arise spontaneously: among the astronomical number of possible states, an artificial neural network settles into a finite few. While in the model of the genome the attractors might represent cell types, it has been suggested that in the networks of the brain they correspond to memories and patterns of behavior.

Move up a level of abstraction and one can think of the economy as a network. Here the units are people, instead of genes or neurons, and the signals they trade consist of money or goods and services. From all these local interactions an economy arises, a superorganism with a repertoire of behaviors. Kauffman, Langton, and some of their more enthusiastic supporters speculate that, along with the genome, neural and economic networks exist on the "edge of chaos," and that in this realm of complexity they can engage in the most powerful information processing.

But it is easy to get carried away. For some scientists at the Santa Fe Institute, these comparisons seem premature. By the early 1990s, the phrase "edge of chaos" had become so replicated a meme that some of the younger scientists could not say it with a straight face; they would smirk and roll their eyes, holding up fingers to signify quotation marks around what was becoming to many a tiresome cliché. How much insight did the notion of an edge of chaos really give into the origin of this elusive quality called complexity? In saying that complexity exists on the edge of chaos, were Kauffman and Langton saying anything truly profound, or just stating the obvious? Cris Moore, a young physicist at the institute, once put it like this: He could see why there is nothing very interesting about a string consisting of all 1s or all 0s. And he could agree that there is nothing very interesting about the set of every possible permutation of 1s and 0s. And he could accept that *somewhere* in between these two extremes interesting, complex patterns occur. But was there anything deeper to the notion than that?

Complexity is a maddeningly slippery concept. In the late 1980s, Langton and Norman Packard tried to get a grip on it by studying the behavior of cellular automata, those checkerboards of cells that generate complex behavior from a few simple local interactions. The cellular automaton is to the study of complexity what *E. coli* or a planarian is to biology—a relatively simple preparation used to open a window on perplexing phenomena. To further simplify matters, computer scientists often study one-dimensional automata, consisting not of a grid but of a horizontal string of cells each of which can be either on or off. How does complex behavior arise from so simple a system? In a one-dimensional cellular automaton, a cell is affected only by the values of its left-hand and right-hand neighbors. If its radius is three, then its "neighborhood" consists of the three cells on either side of it; at each tick of the clock, each cell will observe the values of its neighbors, perform a computation, and, depending on its own value, turn itself either off or on. A rule might say that if the three cells to the left are on and the three cells to the right are off, then the middle cell should always be on. Or if the left-hand sequence is on-off-off and the right-hand sequence is off-on-off, then the middle cell should change value. In an experiment, one decides on the neighborhood, defines a set of rules, and then starts the system with a random pattern of cells that are on or off, 1 or 0. With each tick of the clock the system is updated and a pattern unfolds. If the computer is programmed so that each new generation is placed beneath its predecessor, patterns unscroll like those in Figure 10.

In 1984, the physicist Stephen Wolfram proposed that the menagerie of one-dimensional cellular automata, governed by various rules, can be

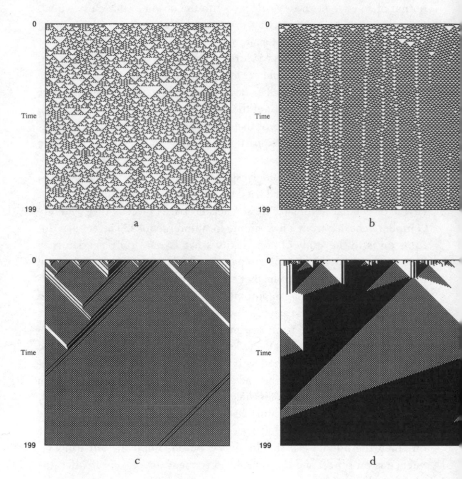

Figures 10 a–d

One-dimensional cellular automata. At time 0, each cellular automaton, programmed according to varying rules, is given a random string of 1s and 0s (cells that are either white or black). At each tick of the clock, a new string is generated and a distinctive pattern unscrolls. Depending on its input, the automaton shown in 10d, discovered by Gacs, Kurdyumov, and Levin, eventually converges on strings that are either all black (as in the example) or all white. Courtesy James P. Crutchfield divided into four classes. Most of them quickly freeze up, no matter what the initial configuration: after flashing through a few patterns they fall into a rut, displaying forever the same string of 1s and 0s—a state cycle of length 1. Others are periodic: some initial configurations will cause them to settle into a longer cycle, churning out the same sequence of patterns over and over again. Both of these kinds of cellular automata are ruled by fairly simple attractors. At the other extreme are automata that generate patterns indistinguishable from randomness. Start them with certain strings of 1s and 0s and they generate what might look like snow on a television set. Though eventually they would have to repeat themselves (the systems are, after all, deterministic and finite), the cycle might take longer than the universe is likely to exist. A slightly different input pattern will send the automata off on an entirely different journey through the space of possible patterns. These automata are, in this sense, chaotic. Most interesting to those in search of laws of complexity was a fourth class of cellular automata that are not fixed, periodic, or chaotic— they produce what seems like complex behavior.

Drawing on Wolfram's taxonomy, Langton suggested that the four types of cellular automata could be arrayed on a scale ranging from orderly to complex to chaotic. To measure where an automaton fit on this progression, he devised a quantity called the lambda parameter. After studying a random sample of cellular automata, he became persuaded that those with intermediate lambda values, on the "edge of chaos," were the ones most likely to be capable of rich and interesting behavior, including computation—the ability to make compressions and detect regularities in the world.

Packard went even further, attempting to show, as Kauffman came to believe, that Darwinian evolution would naturally take cellular automata to the edge of chaos. A cellular automaton can be thought of as a computer. Given an initial string of 1s and 0s (the question), it will transform them, according to its rule table, into an output string (the answer). Packard was intrigued by a cellular automaton rule, discovered by the mathematicans Peter Gacs, G. L. Kurdyumov, and L. A. Levin, which had two basins of attraction. Given a random string of binary numbers, it would eventually converge on one of two patterns, producing a string consisting either of all 1s or all 0s (Figure 10d). It had been noticed that whether the Gacs automaton fell into one basin or the other depended, to some degree, on the density of 1s in the input string. Given a string in which more than half of the bits were 1s, it would say 11111111111; otherwise it would reel off 0s. Although the Gacs rule is far from perfect, it carried out the task with enough reliability that it was tempting to speculate that it was performing a computation.

Packard wanted to see if he could start from scratch and evolve a cellular automaton that would carry out this task of measuring the density of 1s in an input string. If the pattern an automaton churns out can be thought of as its phenotype, then its genotype is the set of rules that generates this behavior. Starting with a random collection of these "genomes," encoded as bit strings, he would have the computer test how well each did at the task and assign it a fitness rating. Those genomes with the lowest fitness would be removed from the pool and the remainders would be altered in one of two ways: (1) by random mutation, in which a bit would be arbitrarily flipped to its opposite value; or (2) by sexual recombination, in which part of one string's genome would be swapped with part of another's. This technique of breeding bit strings, called a genetic algorithm, was invented by John Holland, a University of Michigan computer scientist and member of the Santa Fe Institute's scientific board, and developed over the years by his students, including Melanie Mitchell, a scientist at the institute. Packard reported that as the rules in his primordial soup evolved, their lambda parameters tended toward the intermediate range of Langton's scale, presumably on the edge of chaos.

For a while Packard's experiment was embraced in some quarters as evidence for the existence of a new natural law: the tendency of complex systems to evolve to the edge of chaos. But according to other scientists at the institute, the order Packard thought he saw was an illusion. In 1993, Jim Crutchfield, Melanie Mitchell, and an intern, Peter Hraber, decided to put Langton's theory to their own rigorous test. Trying to replicate Packard's experiment, they found that the lambda values of the cellular automata that prevailed in the evolutionary struggle did not put them on the edge of chaos, if indeed there was such a thing. In fact, according to Langton's own scale, they were deep in the chaotic regime, though their behavior was far from chaotic. Mitchell and her colleagues even found that the Gacs rule itself—the best found thus far for performing the density classification task—was in Langton's chaotic regime. They concluded that it is far from evident that the ability of a cellular automaton to compute has anything to do with lambda.

Langton insists that all Mitchell and her colleagues have shown is that Packard's original experiment was not a very good test of his own ideas. Rules that simply measure the density of 1s in an input string are not particularly complex, he notes, and he says he was surprised that Packard's work seemed to suggest that they lived on the edge of chaos. In any case, Langton says, he never meant lambda as more than a useful tool describing a general tendency among cellular automata rules—not a precise law that would apply without exception to every one. The larger the rule space, the more likely it is that a rule drawn at random will produce simple ordered behavior when lambda is low and random behavior when lambda is high. The rule spaces in the Mitchell experiment, he says, were simply not large enough.

Crutchfield, however, has come to believe that the whole "edge of chaos" scheme is suspect, that the space of cellular automata rules cannot be contorted into a simple continuum between order and chaos, with complexity conveniently tucked in between. And he doubts that there is anything resembling a neat scale, measured by something like the lambda parameter, onto which any system can be assigned a spot.

The problem, he believes, is that in studying cellular automata, Langton, Kauffman, and some others are looking through the wrong eyeglasses, taking concepts from one field and misapplying them to another. In the study of dynamical systems there is a well-established concept called the "onset of chaos" describing, for example, the way the rhythm of a dripping faucet will undergo period-doubling, becoming more and more complex until its patter is almost indistinguishable from randomness. Crutchfield had demonstrated early on that computational power increases at the onset of chaos in dynamical systems. But dynamical systems are continuous, taking on a smooth range of values; cellular automata are digital and discrete. He was dubious that the concepts of dynamical systems theory—period-doubling, chaos, basins of attraction, and so forth—could be applied so easily to cellular automata. In the halls of the institute, he could be heard to complain that in trumpeting the virtues of an *edge* of chaos, Kauffman and Langton were borrowing the earlier notion and bending it out of shape.

Langton is quick to point out that a continuous system can be simulated to any desired precision by a discrete system—this very fact allows us to use digital computers to study, for example, the motion of the solar system. The finer the grain of the simulation, the closer it will mimic a continuous system. A cellular automaton with an infinite number of states would be a continuous system, with the color of a cell varying smoothly through all the grays between white and black. As the number of states is decreased, the grain becomes coarser and the similarity diminishes. But vestiges of continuous behavior remain. Langton argues that even in two-state, black-or-white cellular automata there are traces of phenomena like period-doubling. His edge of chaos, he argues, is simply the remnant of Crutchfield's onset of chaos, a kind of broken symmetry.

Whether a cellular automaton's computational power indeed increases as it approaches a transition between complex and chaotic is still an open question. Other research lends support to the notion that selection pressures can drive evolving systems to a transition between orderly and disorderly behavior. But whether a system is discrete or continuous, one must pause and wonder: Are these qualities we call orderly, random, and complex objective features of the universe, or simply projections of the human mind?

Instead of a simple scale ranging from orderly to complex to chaotic, Crutchfield imagines a hierarchy of computational devices of increasing power and complexity. An observer on one of these rungs, studying a system on a rung above, will ultimately be overwhelmed by its intricacies. The observer might find patterns and hints of order, but can never completely grasp the more complex system; some of its behavior will always seem inexplicable and random. But a more powerful information processor, on a higher rung looking down, will see the same system as deterministic, perhaps even trivial. To some extent, complexity is in the eye of the beholder.

The models we build and the systems we seek to explain can be thought of as occupying different rungs of the ladder. Ultimately the best we can do is to observe the surface behavior of a system; we can never know completely what is going on inside. Then we try to map that behavior onto a mechanism of our own invention—a scientific model. If the structure of the two mechanisms, model and modeled, happens to be similar, then the system in question can be described compactly; it will seem to obey simple laws. If the mesh is imperfect, the same system might seem unfathomable, resisting any simple representation. A good example of this is the reciprocal nature of sine waves and impulse waves (which, because of their infinitely short duration, can be thought of as bits, 1s and 0s). As we saw earlier, it takes an infinite number of impulse waves to perfectly represent a sine wave, and vice versa. An observer who can record a simple melody in the form of sine waves—squiggles in a record groove—will have a fairly compact representation: the undulations on the plastic are similar to the undulations of the sound waves bending the air. But if we record the melody digitally, as a string of bits, the representation will be vastly longer and seemingly more complex.

Whether something appears simple, complex, or random depends on the observer as well as on the observed. Suppose we found a bit string of a performance of a portion of the *St. Matthew Passion*. Without knowledge of where it came from and the coding scheme used by the recording device—and without brains equipped with the filters to recognize the patterns within—the sequence would appear overwhelmingly intricate, indecipherable, perhaps indistinguishable from random, white noise. But if we fed the string into a CD player, programmed to detect just the right patterns, music would pour out.

Or suppose we find, scrawled on a crumpled piece of notebook paper on the floor of a Las Vegas casino, a seemingly patternless sequence of numbers. Because of the context we might assume that this is the random output of a roulette wheel hurriedly scribbled by a player hoping to beat the system. Scrutinizing it, we might even find what appears to be evidence of a bias in the wheel: some numbers that appear more often than others. But if we are told that the paper was dropped by a mathematician who, when he wasn't playing the tables, studied relationships among transcendental numbers, we might tinker with various combinations of sin, cos, tan, and so forth, until we had found a formula that came close to generating the string. The discrepancies would be dismissed as background noise caused by an occasional misfiring neuron or a bad calculator key.

We have at our disposal many different kinds of representations for mapping the world. We can think of them as lenses or filters each of which refracts the world a little differently. Observed through one filter, a system that seems random or chaotic on the surface might show signs of hidden orders—a signal picked out of the noise, complexity lurking within. On the other hand, if we choose an inappropriate filter, a very complex system might seem deceptively simple; everything that doesn't fit into our theory of its behavior is relegated to the category of background noise. As Crutchfield points out, an observer whose computational powers are limited to counting the ratio of 1s and 0s in a string will find no difference between the record of a coin flip and the binary expansion of pi. Both, on average, contain an equal number of 1s and 0s.

Instead of regarding complexity as a fixed platonic essence sitting in the middle of a neat scale, perhaps we should think of it as an ever-changing relationship between observer and observed. The implication, of course, is that there may be no preexisting, canonical order woven into the universe, waiting to be found. The orders we alight upon are, at least in part, human inventions; they depend on the lenses we use. We take a string of data, plot it onto a graph, and fiddle with the scales and the dimensions until, lo and behold, we have found a shape, a curve, an attractor. Of course, some points will lie outside our perceived design, so we call them experimental error, random noise.

This anti-platonic view, Crutchfield insists, is not an excuse for complete relativism, in which one theory is considered as good as another. Think of a theory as consisting of two parts: an algorithm that approximately reproduces the order we have uncovered, and an error string. We always try for the theory that can be expressed with the most compact algorithm (Kepler's laws are more elegant than Ptolemy's), but if we compress too much we will simply generate a longer error string. If we have a spray of points on a graph, we can simply connect the dots, yielding a model of great intricacy. Or we can insist that we see great simplicity: a hyperbola or a parabola. We draw our line and the data that lie outside it are dismissed as random error. Crutchfield agrees that the best theories are the ones that are most compact, but in making our measurement we must include not just the size of the theory but also the size of the error string. An overly simplified theory with a long error string would be just as bad as an overly elaborated theory (a connect-the-dot diagram) with a tiny error string.

When we start to examine just what we mean by complexity we see how elusive the concept really is. In contemplating the mysteries of the Tesuque bingo parlor, we saw how, in the Chaitin/Kolmogorov definition, the complexity of a process is measured by the size of the smallest computer program that will generate it. The sequence 1010011000110000111 is somewhat more complex than the sequence 1111111111111111111 because it is less compressible—it takes a longer algorithm to spit it out. Fine, so far. But as we have seen, under this definition, a completely random string of numbers, like those emitted by a bingo machine, is more complex than, say, the expansion of pi. The only way to generate a random string is with a program that says PRINT followed by the entire sequence. Pi can be generated with a simple algorithm that divides two numbers: the circumference and the diameter of a circle. Under this definition, a random string of one million letters and spaces will be more complex than a novel of equal length. The novel can be compressed by using spelling rules to filter out redundant letters: q is always followed by u; i goes before e except after c. And a random bit string will be more complex than a computer program or a digital image of the Grand Canyon, both of which can be compressed by finding patterns and squeezing out redundancies. The most orderly, compressible things are the least complex; the most disorderly are the most complex. But this defies the intuition that complexity is a tension between predictability and surprise, between order and randomness.

Some scientists have tried to overcome this problem by taking into account the difficulty of generating a bit string. Seth Lloyd and the late Heinz Pagels suggested a measure called thermodynamic depth: the amount of entropy that is dumped into the environment as a by-product of the production process. (Murray Gell-Mann suggests a slight variation: the complexity of something can be roughly measured by how much money it costs to make it.) A cell takes more energy to make than a random broth of the very same molecules. Lloyd concedes that there are limitations to the definition: monkeys banging on typewriters expend as much energy as a novelist.

Gell-Mann prefers a measure that he calls effective complexity. Take a system, the rules of English, for example, and imagine its most compressed representation—in this case, the size of the grammar book or the mental grammar coded within a native speaker's head. The length of this compression, or schema, as Gell-Mann calls it, is a measure of its complexity. Languages with more rules and exceptions to the rules are more complex than languages with shorter grammars. In the biological world, more complex creatures would have larger genetic schemata, genomes encoding a wider range of enzymes and other proteins. Under this definition, a random system would have no schema—it is incompressible—so its complexity would be zero. So far, so good. But effective complexity has its own problems. The genomes of a baboon and a human, Gell-Mann notes, are almost identical. A tiny number of genetic differences seem to account for our ability to communicate with language, to devise or discover mathematical relationships, to build systems to explain the universe.

Taking this into account, Charles Bennett suggests defining complexity using what he calls logical depth: the number of computational steps it takes to generate a structure. Two genomes may be similar in length, but one may contain a short segment that acts as a powerful program, switching on genes that switch on other genes, causing many layers of unfolding. Generating a random string of numbers takes just one step: PRINT 233729 . . . It is logically shallow. Generating a computer program or a novel requires many levels of computational gymnastics.

We think we know complexity when we see it, but we strain at a definition. And how easily we delude ourselves. Brian Arthur, a Stanford University economist and Santa Fe Institute regular, tells of a study in which a psychologist, Julian Feldman, gave his subjects random strings of 1s and 0s and asked them to find patterns. "Subjects would form a hypothesis—'You're following with a 0 and two 1's,' or 'You've begun a progression of two 1s and two 0s'—for some time, and stick to it as long as it predicted well," Arthur wrote. "They allowed for exceptions fairly liberally: 'You just gave me the 1 to throw me off.' But if the pattern-hypothesis performed badly over several predictions they would change or drop it in favor of a different one."

Gell-Mann defined superstition as seeing patterns that aren't really there—and denial as not seeing patterns that really *are* there. Deciding on which side of the line to place some of the Santa Fe Institute's conceptual inventions is not an easy task. We are left to wonder to what extent abstractions like fitness landscapes and scales ranging from order to chaos are features of the universe and to what extent they are attractive inventions—more sophisticated than dividing the world into concentric tetrads of mesas and mountains extending from a pueblo's spirit hole, but projections nonetheless.

Most of us feel certain that the biological world inexorably increases in complexity. At least on this planet, there seems to be a trend toward a greater diversity of creatures made from a greater variety of cell types. One can make strictly Darwinian arguments for why complexity might be likely to increase. There is a survival advantage to processing information, and perhaps more complex creatures are capable of making better maps. If Kauffman and Langton are right, complexity might increase because of internal dynamics driving systems to something called the edge of chaos. But does complexity *necessarily* increase at all? Or does it just appear that way as we gaze through a narrow set of filters?

Measured by sheer mass, the most successful life form of all is still the unicellular microbe. And of all the vertebrates, only a tiny minority, including ourselves, have exhibited increasing brain size. We can tell Just So stories about why the human brain grew so rapidly, but perhaps it was merely a fluke. If Stephen Jay Gould is right, we came frighteningly close to not existing at all.

In *Wonderful Life*, he turns back the clock 570 million years and evokes a time called the Cambrian explosion, when, after more than two billion years of unicellular life, the first multicellular creatures began to appear. We like to think of life starting out simple, with a few species, and then growing more complex, diversifying into the vast menagerie we see around us today. But the fossil evidence indicates that quite the opposite might have occurred. Trapped in the petrified mud of a formation called the Burgess shale in British Columbia is a strange zoo of some of the first multicellular creatures, many so different that they seem to fit into none of our conventional categories.

Biologists classify life into a hierarchy, starting with the most general category, that of kingdom (plant, animal, fungus, and so forth), and descending through the progressively narrower divisions of phylum, class, order, family, genus, and species. Phylum, the level directly beneath kingdom, consists of twenty to thirty categories (there are different ways of carving up the biota), including arthropods (insects, spiders, lobsters), mollusks (clams, snails), annelids (earthworms), echinoderms (starfish, sea urchins, sand dollars), chordates (all vertebrates, including reptiles, fish, mammals, and, of course, *Homo sapiens*). These, as Gould so evocatively puts it, are "the fundamental ground plans of anatomy," "the major trunks of life's tree." For centuries scientists tended to think of these categories as fixed, as though the creatures of the earth were dutifully filling in the squares of an ethereal chart in the platonic skies. After Charles Walcott, head of the Smithsonian Institution, discovered the Burgess shale, he spent the rest of his life forcing ("shoehorning" is Gould's word) the strange creatures trapped in the ancient mud into science's preconceived categories. And what strange creatures there were: Opabinia, with five eyes (at least four of them on stalks) and a tubular nozzle with a claw on the end; Odontogriphus, the "toothed riddle," with its mouth surrounded by dental projections that might have formed the base of tentacles; and, weirdest of all, Hallucigenia, its tubular body studded with seven pairs of thorny spines, matched by seven long tentacles with pincers, and, as if as an afterthought, a cluster of six additional, shorter tentacles on one end. Which end? We do not know. Hallucigenia is a creature so strange that it is not clear which is front or back, up or down.

By the early 1970s, it was becoming clear to a younger generation of scientists that no amount of stretching would accommodate all these animals. In the climax of his book, Gould tells how it slowly dawned on his heroes, Harry Whittington, Simon Conway Morris, and Derek Briggs, that many of the Burgess fauna could be seen as representing entirely new phyla, basic body plans that no longer exist. Gould estimates that half a billion years ago, at the time of the Cambrian explosion, there were fifteen to twenty phyla that we do not find today. There is good reason for the imprecision. Extrapolating from two-dimensional impressions to three-dimensional creatures is a subtle affair, like trying to abstract depth from the flat surface of the sky. While it has been speculated that Hallucigenia's tentacles might represent seven different mouths, Gould reluctantly suggests that what appears to be a wonderfully strange animal may actually be a fragment of an appendage of a less interesting larger creature.

After the Cambrian explosion came the Cambrian decimation, in which the profusion of life was narrowed to the subset occupying the planet today. Why did so many of these early evolutionary experiments fail? In the classical Darwinian view, those creatures that survived were the fittest; they outcompeted those with less efficient body plans. Gould suggests, however, that we are flattering ourselves and our fellow fauna to believe that we deserved to prosper over Hallucigenia or Opabinia. As he and his colleagues scrutinized the Burgess fossils and imagined what life might have been like then, they could see no reason to say that the winners were necessarily fitter than the losers.

Of course, one can always engage in circular reasoning, examining the survivors and making up Just So stories of why certain traits must have led to their domination. But if the tides were turned and creatures like Hallucigenia had given birth to whole phyla, then stories could just as easily be constructed about why they inevitably prospered. Rather, Gould believes, the Cambrian decimation was a lottery. Survivors were picked at random. In fact, toward the end of his book, Gould ventures that if a certain wormlike creature called *Pikaia gracilens* had not chanced to survive the decimation, the phylum Chordata would have perished along with Hallucigenia, closing off the possibility of vertebrates. With a different roll of the dice, different body plans might have survived, including none of those we take for granted. The situation, he says, is not unlike that of the movie *It's a Wonderful Life,* in which a Christmas angel gives a despairing George Bailey the chance to see what his small town would be like if he had never lived.

"The modern order was not guaranteed by basic laws (natural selection, mechanical superiority in anatomical design), or even by lower-level generalities of ecology and evolutionary theory," Gould writes. "The modern order is largely a product of contingency. Like Bedford Falls with George Bailey, life had a sensible and resolvable history, generally pleasing to us since we did manage to arise, just a geological minute ago. But, like Pottersville without George Bailey, any replay, altered by an apparently insignificant jot or tittle at the outset, would have yielded an equally sensible and resolvable outcome of entirely different form, but most displeasing to our vanity in the absence of self-conscious life. . . .

"[T]he Burgess not only reverses our general ideas about the source of pattern—it also fills us with a new kind of amazement (also a *frisson* for the improbability of the event) at the fact that humans ever evolved at all. We came *this close* (put your thumb a millimeter away from your index finger), thousands and thousands of times, to erasure by the veering of history down another sensible channel. Replay the tape a million times from a Burgess beginning, and I doubt that anything like *Homo sapiens* would ever evolve again. It is, indeed, a wonderful life."

But if not us, would some other creature have evolved until its nervous system became so complex that it developed self-awareness and consciousness? Gould has his doubts. He concedes that the development and evolution of life were not entirely a random walk, that the emergence of some kind of life was "virtually inevitable, given the chemical composition of early oceans and atmospheres, and the physical principles of self-organizing systems." And once life was under way, he allows that physical constraints—"rules of construction and good design" based on laws of surfaces and volumes and so forth—would channel evolution in certain directions. He believes "bilateral symmetry can be expected in mobile organisms built by cellular division." But these "channels are so broad relative to the details that fascinate us!" The question, he says, is where to draw the line between these background laws and the details. Gould puts the modern phyla, including vertebrates, among the details and concludes: "Whether the evolutionary origin of self-conscious intelligence in any form lies above or below the boundary, I simply do not know. All we can say is that our planet has never come close a second time."

And, he reminds us, there is more than one way that a creature might solve problems presented by the "rules of construction and good design." For example, as creatures grow larger, they have to ensure that none of their cells are too far from the surface, the source of air and other nutrients. Volume (length cubed) increases faster than surface area (length squared), so cells must constantly struggle to keep from being smothered. The organisms in our world solve this problem by involution, increasing internal complexity: lungs are lined with little sacs called alveoli, intestines with villi. One hundred million years before the Burgess explosion, Gould tells us, there existed a radically different kind of multicellular creature that solved the volume-to-surface problem in quite another way. Instead of increasing internal complexity, creatures of the fauna called Ediacara, named for the location in Australia where their fossils were first found, grew larger by spreading themselves into thin sheets and ribbons and pancakes, always ensuring that their cells were not too far from the environment. The Ediacara fauna were traditionally seen as precursors that later evolved into some of the involuted creatures of the Burgess shale, the arthropods and annelid worms. But Gould tells of an alternate interpretation: maybe Ediacara was a failed experiment, unrelated to creatures today. Because of some random event, these creatures were killed off before they spread throughout the world, filling up the niches created by multicellularity. What if the dice had fallen a different way? If the tape were replayed and Ediacara prevailed instead of the Burgess fauna, Gould says, creatures with internal organs probably never would have developed—no hearts, no lungs, no livers, and, most important, no brains.

But if multicellularity and internally convoluted animals did appear, wouldn't consciousness be inevitable? Gould asks us to consider the dinosaurs. We like to think that mammals, with their increasing intelligence, won over the giant reptiles, David over Goliath. In fact, for most of the reign of the dinosaurs, mammals were small creatures surviving because they could burrow beneath the feet of the lumbering lizards and hide. They did not begin developing bigger bodies and bigger brains *until* some chance event killed off the dinosaurs, vacating the niche for large vertebrates. Currently, the favorite story on the death of the dinosaurs is that an extraterrestrial impact, perhaps in the Yucatán, severely disrupted weather patterns, sending the climate out of kilter. At a lecture at the Santa Fe Institute, a champion of this theory, Walter Alvarez, estimated that (depending on whether the meteor or comet was traveling with or against the motion of the earth) a difference of minutes or seconds might have caused it to land in the ocean, muffling its impact. The dinosaurs might have survived and mammals would have remained a nuisance underfoot. He then startled the audience with an artist's rendering of a reptilian humanoid, the imagined fruit of continued dinosaur development—the *Star Trek* view of evolution. But in *Wonderful Life,* Gould presents a different possibility. There is no evidence that during the reign of the dinosaurs their brains were growing larger. They seemed comfortable in their niche, with little reason to develop increased intelligence. "This situation prevailed for a hundred million years; why not for sixty million more?" he asks. No extraterrestrial impact, no evolution of consciousness. "In an entirely literal sense, we owe our existence, as large and reasoning animals, to our lucky stars." And so may we someday owe them our extinction.

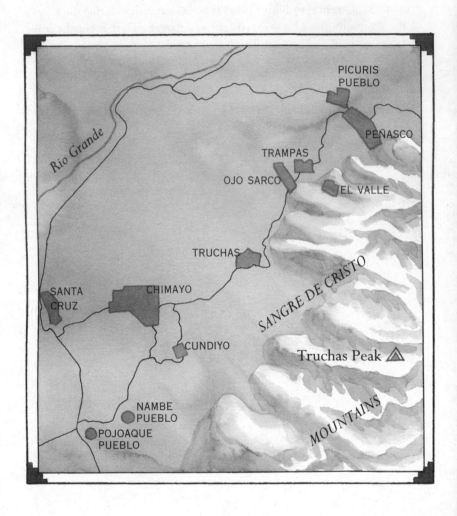

THE LEAP INTO THE UNKNOWN

Every year during Holy Week, the east shoulder of the highway that heads north from Santa Fe becomes a footpath, as pilgrims from all over the countryside begin their annual trek to the village of Chimayo, reputed to be a place of miraculous healing   the Lourdes of America. Most of the trekkers seem unexotic enough, no different from the people you might see shopping on a Saturday at the Villa Linda Mall south of Santa Fe or at the Wal-Mart on the city's commercial thoroughfare, Cerrillos Road. To warn motorists of their presence, some wear bright Day-Glo colors and carry flashlights or green photochemical beacons issued by the state police. Only occasionally does one see a sign that this is more than an afternoon's outing or a walkathon to raise money for some disease, that a religious procession is under way. Some groups of walkers carry the embroidered standards of their church or religious organization, hoisting them like sails against the spring breezes. Others hold crucifixes in front of them. And, every so often, one will see a pilgrim carrying over his shoulder a large wooden cross, as though in imitation of Christ dragging himself up Calvary.

At first this flow of humanity is but a trickle. But as Good Friday approaches, the trickle becomes a flood. Beginning in Santa Fe, or points along the way, the pilgrims walk north, past the Santa Fe Opera, which in a few months will begin another season of music under the stars; past Trader Jack's Flea Market with its selection of imitation Indian arts and New Age paraphernalia; past Tesuque pueblo and its bingo hall, where the camel gazes upon the procession with its contemplative stare.

Fifteen miles from Santa Fe, the pilgrims reach Pojoaque pueblo with its shopping center plaza. Here the road splits in three. To the left is the route to San Ildefonso pueblo and Los Alamos, the city of science. Straight ahead is the road that follows the Rio Grande northward to Taos. Ignoring these two alternatives, the pilgrims take the right-hand fork, a winding two-lane road that leads up into the Sangre de Cristo Mountains and another of New Mexico's almost insular worlds. For the first few miles, the narrow highway follows the Nambe River, on its way down from the frigid headwaters at the base of Lake Peak, the eastern boundary of the Tewa world. Just before the entrance to Nambe pueblo, the road leaves the green river valley, veering northward through a red eroded landscape that, except for the white wooden crosses on the mesas and hilltops, looks a lot like the *Voyager* photographs of Mars. Passing through this otherworldly terrain, the pilgrims finally reach a point, some twenty-five miles from the beginning of their journey, where they can pause and look down a steep hill at their destination: another green valley, where lies the village of Chimayo.

Marching down the final turns of the highway, they head for a small adobe chapel with a tin roof and twin steeples called El Santuario de Chimayo, a place where miracles are said to occur. Filing past the gravestones in the courtyard, they enter the double doors leading to the nave and sanctuary, adorned with the ornate wooden panels (called *retablos*) and the statues (*bultos*) carved by the mountain's *santeros* to honor Mary and the saints. Pausing near the altar, with its towering crucifix, the visitors each cross themselves and whisper a prayer. Then they turn left, passing through the narrow doorway leading to the sacristy, where a collection of the church's relics is displayed.

It is in a dark, tiny side room just off the sacristy that their ultimate destination, the object of their veneration, resides. Through an entrance so small it might be the mouth of a cave, the pilgrims enter one by one. Inside, carved into the floor, is the shallow earthen pit that has given Chimayo its fame. Crouching before it to say another quick prayer, the visitors reach down for a pinch of the miraculous soil said to heal all wounds. Some rub the dirt on sore limbs or reach discreetly inside their shirts or blouses to apply it to the chest, or they take it home to mix it with water or food to treat internal ailments. All around the sacristy hang the crutches and braces of those said to have been cured. Others have left letters, telling how the holy dirt healed an ulcer or destroyed a tumor that radiation or chemotherapy was unable to subdue. By the time Good Friday has arrived, the visitors number in the thousands and the pit is refilled again and again with soil carried from a special place down the hill by the Santa Cruz River.

Far away, on the other side of the Rio Grande Valley, are the laboratories of Los Alamos, populated by a people driven by the search for natural law. If they believe in a God at all, it is likely to be one who invented the laws, established the symmetries, then left the universe to unfold on its own, mathematically. Faced with what is said to be a miracle, these scientists will analyze it, search for a way to dispute it or explain it away, absorbing it within the fabric of laws they swear allegiance to. Beginning in the late 1970s, a small group of scientists at Los Alamos trained their most sophisticated tools of analysis on the Shroud of Turin as part of a worldwide effort to prove its authenticity. Though some of the researchers were Roman Catholics who hoped to prove that the ancient relic was really Jesus' burial shroud, they were ultimately forced to bow to the verdict of science: carbon-14 dating and other tests indicated that the cloth, venerated for centuries, was a medieval forgery.

The simplest explanation for Chimayo's fame is the one a biomedical researcher might consider the least likely: that the dirt really has the power to cure so wide a spectrum of illnesses. Surely, the skeptics say, a careful analysis of the data would show no cures that could not be attributed to a statistical fluke, the placebo effect, or the delusions of mass psychology. If people believe in the miracle of Chimayo, then any chance improvement in their health will be attributed to the power of the holy dirt. Any failure will be explained away. With a different chain of historical circumstances, it is easy to suppose, people might flock instead to some nearby town—Cundiyo or Cuyamungue—to venerate an oddly shaped rock or a legendary tree. Those of the secular faith seek a reason for Chimayo in the rolls of the dice, the contingencies of history.

But no investigation, no appeal to scientific law, will dislodge the miracle of Chimayo from the hearts of so many New Mexicans. Chimayo's Santuario is a place for those who believe that the laws, like anything else the Almighty created, can be suspended, that extraordinary things can happen which only the faithful can understand.

Some say that the miracle of Chimayo predates the arrival of the Spanish in the sixteenth century, that before the first Franciscan priest set foot in the valley, the Tewa Indians came to Chimayo to be cured. According to native mythology, when the Hero Twins killed one of the monsters that stalked the early earth, fire and hot water rose from the ground in several places: Black Mesa at San Ildefonso, Cabezón Peak on the Navajo reservation, and Chimayo—Tsimayo, as the Tewa called it, "the place of good obsidian." The fires subsided, leaving a pool, which evaporated to form dust with medicinal powers. By building a sanctuary at the site, the Catholics adopted the Indians' ancient magic as their own.

The descendants of the Spanish colonizers tell a different story. They say the site was discovered in 1813, during Holy Week, by Bernardo Abeyta, a leader of the Penitentes who was sitting in the hills, whipping himself in penance for his sins, when he saw a light shining from the ground. Digging down into the bank of the Santa Cruz River, he found a glowing crucifix. Rejoicing, the people of Chimayo formed a procession, carrying the cross to the parish church, down in the village of Santa Cruz, and placed it on the altar. But the next day the cross was gone. In defiance of the rules and regulations of natural law, it had mysteriously transported itself back to the hole in Chimayo where it had been found. Another procession was formed to return the cross to Santa Cruz, but again it levitated back to its original spot. After the third try, the legend goes, the people understood what the cross was trying to tell them, that it belonged in Chimayo. Abeyta built a chapel for the cross and petitioned the archdiocese for permission to build a church. For reasons that remain a mystery, he specified that the church, today's Santuario, would venerate an image of Christ he referred to as Our Lord of Esquipulas.

To this day, no one really knows why Abeyta chose the name. Esquipulas is a Mayan village in Guatemala, some two thousand miles away. A basilica built there in 1759 honors a figure called the Black Christ, a crucifix carved from dark balsam wood, perhaps in deference to the brown-skinned natives whose gods the Catholics were trying to subdue. Esquipulas is also venerated as a place of healing, and every year, on January 15, people from the countryside make a pilgrimage to the town. Many of them carry away small tablets of clay, stamped with holy images, which they eat or dissolve in water to cure afflictions and disease.

Confronted with this weird coincidence, the mind tries to impose a sensible story, to explain it within the skein of cause and effect. The odds are vanishingly small that two remote villages where the curative powers of the earth are venerated would both happen upon the name Esquipulas. And nothing in science allows for apparitions—a real Esquipulas, saint or spirit, manifesting himself in different parts of the world. And so we seek a channel through which information might have flowed between two parts of the planet. Esquipulas was revered in parts of Mexico. Perhaps Bernardo Abeyta had visited one of these towns, or even the distant Guatemalan village, or heard about them from a traveler or a priest. Perhaps he also knew of the Tewa legend about the magic of Chimayo and was inspired to link the two stories together. Or was it the other way around? Perhaps the pueblo Indians later heard the story of Abeyta, retroactively absorbing it into their own myths. According to a legend recorded at Isleta pueblo by the anthropologist Elsie Clews Parsons, it was not a Spaniard but an Indian who discovered a little head sticking from the ground in Chimayo. He took it to Santa Fe, not Santa Cruz, but it kept returning to the spot by the river. Parsons's informant—whose name, coincidentally, was also Abeyta—called the head Escápula.

How adept we are at weaving stories that explain the world according to what we hold most dear. The legend of Chimayo is further confused by the later emergence of a rival to Esquipulas: the Santo Niño de Atocha, a manifestation of the Holy Child said to walk the valley at night healing sick children. In the middle of the nineteenth century, a chapel to the Santo Niño was built near the Santuario, luring away some of the visitors drawn by the Esquipulas legend. The Santo Niño is revered in many parts of Mexico, particularly at a shrine in Fresnillo in the state of Zacatecas. The new chapel in Chimayo became so popular, with pilgrims bringing gifts of baby shoes to replace the ones said to be worn out by the infant on his nocturnal wanderings, that the Abeyta family obtained its own Holy Child and placed it on an altar next to the pit of healing dirt. Before long the legend grew that the small figure had been discovered by a little girl and her father, who heard a bell ringing underground. They dug a hole and found the Holy Child, who had the power to cure. Over the years, the legends of Esquipulas and the Santo Niño have become so entangled that it is hard to tell them apart. Today most of the pilgrims who gather samples of the miraculous soil attribute its powers to the Holy Child.

In the never-ending drive to bring order to the world, humanity's different tribes are constantly bumping into one another, encountering alien systems, webs of explanation strung together according to different assumptions. Sometimes one tribe tries to eliminate the competition by wiping out the infidels, waging holy war. More often people simply try to absorb the other religion into their own. Did the Spanish co-opt the Chimayo legend from the Tewa? Or was it the Tewa, so adept at absorbing new gods into their ever-expanding pantheon, who adopted the Spaniards' magic? We'll probably never know. By asking this kind of question we betray our own attempt to absorb the Chimayo legend into a framework in which rational explanations must always prevail. Faced with the mysterious, we strain to impose a story that makes sense of it, in terms of what we consider fundamental—whether it is the power of the Trinity, the power of the Tewa Cloud Beings, or the power of natural law.

As one leaves the Chimayo Valley and heads higher into the mountains, closer to the stars, the twentieth century with its smug certainties is quickly left behind. Except for the dozen years between the Pueblo Revolt and the European Reconquest, Spain (and later Mexico) dominated this land until the United States seized it in 1846. To this day, some of the older people in the villages of the Sangre de Cristos speak a dialect of Spanish with echoes centuries old. It is here that some of the last remnants of Los Hermanos Penitentes, the Penitent Brothers, continue to celebrate Holy Week by reliving the suffering of their savior. Closeted away night and day in the adobe chapels called moradas, they carry out rituals, many of them secret, that have been handed down for two centuries, perhaps many more.

Driving higher and higher, one passes the turnoff to Cordova, famous for its woodcarvers, and an old cemetery filled with colorfully decorated graves. The narrow, twisting road is itself marked with crosses where drivers have taken a turn too fast and plunged over the edge. Several miles beyond Chimayo, at the edge of a high mountain meadow, is the village of Las Truchas. With the thirteen-thousand-foot Truchas peaks rising behind it and the Rio Grande Valley spreading out down below, the village sits in one of the most beautiful settings in northern New Mexico. But the location was not chosen for its grandeur. In the mid-1700s, the Spanish government decided that a buffer zone was needed to protect the settlements in the Santa Cruz Valley, and beyond that, Santa Fe, from raids by Comanches from the eastern plains. Offering grants of free land, the government recruited poor families, who must have felt they had little to lose, and established the outpost settlements of Truchas and nearby Las Trampas.

The villagers, considered so expendable by their government, fared little better with the Catholic Church. By this time the Pueblo Revolt had come and gone. But the Church never regained the full strength it had before the uprising. The few priests in the region made only sporadic visits to the hinterland. In this spiritual vacuum, the seeds of Roman Catholicism grew in strange and unruly ways. Lacking finely wrought relics like those in Santa Fe, the people of the Sangre de Cristos carved their own. Spanish Catholicism had long held a fascination for the macabre. The local *santeros* seemed to compete with one another to magnify the suffering of Christ with the bloodiest of crucifixes.

By the early 1800s, Penitentes had appeared throughout northern New Mexico. Filling in for the priests who so rarely came, the Hermanos, as they are also called, prayed for the sick and officiated at wakes and burials, keeping the embers of the religion alive. Emulating Christ, the Penitentes quietly and sometimes anonymously performed acts of charity. But they especially took to heart the story of Christ's suffering at Calvary. During Lent and Holy Week they did penance by flaying themselves and each other with yucca whips; some walked with branches of cactus tied to their backs. They crawled on their knees in long processions and reenacted the march to Calvary carrying crosses so heavy that the wood lacerated their shoulders. On Good Friday they reenacted the Crucifixion, tying one of the brothers to a cross on the top of a hillside. When he passed out he was taken down and revived, though tradition has it that occasionally a brother would die on the cross, assuring himself a place in heaven. In trying to piece together the story of the Penitentes, it is hard to separate historical fact from the embroideries of legend. According to local folklore, the brothers were sometimes joined in their processions by skeletons in white trousers and black hoods, lashing themselves and the backs of the not-yet-departed.

At the height of the Penitentes' influence in the mid- to late 1800s, most of the men in many northern New Mexico villages belonged to the brotherhood. In secret meetings in the rustic chapels called moradas, the *Sangrador,* or bloodletter, would use sharp pieces of flint to mark initiates with the seal of the brotherhood—three incisions on the back in various configurations. The *Celador,* or warden, administered the sacramental punishments. His assistant, the *Coadjutor,* was in charge of washing the whips and treating the brothers' wounds.

Where did these practices come from? What is the source of the cultural memes that implanted themselves so firmly in the minds of these people, causing them to worship so fervently? Again, we the rationalists are faced with the problem of untangling the circuit of ideas. Over the years, historians have offered various interpretations. Could the Penitentes be direct descendants of the flagellant cults that spread through Europe during the plague years of medieval times? Have the rituals been passed down, generation by generation, for half a millennium? Or were they borrowed more recently, implanted perhaps by a member of an eighteenth- or nineteenth-century flagellant society who traveled from Spain to the New World? Some historians believe the Penitentes are an offshoot of a Catholic lay group called the Third Order of St. Francis, which dates from the thirteenth century and was popular in New Mexico after the Spanish Conquest. There have also been suggestions that some of the rites were borrowed from the Tewa, who secretly practiced flagellation in kivas instead of moradas. Of course, the Tewa could have been influenced by the Penitentes. Or we might have a case of parallel evolution. Perhaps there is something about the human spirit that, when times are hard, naturally seeks deliverance in self-inflicted pain. There is so much about this land and its tangle of beliefs that we will probably never understand.

The Penitentes have seen no need to tell the rest of the world their own origin theories. And so historians rely on documents, scraps of evidence from which to re-create another world. Wherever the brotherhood came from, mentions of it do not begin to appear in church records until the early nineteenth century. While some historians take this as evidence that the brotherhood is a recent transplant, perhaps coming from Mexico or Guatemala, others note that penitential sacrifice was so common among Spanish Catholics in the previous century that church officials might not have considered the practice worth remarking on.

Though the ecclesiastical authorities in Santa Fe made public pronouncements against the group and occasionally tried to suppress it, they generally seemed to tolerate the Hermanos, encouraging them to do their penances in private and to avoid inflicting upon themselves serious bodily harm. In 1947, as a reward for their discretion, the Archdiocese of Santa Fe granted official recognition to the chapters of the brotherhood that agreed to abide by the Church's rules. The heterodoxy had been largely absorbed. By this time all but a few of the Penitentes' rituals were performed inside the morada or late at night in remote locales. Today, only the few brothers who are left know what goes on in the moradas on the long nights of Holy Week.

On the Wednesday night of one Holy Week, not too long ago, as the throngs of pilgrims continued to arrive in Chimayo, up the mountain in Truchas half a dozen of the remaining members of the local brotherhood were inside the morada, preparing to celebrate the Last Supper. Holy Week had come early that year. Flurries of snow were still blowing through the arroyos as smoke flowed from the chimney of the morada, a long, plain adobe building with a simple cross mounted above the door. The morada is perched on the edge of town, with a panoramic view of northern New Mexico. From the doorway one can look east and see the Truchas peaks towering close by. Far to the south the transmitting towers on the Sandias blink their red signals. To the west the lights of the Rio Grande settlements shimmer in the valley below, the more distant nebula of Los Alamos hovering in the mountains above them. In the daylight one can stand near the morada and see San Ildefonso's Black Mesa and Chicoma Peak, the western holy mountain.

On most nights Truchas is silent, except for the distant sound of barking dogs. But on this night, from inside the morada comes the harmonious sound of chanting, smooth and relaxing, almost hypnotic, like that which might be heard outside a medieval monastery. With different words and rhythms the sounds might be coming from a kiva. With life always verging on confusion, we find comfort in the repetition of familiar sounds.

Inside the morada, three of the Hermanos are on their knees before an altar reciting the rosary. Kneeling in the center is the Hermano Mayor, or head brother, Leroy Vigil, a soft-spoken, rugged-looking man who works by day at Los Alamos for the company that maintains the laboratory's various mechanical systems; one of his daughters, Della Ullibari, works in the publications office at the Santa Fe Institute. But tonight these outposts of the scientific quest seem worlds away. The simple chapel, decorated with floral wallpaper, is heated with a woodstove. Behind the altar is a traditional northern New Mexican *retablo,* the squares painted with images of the Virgin and saints. To the left and right, small crucifixes adorn the walls. And at the focus of the room, towering above the kneeling brothers, is a tall statue of Christ in purple robes, his head gashed and bleeding from the crown of thorns. A few visitors sit on folding chairs, watching and sometimes praying.

When the service ends, the brothers kiss the floor in front of them, and the guests are invited to the kitchen for the Last Supper. As they sit at a long wooden table, several women from the village serve dinner—plates of red chili, beans, corn, chicken patties, orange drink, coffee. The brothers stand behind the table, holding up two large wooden crosses as they sing the Spanish hymns called *alabados*. When the dinner is over, they thank their guests for helping them celebrate the night when Jesus bade farewell to his disciples.

For the rest of Holy Week the six brothers spend day and night inside the morada, emerging for the occasional ceremonies outsiders are allowed to see. What goes on the rest of the time must be left to the imagination. The wives and families of the Hermanos have learned not to ask.

On the morning of Good Friday, the Penitentes leave the morada to help the village commemorate what for many is the most important day of the year. This is the morning of the Encuentro, the "Encounter," when Christ met Mary on the way to the Crucifixion. Praying and singing, the brothers slowly carry the Cristo, blindfolded, up the dirt road past Tafoya's General Store to a little church called the Mission of the Holy Rosary, which sits near the center of town. As they gradually make their way through the village, a group of women, members of the local Carmelite Society, begin their own procession, carrying a figurine of the Madonna, dressed black in mourning, out of the church. The men and the women meet in the small courtyard, and Christ and Mary are held against each other in an embrace. Then, one by one, the villagers kneel to kiss Christ's gown, then Mary's.

In the early afternoon, the Penitentes are back in the morada performing the Via Crucis, a version of the traditional stations of the cross. In the past the fourteen stations were often represented by little crosses leading from the morada door to a large cross, the Calvario, on a nearby hill. The Penitentes would visit each station, one by one, reliving in their songs Christ's condemnation by Pilate, his trip up the hillside carrying the cross, his first fall, his meeting with his mother, . . . his second fall, his third fall, his crucifixion. As they cycled through this familiar pattern, some of the brothers would whip themselves in sorrow for the trouble mankind's sins brought to the Holy Redeemer.

Today, in Truchas, the devotion is far more subdued, taking place entirely inside the morada, visitors coming and going throughout the long service. Afterward, some of them drive down the hill to Chimayo, which, by this time, is a colorful traffic jam of people. Or they spend the day with their families, waiting for nighttime, the climax of the week.

At about ten in the evening, the Hermanos make another procession from the morada to the church, carrying again the statue of Jesus. One brother holds a lantern, another a flashlight to read the words to the *alabados*. As they sing, one plays a flute while another occasionally spins a noisemaker called a *matraca*. Filing into the church, they place the Cristo by the altar. Then they light a row of thirteen candles along the altar rail—one for Christ and each of the disciples. For an hour or so the brothers kneel on the hard floor singing the rosary. Then the service, called Tinieblas, begins.

Tinieblas, which derives from a traditional Catholic service called Tenebrae, means "darkness" or "confusion." As the evening progresses, it becomes clear how the ceremony got its name. The old church fills with the mournful sound of the Penitentes' *alabados*. When they reach the end of the first verse, one of them swings the *matraca* and a candle at the end of the row is put out. The first apostle has betrayed Christ. At the end of the next verse, a candle from the opposite end of the row is extinguished. Another betrayal. The progression continues, a candle from one side and then the other extinguished, the darkness converging on the single candle burning in the center, Christ himself. Finally it is extinguished—the Savior's life gone in a wisp of smoke—and the rest of the church lights are turned off.

Hidden by the darkness, several of the Penitentes retreat into a side room. The *alabados* continue. Prayers are offered for the recently departed. And then, suddenly, pandemonium breaks out: the rasping sound of the *matraca*, the shrill cry of flutes, the stamping of feet, what sounds like a box of glass being shaken—too many sounds for the mind to sort out. In the past, it is said, the Penitentes would whip themselves during the confusion and cry out in pain. This is the hour of darkness after Christ's death when, Matthew tells us, "the curtain of the temple was torn in two, from top to bottom; and the earth shook, and the rocks were split; the tombs also were opened, and many bodies of the saints who had fallen asleep were raised. . . ."

For these few minutes, the world seems devoid of order—be it that of man or God. But faith ultimately triumphs. Chaos is vanquished. The lights inside the church go on again. More prayers are offered for the dead, and finally, near midnight, the service ends. With their heads bowed, the Penitentes walk backward out of the church and return to the morada. On Easter Sunday they will attend a traditional mass with their families. They have done their part to keep alive the memory of the suffering and death of Christ. The celebration of the resurrection is left to the Church and its authorities.

Many of those who study the origins of Christianity believe it is no coincidence that Holy Week occurs near the time of the spring equinox, when night has become as long as day again, and the sun is halfway home from its journey south. The world emerges from darkness, the Son emerges from the grave—and soon the seeds will emerge from the darkness underground, as the people, in pueblo myth, came up from the underworld.

All over northern New Mexico, dancers file out of the kivas to celebrate this repetition of the heavenly cycles. Just down the road in Nambe, Truchas's Tewa neighbors commemorate the Easter holiday with the precise repetitions of the bow and arrow dance. Across the valley, the men and women of San Ildefonso perform the basket dance. The celebration of light and the emergence from darkness seems to be hardwired into our brains by the cycles of the sun, by the experience of birth itself. The specific ways the celebration develops are molded by the contingencies of history. But beneath it all is this longing—not only for light but for pattern, for the predictable return of the sun. Marooned on this planet beneath an insentient sky, we find solace in the belief that behind the world's veil of confusion is a detectable order. Beyond the unfeeling stars is an intelligence, whether we think of it as an intervening god or as the mathematical laws of physics. The deepest mystery of all, impenetrable to both science and religion, is how we fit into the scheme.

New Mexico's Hispanic population is overwhelmingly Catholic. But since the American occupation, small bands of Protestants have tried to compete with the Catholic Church for the souls of the people of the Sangre de Cristos, offering yet another way of carving up the world. In Truchas, just beyond the Mission of the Holy Rosary, is a Presbyterian church. And beyond that is the ramshackle little chapel, abandoned in recent years, that was described at the beginning of this book: Templo Sion, an outpost of the Asambleados de Dios, the Assembly of God. Rising in counterpoint to the solemn intonations of the Hermanos were the enthusiastic cries of the born-again. Like other fundamentalists, members of the Assembly of God, a group founded in the Midwest around the turn of the century, believe in the literal truth of the Bible and in the imminent return of Christ. They are also pentecostals, who distinguish themselves from other fundamentalists by their belief in glossolalia, the speaking in tongues. In their most intense moments of fervor, worshippers spew forth seemingly random syllables said to encode messages from the realm of God. For decades the Allelujahs, as some of their neighbors mockingly call them, have proselytized in the Sangre de Cristos, trying to harness the fire of the Catholic faith and channel it in other directions. Even a few Penitentes, at odds with the Mother Church, have joined the pentecostals. Though Templo Sion is now boarded up, down the road between Chimayo and Santa Cruz is another church of the Asambleados de Dios called Templo Calvario, its dirt parking lot jammed on Easter Sunday with cars.

The most ardent fundamentalists believe the world in which they live is under siege. The Catholic Church is an agent of Satan, the "Whore of Babylon" described in the Book of Revelation. And secular culture, including much of the science that goes on in Santa Fe and Los Alamos, is part of the godless empire described in Revelation as the kingdom of the Antichrist. When the prophecies are fulfilled, the fundamentalists teach, Christ will make his Second Coming and destroy the kingdom of evils. But, as the Book of Matthew teaches, Jesus will not return until the words of the gospel have been brought to every man, woman, and child on earth. It is to ensure this mass dissemination of information that the fundamentalist churches send their missionaries all over the earth, including areas that are strongly Roman Catholic.

Throughout Latin America, the Assembly of God is challenging the hierarchy of Rome with this alternate brand of salvation. In northern New Mexico these pentacostals are enticing a people oppressed by poverty with the assurance that their hardships are not random and meaningless—they are foretold in Revelation and in the books of Matthew, Daniel, and Ezekiel. The occupation by the Anglos, the onslaught of secularism, the prosperity of Santa Fe compared with the poverty of Truchas and Chimayo—all these injustices are part of the master plan.

Satan has lured mankind into building its Towers of Babel, elevating the creations of people over the creation of God. But to the fundamentalists there is no such thing as human progress. What the Bible foretells is inevitable decay. Since Adam and Eve ate of the fruit of the tree of knowledge, mankind must be punished. Its creations must crumble to make way for the Savior's return. Salvation will come not from mankind's science but from Christ himself when he reestablishes his reign. In the meantime, civilization's unraveling is something not to be dreaded and fought against but welcomed with open arms. Christ is about to return to the planet; the suffering is all for a purpose and it is about to end.

As the year 2000 approaches, millenarian sentiments are flourishing all over the world. In the swirl of events—the downfall of the Soviet Union, the twists and turns of Middle East diplomacy—the fundamentalists believe they see the prophecies being fulfilled. Behind the chaos are the unmistakable hints of a celestial agenda. Once all the pieces click into place, Christ will stage his return, beaming believers up to heaven in a mass ascent called the Rapture, which for everyone else will be followed by a seven-year period of hell on earth called the Tribulation. And then, the Battle of Armageddon, also known as World War III. Some of the horrors of this time are described in Revelation: a plague of locusts with breastplates of iron (helicopters, according to some interpretations); a great star shooting from the sky and poisoning a third of the earth's waters (a nuclear missile?). But the forces of good will triumph and the new millennium will begin.

Historians tell us that many parts of the Bible were written to comfort the people of ancient times. The prophecies of Daniel were written for Jews persecuted by the Syrians; Revelation for Christians who suffered under Emperor Nero. In Matthew, when Jesus says his return will be presaged by earthquakes, famines, wars, and "rumors of wars," he seems to be speaking to his own disciples, promising to return within their lifetimes. But to the fundamentalists, the Bible is not a historical document. It is the blueprint of creation. It is here one should look for timeless laws.

Scientists scrutinize the impressions our senses receive from Nature, hoping to find patterns. For decades, fundamentalist scholars have scrutinized the verses of the Bible, certain they are unearthing nothing less than a description of the future of the planet. The result of these hermeneutical excavations is elaborate charts, diagrams of history, both past and future. Bible verses are correlated with Bible verses forming networks as dense with connections as the wiring on a computer chip. The best-known prophecies were mined with great effort by the religious scholar C. I. Scofield, who footnoted and footnoted every chapter of the Bible until his commentaries were as long as the Scriptures themselves. Fundamentalists take them as literally as they do Holy Writ. First published in 1909, *The Scofield Reference Bible* forms the basis for many of the predictions in Hal Lindsey's best-selling apocalyptic thriller *The Late*

Great Planet Earth, or in Pat Robertson's television commentaries on the Christian Broadcasting Network, in which events in the news are said to presage the destruction of man's ephemeral creations. Fundamentalists reject the modern view that there is something random or contingent about history. It can be thought of as the output of a great computer, whose software is laid out in the Bible. Once you have deciphered this code, everything can be known. The result is an all-encompassing system in which there are no accidents. Everything unfolds according to a plan.

As we learn from the particle physicists, if we ascend to a higher level of abstraction, things that seem different on the surface suddenly appear as manifestations of a deeper unity. Viewed from on high, fundamentalism can be seen as a manifestation of the rational spirit—this drive for prediction, this search for buried orders. But the structures it builds serve a very different end from those of science. If the cosmologists find a theory of the initial conditions, a quantum wave equation for the universe that seems to require no outside mover, then so much the worse for the Almighty. To the fundamentalists, God and the Bible are the foundation and any science must be built upon this rock, interpreted within this framework. In seeking rational explanations for miracles, science tries to absorb religion into its conceptual mesh. The fundamentalists try to turn the tables. Rather than reject science, they try to absorb it into their own system, like the Tewa absorbing Catholicism or the cult of Esquipulas absorbing the legend of the Santo Niño.

In the days before Einstein and Planck, when Newton's laws reigned absolute, even the most fervent evangelicals looked to science to justify their most heartfelt beliefs. Copernicus, Galileo, Kepler, and Newton delineated a universe that seemed to run as smoothly as an elegant machine. God was the engineer of these celestial clockworks. Though some Protestants lamented that the earth was no longer in the center of the Almighty's gaze, science's new cosmology seemed to stand as a formidable challenge to the Catholic Church's version of things.

In the Catholics' universe, arranged according to what has been called the Great Chain of Being, God sat at the top of a celestial ladder with mankind at the bottom. On the rungs in between were the various grades of angels. Then came the Pope, the cardinals, the archbishops, the bishops, the priests. God ruled by fiat, passing down his decrees. Protestantism, on the other hand, was driven by the belief that people could discern God's rules for themselves, without the intercession of the saints or the help of ecclesiastical authorities. The rules of the universe were not promulgated continuously by decree, as in some celestial corporation; they had been wired in at the beginning by the Great Designer.

But then came evolution, relativity, and quantum theory. According to these new sciences, the beautiful biological orders came not from a Great Designer but from random mutation and selection. Space and time were not absolute but relative; the behavior of subatomic particles could only be described statistically. Worse still, the fundamentalists believed, the world described by Einstein, Planck, Heisenberg, and Bohr was no longer simple, discernible to the common man. Now physics could only be understood with abstruse mathematics and imaginary spaces. The scientists, it was feared, were becoming the new priests. About the only thing to come out of post-Newtonian science that the fundamentalists eagerly embraced was the second law of thermodynamics with its image of inevitable decline.

To try to recapture the science they believed was stolen from them, the fundamentalists began to develop their own system—an alternative science that rejects the notion that chance plays an important role in human affairs. Depending on one's assumptions, data can always be cast in different patterns. Shake up the kaleidoscope and evidence that life and the universe evolved over eons can suddenly appear to support the notion that everything was created in seven days.

Radioactive dating studies can be reinterpreted to show that the earth is not four billion years old, as geologists believe, but six thousand years—an age that can be obtained by adding up the biblical generations before Christ. And since six thousand years is hardly enough time for a universe with a diameter of ten to twenty billion light-years to have formed, practitioners of what has come to be called "creation science" try to reinterpret astronomical data, to recalibrate the Hubble constant or show that there really is no upper limit to the speed of light. Some creationists, as certain as Einstein that God does not play dice, have strained to develop a deterministic model of the atom.

A few fundamentalists even interpret certain Bible passages (Ecclesiastes 1:5 says, "The sun rises and the sun goes down") as requiring a geocentric universe. To put the earth back at the focus of the Creator's attention, one creationist scholar has written a paper reinterpreting Michelson and Morley's landmark experiment. After the two scientists found that their perpendicular light beams moved at the same speed, it was eventually taken as evidence that there was no aether and that the velocity of light was constant. As Fitzgerald and Lorentz suggested, the effect was caused because objects shrink in the direction of their motion, so the light beam moving with the earth would traverse a slightly smaller distance than the beam moving perpendicularly. And Einstein went on to develop his special theory of relativity in which moving objects shrink and their time slows, always by the exact amount necessary to preserve the speed of light as a constant. As some creationists see it, there is a simpler, more elegant answer: Michelson and Morley's light beams moved at the same velocity because the earth is standing still, at the center of creation, just as the Bible requires.

To all but a true believer, creationism is pure superstition—all these contortions, these most elaborate of epicycles, to get the universe to conform to the words of an ancient text. In science, it is said, nothing is sacrosanct; the most beloved doctrine can be overthrown by contradictory data. It is because of this willingness to put its feet to reality's fires that science, like no other system of belief, has given us so much power to predict and control nature. In this way it is unlike any religion or philosophy.

But as the philosophers tell us, the empirical method is not always as straightforward as it seems. A powerful theory can take on a life of its own, rooting itself so deeply in the mind that it becomes almost invulnerable to reality's challenges. Threatened with data it cannot at first explain, a theory is adjusted; elaborations are added to elaborations until the structure sits more comfortably on its platform or becomes so unwieldy it can no longer stand. Here the drive is not to preserve God at the center of creation but to preserve—as long as possible—a human invention that has served us so well in our neighborhood of the galaxy that we cannot help but hope it will turn out to be of universal import.

With the powers of compression molded into our brains by evolution, we find laws that not only apply to falling objects but can be used to predict the motion of the planets. Newton's laws do such a wonderful job of describing the motion of things nearby that it seems unthinkable that they would not apply (with a relativistic twist and a hefty dose of dark matter) to everything in the universe. To cosmologists, the big bang theory explains so much that it has become almost as fundamental as Newton's laws. When faced with discrepancies—galactic structures so large that gravity could not possibly have formed them in the allotted eons—we suppose there must be even more of this invisible stuff.

The incredible smoothness of the background radiation and signs that the universe is "flat"—poised on a knife edge between being open (expanding forever) or closed (eventually collapsing in on itself)—seem like coincidences too good to be true. Some Christians take phenomena like smoothness and flatness as miracles. Better than the holy dirt of Chimayo, they provide evidence of a willful God, a master craftsman. But the point of science is to expel miracles, to explain the world through natural law. And so some cosmologists take a leap of faith in a different direction and embrace the doctrine of cosmological inflation: In the beginning the universe all of a sudden started wildly expanding, smoothing and flattening space as it grew. They find hypothetical mechanisms within particle physics to explain how this antigravity might have occurred. And if the inflationary version of the big bang theory supports the notion that 99 percent of the universe is made of dark matter—exotic particles we have never detected—then so be it. A primary justification for these artful inventions is that they give rise to the kind of genesis science instinctively seeks: one that required no God to set the knobs precisely, one that is not miraculous—too good to be true.

Some particle physicists find it aesthetically repellent that there seem to be four fundamental forces, each with its own peculiarities. So they posit a perfect netherworld with one force, a symmetry that was randomly shattered, giving rise to our imperfect universe. They recoil at the zoo of subatomic particles with their seemingly arbitrary masses, so they posit that in the netherworld all particles—the fermions and the force-carrying bosons—were one perfectly massless particle, another symmetry that was shattered. When the attempts at bringing order to the subatomic world require the invention of abstract qualities like strangeness and charm that manifest themselves only in imaginary mathematical spaces, then our wonder only increases.

We perform these heroic feats of imagination to preserve what we believe in our heart of hearts to be true: that the universe is ultimately as symmetrical as the music we make, the diamonds we carve, something harmonious to the human mind. But when we look out on the messiness of creation, when we consider our limited vantage point, our blindered senses and brains, then this belief that we can penetrate the veil of contingency and happenstance and behold a crystalline perfection begins to seem like the deepest of faiths.

In the beginning was a world of mathematical purity that shattered to give birth to the world in which we find ourselves. How is this belief so different from the Fall from the Garden of Eden, or the emergence of the Tewa from the heavenly underworld? We, the pattern finders, the pattern makers, instinctively long for symmetries. Rather than let ourselves be overwhelmed by the messiness, the randomness, the unruliness that so often prevails, we construct our creation myths, we dream of a time when order prevailed.

Over centuries we've stitched together a patchwork of explanation: laws of physics, chemistry, biology, geology—little windows on the world, clusters of concepts each of which makes sense of a small portion of reality. Then, driven by this talent, this compulsion to seek out regularities, we find patterns among the patterns, clusters of clusters. And from this higher level of abstraction we think we behold clusters of these metaclusters. Not satisfied with the power of general relativity to explain the heavens and quantum theory to explain the subatomic realm, science dreams of uniting them into a theory of everything. Rung by rung, we climb up the ladder of abstraction to a vantage point from which everything is one—ruled by a higher symmetry. But it is a daring leap of faith to believe that all can be unified, that the brains that evolution gave us to steer our way through earthly labyrinths are equipped to perform this ultimate abstraction.

We are endowed by nature with this marvelous drive to find order. But we constantly bump up against our limits. Just as a frog can only see objects that move across its visual field with certain motions, so are we aware of only a tiny part of the electromagnetic spectrum. But we assume that we can supplement our senses with our minds and with our mathematics. We theorize about frequencies beyond our horizon, the invisible rays of infrared and ultraviolet light, of gamma and radio waves, and we build instruments to detect them. Then we weave stories about how these hidden worlds must be. When we fail to find symmetry in the world around us, we imagine extra dimensions, higher vantage points from which the world will regain its perfection. But for all our efforts, the whole truth will always elude us. Try as we might, we will never succeed in squeezing the immensity of creation into our tiny heads.

Gödel, after all, proved that mathematics itself has its limits. In his famous incompleteness theorem, he showed that no logical system can be used to prove its own consistency. To do so, one has to step out of the system, pop up a level, and study it from a higher vantage point, using more powerful mathematical tools. But proving the consistency of that system requires popping up another level, and so on, ad infinitum. Again, it seems, there is no highest vantage point, no ultimate abstraction. We are part of the very universe we are trying to understand. We are trapped inside the system. There is no Archimedean point on which to stand and behold all of creation.

Wittgenstein saw this in a more general way. In our search for order, we use language, verbal or mathematical, to make theories, representations—"pictures" of the world. But we have no way of explaining how it is that our theories are capable of describing what we take to be reality. That would require another theory, but then we would have to explain how that theory is capable of explaining the first theory—and that would require a third theory. And on and on. To keep from falling down this rabbit hole, we have to stop at some point, dig in our heels, and simply declare that our representation is valid. You cannot have a theory of representation—you can only represent. Any effort to explain the world must begin with a leap of faith.

We dream of a reason that is transcendent. But we are matter-bound. We build these great towers of abstraction, but ultimately they all rest on a platform of belief, the postulates that we must accept as true because there is no way ever to prove them. Unlike the scientists in Los Alamos or even the fundamentalist scholars, the Penitentes are content to surrender to the mystery, to accept with resignation that there is so much that can never be known. We can wonder at the rolls of the dice that led to the story of the miracle of Chimayo, or to a small group of men kneeling before the Cristo in the morada in Truchas. But to them their faith needs no explaining. It is something that has always been.

CONCLUSION:

_____

THE RUINS OF LOS ALAMOS

There is not much left of the village of Otowi, just an undulating landscape of mounds and depressions that the eyes of an archaeologist can identify as toppled walls and buried kivas, symmetry that is all but washed away. Scattered among these ruins are the shards of white pottery with black geometric designs so familiar to hikers on the Pajarito Plateau—more patterns left in pieces, a jigsaw puzzle that will never be completely reassembled. Lying halfway between San Ildefonso and Los Alamos, Otowi—just half a millennium old—stands as one more reminder of the fragility of the webs we weave.

Unlike the neighboring ruins of Tsankawi, built high on a mesa top, Otowi sat low to the ground, at the bottom of a ravine between two mesas. If a villager wanted to gain a higher vantage point, he could scramble across the arroyo that runs nearby, then climb up a rock slide to the base of Otowi Mesa, honeycombed with cliff dwellings. Here, high above the ground, the villagers left strange markings that we puzzle over today. Scratched into the walls and ceilings of the shallow caves are spirals, zigzags, concentric circles, checkerboards—the same kinds of simple designs one sees all over New Mexico.

Except for the most obvious examples—stick figures of people and birds and deer—we have little idea what these patterns mean. Whatever message was intended by these long-dead scriveners has been lost amid the noise. And yet we instantly recognize these markings as patterns, signs of an intelligence akin to our own—the bare bones of a signal arcing across the centuries. Though we will never know precisely what the Anasazi of Otowi were trying to say, we can feel the presence of other minds at work, kindred spirits engaging in the human game of playing with patterns. These people, too, were fascinated by symmetries. Their brains resonated to the repetition of simple patterns: V's repeated to form a zigzag, U's linked to produce an almost sinusoidal wave; right angles ascending to form a staircase, a pattern later echoed on Catholic churches throughout the Southwest; lefthand staircases mirrored by righthand staircases, forming the outline of a tower and demonstrating the concept of bilateral symmetry; concentric circles converging on their logical limit, the point—visual rhythms as mesmerizing as the chanting of a song or the cycles of a dance.

These people, about whom we know so little, had stumbled upon the mysterious power of representation. With half a dozen crude strokes cut into a cliff face with an obsidian knife, one can make a figure that looks like a human face: a circle with four marks standing for eyes, nose, and mouth. It doesn't matter that faces are never really circular or that facial features do not really look like dots and lines. As long as the marks are arranged in the same rough configuration one finds on living, breathing faces, the marks prod the mind with a feeling of recognition. Half a millennium later, through all the weathering, all the winters, the brain still responds.

Once people began scratching figures on cliffsides, it must have quickly become apparent that no two representations need to be precisely alike. The eyes can be closer together or farther apart. The nose can be evoked by a vertical mark or a dot, the mouth by a horizontal line or a circle. Within the brain's mental spaces, each of these configurations sits on the slopes of the same basin of attraction—this category we call face. Other neurological filters allow us to recognize circles and spirals. Viewed at the finest grain, each of these simple patterns—scratched into the hardened volcanic ash—is different, just as two seemingly identical letters in a newspaper, magnified a hundred times, are utterly unique.

Yet by ignoring a lot of detail—by coarse-graining—we can sense the similarity, the underlying theme. A signal can be detected amid the noise. No two spirals are ever precisely the same. By lumping together a wide range of approximations, we create this concept, spiral.

The next lesson a culture learns might be this: A marking need not bear any resemblance whatsoever to the meaning it is intended to evoke. It might be decided, by convention, that a left-hand spiral stands for evil and darkness, a right-hand spiral for goodness and light. Or one might stand for summer, the other for winter. The face is a representation; we recognize it because of its crude resemblance to the real thing. But the spiral is a symbol; its meaning must be assigned arbitrarily. It is a frozen accident.

Over the centuries the difference between representation and symbol tends to fade away. Once a meaning is established, the pictures we draw can become so stretched and distorted that their originators would barely recognize them. What begins as a representation, a crude picture, becomes stylized and abstracted until—as in the elaborate Chinese and Japanese ideographs called kanji—no more than a hint of the original image remains. But as long as we can distinguish one image from another, the meanings can be preserved.

Once representations cross the line into symbolism, there is no way to pull the meaning from the image itself. A circle divided into four quadrants might mean one of four things, depending on which section is marked with a dot. Since the meaning is not intrinsic in the markings, we must have a codebook to understand. For the Anasazi, any codebook that might have existed was apparently printed nowhere but in their memories, the delicate patterns of neural connections impressed on one generation after another—information that has been lost somewhere along the way, or sequestered among those with keys to the kiva.

Up the hill from Otowi, in Los Alamos, is an outpost of a civilization that has become so adept at the magic of symbolism that it has made the next great leap: symbols can be symbolized by other symbols. Here too the connection can be perfectly arbitrary. By using numbers to stand for letters, punctuation marks, and spaces, a book can be converted into a long number. Then this number can be converted again into the simplest possible code—one that recognizes nothing more sophisticated than presence or absence, 1 or 0. This is exactly what is done when a text is digitized and stored on a computer disk. The recorded knowledge of a civilization—every book in the British Library or the Library of Congress, every musical recording, every videotape and film—could conceivably be converted into a horrendously long number, bytes and bytes of information.

Suppose we wanted to transmit this knowledge, everything we had ever learned, to another world. First we would want to make the representation as compact as possible. By squeezing out redundancies we could compress the number so that it would occupy smaller and smaller spaces. In fact, if we are adept enough we can represent the number in a manner that requires almost no space whatsoever. We simply take the long string of digits and put a decimal point in front of it so that it becomes a fraction between 0 and 1, a mere point on a line. Then we choose a smooth stick and declare one end 0 and the other end 1. Measuring carefully, we make a notch in the stick—a point on the continuum representing the number. All of our history, our philosophy, our music, our art, our science—everything we know would be implicit in that single mark. To retrieve the world's knowledge, one would measure the distance of the notch from the end of the stick, then convert the number back into the books, the music, the images.

The success of the scheme would depend on the fineness of the mark and the exactness of the measurement. The slightest imprecision would cause whole Libraries of Alexandria to burn. And though we think of the stick as a continuum, quantum graininess might put a fundamental limit on just how fine the mark could be. But long before one approached these constrictions, it seems, an inconceivable amount of information could be recorded.

Suppose the medicine men of Otowi had discovered this trick. Suppose, contrary to all evidence, that they had developed a written language, a number system, and tools of enough precision to encode a single book of sacred knowledge into the notch of a prayer stick—the very book, perhaps, that explains what the symbols on the rock walls mean. And suppose a hiker, exploring one day in the caves above Otowi, found the stick. Could the knowledge be recovered?

First of all, we wouldn't know which end of the stick to measure from. Even if the ends were marked, how would we read the symbols? Since there are only two choices, we could measure from each end and compare the results. Perhaps a statistical analysis would show that one string was filled with regularities—patterns—while the other was largely random. But perhaps the message was so compressed that, read either way, it would appear random. We would have to decode both strings.

In any case, our next task would be to convert the number into something that could be translated back into language. We might assume the knowledge was coded using a base-ten number system, because these were fellow humans with ten fingers. But some cultures counted by the spaces between their fingers, base eight, or by the three segments of each of the four fingers on a single hand, base twelve. If that was the only problem, we could again try various possibilities, seeing which made sense and which nonsense. The insurmountable barrier would come when we tried to convert the numbers into a message. The only way we could know which numbers stood for which symbols is to look them up in a codebook. The codebook could be included somewhere in the string along with the sacred text, but how would we distinguish it from the rest of the message? How would we decode the codebook? Shades of Wittgenstein: when we try to describe how a representation represents, we fall into an infinite regress.

Without some kind of context, this long, long number left by our ancestors would be as impenetrable as a marble-smooth cliff face—as silent as the digital representation of the *St. Matthew Passion* without the machinery to convert it into song. How would we get a foothold? Unless we were lucky enough to find a Rosetta stone hidden in another cave somewhere, the number by itself would be meaningless. And even if we had the codebook, it is not clear how much of the message we deciphered would be meaningful without some knowledge of the culture of the people who encoded it. Since these were fellow humans, surely there would be things we had in common; certain assumptions could be made. But much of the meaning would depend on conventions. And many of these would be frozen accidents. As Wittgenstein wrote, if a lion could talk we could not understand him because we do not live in his world.

From the caves of Otowi, one can look across the ravine at the highway, carved into the side of another cliff, and follow it up to the laboratories of Los Alamos, accumulators and manufacturers of so much information. Just as the Anasazi left messages for us to ponder centuries later, scientists involved in the search for extraterrestrial intelligence, or Seti, send messages that they hope will make sense to creatures in other worlds. We hurl our prayer sticks into space hoping they will be intercepted. We broadcast radio signals into the aether; we etch marks in aluminum tablets and ferry them on space probes beyond the solar system. The messages scientists have sent over the years include stick figures of men and women, not so different from Anasazi rock art, and diagrams of the DNA double helix and of our solar system, accompanied by a star map giving our location in relation to a number of nearby pulsars, the rhythm of their blinking identified in binary code using the frequency of the hydrogen atom as the unit of time. Interceptors of one of our Voyager craft would be the proud possessors of a recording of Glenn Gould playing a Bach prelude. This is part of an album of earthly sounds and images whose liner notes include a diagram showing how to set up an appropriate record player, using an enclosed stylus and cartridge. Surely, we want to believe, there are universals, common denominators that might be recognized by intelligent life forms everywhere. We hope that our view of biological evolution and perhaps even quantum theory (à la Zurek, Hartle, and Gell-Mann) imply that the rise of information gatherers is inevitable. They too might conceive of the universe in terms of time and space and so be mesmerized by rhythm and pattern. They might feel stirrings of recognition when they saw at least some of our marks.

But there are good reasons not to get carried away. As we have seen, there are those who argue that this intelligence we pride ourselves in, this compulsion to find order, is no more than a quirk of evolution, an accident of biological history. And even if information gatherers abound, are we justified in supposing that they carve up the world the way our scientists do? Expecting galactic neighbors to recognize our signals as signals and figure out what they mean may be as hopelessly unrealistic as trying to decode a notch in a prayer stick. If this is so, then we can forget about sending double helices or diagrams of hydrogen atoms to other worlds— we might just as well send the pages of the Internal Revenue Service Code.

Even to a culture that amused itself with the material vibrations we call music, how meaningful would Gould playing Bach be without hundreds of years of musical history? One of the architects of the Seti project, the astronomer Frank Drake, concedes the possibility that the alien audience might not have invented/discovered music, or that it might perversely prefer to listen to the rhythm of the video images, but even that is assuming a very substantial expanse of common ground. Signals are worthless without a context. By now, millions of people have had the eerie experience of "downloading" a computer program over the telephone line. The bits of the code are transmitted in the form of tones, and when the message has been received a new icon appears on the screen.

Double-click to activate it, and, lo and behold, one has a new calculating machine. But the signal itself cannot calculate. If someone had tapped the line and captured the modem tones, he would not have all the information necessary to reconstruct the calculator. First of all, there weren't actually 1s and 0s in the phone line but rather tones arbitrarily assigned these values. One must already have the code. And even then the digital message would be useless without the context—the computer on which it was written to run. A Macintosh program won't run on an IBM.

Even if the meaning of our messages is indecipherable, we can still hold out hope that anyone or anything that we would consider intelligent—an information gatherer—would at least recognize the existence of patterns: regularities unlikely to have occurred by chance. As we ponder what about our knowledge is universal and what is contingent, we hope we can assume that symmetry itself is more than a human construct. Surely, if there is something fundamental in the universe besides matter and energy it is this thing we call pattern or form. Our science, our mathematics, our languages all are patterns of patterns. But where, in a material world, can something so seemingly ethereal as pattern exist?

The platonists have a ready answer: the patterns—shapes, numbers, symmetries, concepts—come before all else. They exist independently in a separate realm of pure idea. The material world is simply their shadow. When we do science or mathematics, we are reading the mind of God. Surely, the platonists might argue, there is something deep about the fact that the initial conditions of the big bang were just so that they gave rise to conscious beings with the power to understand creation. Unlike so many evolutionary biologists, they are likely to see consciousness as fundamental, not as an artifact, a contingency of evolution. They have little doubt that there must be other vessels of consciousness discovering the same mathematical laws. As evidence for this mystical belief, they point to how our brains seem to resonate with the very fabric of nature, using this seemingly natural faculty called mathematics to discover laws that apply throughout the universe, throughout time.

Of course, we do not really know that the truths are truly universal. Rather, we assume that the laws are the same everywhere and then interpret our observations accordingly (adding dark matter to make sure the galaxies spin right). Nor can we be sure just how deep or shallow this understanding really is. We have no standards to measure by. The scientists of Los Alamos divined enough about nature to produce a nuclear bomb. But perhaps to a higher intelligence, nuclear fission is the equivalent of fire. Deeper laws might be denied us because of the limits of our neural systems, the peculiarities of the mesh with which we categorize the world.

At the far extreme from the platonists and their mathematical god are the cultural constructivists, who hold that science and mathematics are human inventions—a set of conventions, or frozen accidents, as utterly contingent on the surrounding culture as is, say, British common law. Science, in this view, is a cultural construction built and enforced by the segments of society that wield the greatest power. Mathematics and science seem so effective because the high priests who practice these arts define the standards by which success is measured. A whole literary and philosophical movement has arisen to "deconstruct" these hidden ideologies, unearth ulterior motives so buried that they are all but indiscernible.

If the platonists are the true believers, then the constructivists are the village atheists. Can we find a middle ground between these two extremes—a way to separate the patterns we stamp on reality from the patterns that reality stamps on our minds?

Perhaps the patterns we discern are neither universal nor arbitrary, but the result of the intersection between our nervous systems and some kind of real world. Borrowing from Kant, we might begin to explain the seeming effectiveness of science and mathematics like this: We have these wired-in neurological filters (Kant called them the "a priori") constantly sifting the barrage of sensory impressions. Then it is the result of the sifting that we study. All else is ignored. Thus the patterns we find tell us as much about the filters as about the filtered. There is something a bit tautological about this. Perhaps when the phenomena of the universe seem to obey mathematical laws, it is because one product of the nervous system—mathematics—is recognizing another: the filtered sensory impressions. We are seeing the shadows of our own brains.

But one can entertain this view without being a solipsist. As long as we are willing to grant the existence of a real world out there, there is no reason to think that our mathematics and our representation schemes are completely arbitrary. The brain evolved in the world. It is molded by the environment. Thus we can expect our mental representations to bear some useful relationship to at least this tiny corner of a single galaxy.

By embracing this view, we abandon the idea of mind as a clear pool of water, passively reflecting truths that lie beyond the senses. The symmetries that so enchant us may be no more than good tools—compact ways for brains to store information. There need not be a true, platonic circle. The concept "circle" could be just a compression, an artifact of the brain's ability to coarse-grain. If we approach the world with too fine a grain we will have to catalog every "circle" we see as a separate object, delineating every point. By taking a kind of average and pretending that there is an ideal called "circle," we can just note the deviations. The brain is finite, so it must coarse-grain, sort things into categories as though there were ideal exemplars. Without this ability to recognize the approximations we call sameness, we would not long survive in the world.

Looked at this way, numbers, equations, and physical laws are neither ethereal objects in a platonic phantom zone nor cultural inventions like chess, but simply patterns of information—compressions—generated by an observer coming into contact with the world. They are configurations of bits, the simplest possible discrimination that can be made: a point in space is occupied or unoccupied, a quality is present or absent. And if information is physical, consisting of matter and energy, then we have come a long way toward exorcising Phaedrus's ghosts. The laws of physics are compressions made by information gatherers. They are stored in the forms of markings—in books, on magnetic tapes, in the brain. They are part of the physical world.

If there are other information gatherers in the universe, they may be making different compressions, carving up the universe in their own way. They may even have discovered entirely different ways of processing information. Some physicists believe it is possible to make quantum computers in which vast numbers of superposed calculations would interfere with one another—canceling and reinforcing until, like all the possible paths of a photon bouncing off a mirror, they yielded a single trajectory, the answer to an otherwise intractable problem. Harnessing the strange properties of quantum mechanics, they might cut to the heart of matters beyond mere brains and Turing machines. But even if different beings with different powers weave different patterns, and marvel at different symmetries, they would share the ability to make binary distinctions. That much, at least, would not be an artifact. They would not mark their bits with 1s and 0s but they would have discovered the notion of information, this most basic of all distinctions. If anything is universal, then perhaps it is this.

Or could information also be an artifact, another of our projections? After all, there are not really any 1s or 0s inside a digital computer, just voltages that we chop up by arbitrarily drawing a line and declaring everything below it 0, everything above it 1. Everything going on in the machine could conceivably be described in terms of continuous currents of electricity without recourse to this notion of information. Similarly, the operation of the eye or even the whole nervous system could be described entirely in terms of biochemistry with no talk whatsoever of signals. The replication of DNA and the transcription of proteins could be described in painstaking molecular detail without the convenience of symbols—G, C, A, and T. By ascending to a higher level of abstraction and analyzing nervous systems and cells in terms of information we are able to get a better grasp of the complexities unfolding. And we can use the idea of information to build devices like digital computers. But from this perspective, bits seem like human contrivances, an interpretation imposed by a mind.

If the human race were wiped from the earth, the computers would keep going until the energy supplies ran down. But without an interpreter, could they really be said to be processing something called information?

Sitting above the fallen walls of Otowi, one can imagine a time when its sister city, Los Alamos, has become one more lifeless ruin on the Pajarito Plateau, another ancient village for archaeologists to unearth. Basket Maker I to III and Pueblo I to V are followed by Information Gatherer I and II.

The weapons laboratories of Los Alamos stand as a reminder that our very power as pattern finders can work against us, that it is possible to discern enough of the universe's underlying order to tap energy so powerful that it can destroy its discoverers or slowly poison them with its waste. Perhaps our world will be leveled by nuclear blasts and fallout. Or, more eerily, by an invincible toxin or virus or a climatic change, a silent killer that will eliminate all the people, all the interpreters, while leaving our creations intact.

Something very much like this seems to have happened in David Markson's disturbing novel *Wittgenstein's Mistress,* though here there are not even dead bodies lying around. For reasons left unexplained, a woman— a painter, we learn—is the last person left on the planet, as though everyone else has simply vanished or been raptured away. The radios she switches on are silent. She can drink from the rivers, the Po, the Mississippi, the Seine.

This lone survivor drives from place to place in abandoned cars, listening to whatever tape cassette happened to be playing when the driver disappeared and the engine died. She crosses oceans in boats, visiting the great art museums of the world. She lives in the Louvre, the Tate, the Metropolitan, where she burns picture frames to stay warm, shooting out a skylight for ventilation. She is careful not to disturb the paintings, even though there is no one else left to see them.

As she travels she carries in her head a load of cultural baggage—a smattering of classical history, anecdotes about famous painters and philosophers, scraps of Vivaldi and Joan Baez. The only words now being produced on earth are those in the interior monologue running inside her brain. But all of this is imperfectly recorded and deteriorating rapidly. She finds herself confusing the Aegean and the Adriatic, Michelangelo and Leonardo da Vinci. Is the Hermitage in Moscow or St. Petersburg? The distinctions are blurring together, the bits are being dissipated. Soon, this last repository of culture will be gone. The paintings will truly lose their frames. They won't be art anymore, just oily pigments smeared on patches of canvas.

Recordings that will never be played, books that will never be read— the only way we can believe any of this detritus will remain meaningful is to imagine the possibility of future visitors, from another world, who hold enough in common with the human race to enable them to reconstruct the codebook, recognize messages as messages, art as art.

What would these creatures make of our science? Aliens trying to decode our records might recognize what seemed to be deliberate patterns in the markings of ink on pages or the fluctuating magnetic fields of computer disks (though, again, if the information had been highly compressed, it would be harder and harder to distinguish from randomness). If they persisted, would they find truths to marvel at, signs of kindred minds? Or would they even recognize the books and tapes as things that might be worth analyzing? One can't go around measuring every notch on every stick.

If we are alone in the universe, then our civilization is like Wittgenstein's mistress. When we are gone, the meaning will disintegrate. For a while the computers will continue to run, with no one to make sense of the output. But without hordes of Maxwellian demons building and maintaining little pockets of order, entropy will overcome. The batteries will disintegrate, positive and negative charges becoming uselessly intermingled. Fuel tanks will rust, their contents leaking irretrievably into the ground. The patterns of magnetism etched on the tapes will break down, the distinctions lost—was this spot magnetized or unmagnetized? The paintings will slowly disappear, along with the texts that explain them and the texts that explain the texts.

Sometimes the intelligence of our species seems like a tiny flame flickering on the periphery of a vast blackness, trying to illuminate the void. Who gave us this burden? Will anyone or anything beyond our celestial campsite ever care? If this web is just something we are spinning for our own amusement, it will die along with its creators.

But when one thinks of the hunger for order that drove the Anasazi cliff dwellers to etch their crude symmetries on rock walls, that propels the dancers of San Ildefonso in their intricate choreography or the brothers of Truchas through the haunting verses of the *alabados* and the fourteen stations of the cross; when one thinks of the burning curiosity that drives the physicists on the mesa top in Los Alamos ever higher into the dizzying altitudes of abstraction—it is hard not to believe that we are all participating in something universal, something holy, that the pageant must unfold beyond our planet. Perhaps we are merely one among a myriad of players—gathering bits, abstracting concepts, building great edifices of theory, these mathematical Towers of Babel, that reach higher and higher above the plains. High enough, perhaps, to make out, just barely, the rhythm of other dancers, the flickering of other fires.

ACKNOWLEDGMENTS

From the moment I returned to New Mexico, I was struck by the generosity of the people I had come to write about. Several scientists agreed to meet with me regularly to talk about the many subjects threaded through this book. During hikes in the Sangre de Cristos, including an ascension of Santa Fe Baldy, Melanie Mitchell spoke about cellular automata and computer science. Seth Lloyd discussed the curious links between information and quantum mechanics over occasional dinners and coffee at the Aztec Cafe and Downtown Subscription. John Casti, my neighbor on San Acacio Street, was an important source of information on the problems of prediction in science and a generous lender of books from his extensive library. Over some memorable dinners, Walter Fontana offered his characteristically original ideas on biology, semiotics, and the philosophy of science. James Crutchfield talked about complexity on two hikes to Atalaya Peak (and another to the far reaches of New York's East Village). Stuart Kauffman held forth on the origins of order in a year-long running conversation whose venues included the patio restaurant at La Posada, the Cloudcliff Bakery, the meadows of Puerto Nambe, and

Penitente Peak. Doyne Farmer and Norman Packard met with me several times to talk about a range of subjects including the origin of life, the predictability of roulette, and chaos in the financial markets. I benefited enormously from a series of superb lectures by Harold Morowitz at the Santa Fe Institute called "Beginning Biology from an Advanced Point of View." John Miller was my consultant on economics, and Alfonso Ortiz discussed his work on Native American anthropology. Della Vigil Ulibarri and her family kindly invited me to Truchas for an unforgettable Good Friday.

Along the way, I also had enlightening conversations with many other denizens and visitors to Santa Fe, including Bryan Arthur, Leo Buss, Carlton Caves, Gregory Chaitin, Paul and Patricia Churchland, Betsy Corcoran, George Cowan, Jean Czerlinski, Murray Gell-Mann, Brian Goodwin, Jonathan Haas, Steven Harnad, James Hartle, John Holland, Atlee Jackson, Erica Jen, Tim Kohler, Rolf Landauer, Christopher Langton, Kristian Lindgren, Gunther Mahler, Alexis Manaster-Ramer, Pamela McCorduck, Richard Palmer, Thea Philliou, Julie Pullen, Tom Ray, Bruce Sawhill, Joseph Traub, Marta Weigle, David Wolpert, Michael Zeilick, Wojciech Zurek, and many others.

In writing the book, I constantly felt the presence of two apparitions hovering over my shoulder: the Scientist, always calling for more precision and rigor, and the General Reader, begging for simplification and willing to settle sometimes for an illuminating metaphor. Hoping against hope to satisfy both, I gave out pieces of the manuscript to a number of volunteer readers, who offered many helpful suggestions. Belonging to the first category were Charles Bennett, Emory Bunn, John Casti, James Crutchfield, Walter Fontana, Murray Gell-Mann, James Hartle, Stuart Kauffman, Rolf Landauer, Christopher Langton, Seth Lloyd, Melanie Mitchell, Harold Morowitz, Alfonso Ortiz, Norman Packard, Julie Pullen, Aephraim Steinberg, Marta Weigle, and Wojciech Zurek. Belonging to the second category were Malcolm Browne, Timothy Ferris, Richard Freedman, Joseph E. Johnson, Alan Lappin, Douglas Maret, Nancy Maret, and Robert Wright. I thank them all for an enormous amount of free labor. During these readings, the book has been in constant flux. The final gel is my responsibility alone. I've worked hard to weed out errors, but some doubtless remain. I'd appreciate hearing from readers through E-mail (johnson@nytimes.com or johnson@santafe.edu), the Worldwide Web (http://www.santafe.edu/~johnson/), or my American publisher, Alfred A. Knopf.

The workshops, conferences, and lectures at the Santa Fe Institute were an indispensable part of my research. I would like to thank the people who helped me find my way around the place, particularly Ginger Richardson, Mike Simmons, Andi Sutherland, and Susan Wider.

At *The New York Times,* I would like to thank Dan Lewis, Mike Leahy, and John Lee for supporting my request for a leave of absence, and Max Frankel and Joseph Lelyveld for granting it.

At Knopf, I'd like to thank the production editor, Melvin Rosenthal, and the copy editor, Edward Johnson, for their intelligent touch.

As always, I thank my editor, Jon Segal, and my agent, Esther Newberg, for being such talented and patient people.

NOTES

Unless otherwise indicated, all quotations come from interviews and conversations that took place between 1992 and 1994 in northern New Mexico. Full citations to books are given in the bibliography, which is divided into two parts: Books About Science and Philosophy, and Books About New Mexico and the Southwest.

The epigraph is from Franc Johnson Newcomb, *Navaho Folk Tales* (Santa Fe: Museum of Navaho Ceremonial Art, 1967), p. 83; quoted in Trudy Griffin-Pierce, *Earth Is My Mother, Sky Is My Father: Space, Time, and Astronomy in Navajo Sandpainting*, p. 142.

Introduction: Kivas, Moradas, and the Secrets of the Nuclear Age

PAGE

3 To get a sense of the dense layers of history and geology that distinguish New Mexico, there is little more enjoyable than driving its roads and highways in the company of two books and an unusually fine map: *New Mexico Place Names: A Geographical Dictionary*, prepared in the 1930s by T. M. Pearce and the New Mexico Writers' Project; Halka Chronic's *Roadside Geology of New Mexico;* and *The Highroad*

Large Scale Map of North Central New Mexico, 4th ed., published by Highroad Publications in Albuquerque. Highroad maps are also available for Santa Fe and other parts of New Mexico.

There is endless debate over whether, on a day trip from Santa Fe to Taos, it is better to take the High Road, through the old villages of the Sangre de Cristo Mountains, and return by the Low Road that runs along the Rio Grande, or vice versa. Starting out on the mountain route gives you the sense of leaving modern life behind as you ascend into an earlier, more peaceful time. And it has this advantage: on the return trip, the sun is low in the sky, illuminating the plains south of Taos to a golden sheen and highlighting the rough texture of the Rio Grande Box—a spectacular way to end a day of exploration. On the other hand, the low route will get you to Taos faster, and on the leisurely return, the descent from Truchas to Chimayo offers stunning views of the Rio Grande Valley, Black Mesa, and the kingdom of the Tewa.

PART 1. FOUR MAGIC MOUNTAINS

Maxwell's quote can be found in Gerald Edelman's *Bright Air, Brilliant Fire,* p. 16.

Chapter 1: Phaedrus's Ghosts

11 Two ideal spots for watching the sunset are the Cross of the Martyrs, on a hill just north of downtown Santa Fe, and the Bell Tower bar at La Fonda, across from the southeast corner of the plaza.

13 There are many good histories of New Mexico. I relied on two recently published ones: Ramón A. Gutiérrez's *When Jesus Came, the Corn Mothers Went Away* and Joe S. Sando's *Pueblo Nations.* Seminal work that all historians depend on has been done by Marc Simmons and Fray Angelico Chavez.

Trudy Griffin-Pearce's beautiful memoir and study *Earth Is My Mother, Sky Is My Father* describes how the Navajos carved up the sky.

15 Alfonso Ortiz's *The Tewa World* elegantly describes the intricate world view of the Tewa Indians. For further references, see my notes to "San Ildefonso Interlude."

17 Some of the ideas of information physics are explored in *Complexity, Entropy and the Physics of Information,* edited by Wojciech Zurek.

18 Chaos and nonlinear dynamics are the subject of James Gleick's *Chaos,* David Ruelle's *Chance and Chaos,* and Ian Stewart's *Does God Play Dice?*

For a sampling of the Santa Fe Institute's world view, I suggest two anthologies of scientific papers: *Emergent Computation,* edited by Stephanie Forrest, and *Emerging Syntheses in Science,* edited by David Pines. Murray Gell-Mann's engaging book *The Quark and the Jaguar* describes his own ideas on complexity. Two journalistic accounts of the institute, both entitled *Complexity,* also evoke the spirit of the place; one is by Mitchell Waldrop, the other by Roger Lewin. Also recommended is Steven Levy's *Artificial Life.*

19 The classic study of the Penitentes is Marta Weigle's *Brothers of Light, Brothers of Blood.* Life in the Sangre de Cristos is eloquently captured in words and images in *River of Traps,* by William deBuys and Alex Harris, a writer and a photographer who lived in the village of El Valle, and in several chapters of deBuys's excellent *Enchantment and Exploitation.* For more references, see the notes for "Truchas Interlude."

20 For an interesting analysis of the Fiesta, see Ronald Grimes, *Symbol and Conquest: Public Ritual and Drama in Santa Fe, New Mexico.*

24 The Robert Pirsig quote is on pages 38–39 of the Morrow edition of *Zen and the Art of Motorcycle Maintenance.*

27 Tsankawi Mesa is an unattached outpost of Bandelier National Monument. It can be reached by driving from Santa Fe toward Los Alamos. Beyond San Ildefonso, the highway climbs to the top of the Pajarito Plateau and forks. Take Highway 4 toward Bandelier. Less than a mile from the interchange, watch for the parking lot for Tsankawi on the left side. (The main monument is still twelve miles farther down the road.)

28 Von Däniken's *Chariots of the Gods? Unsolved Mysteries of the Past,* translated by Michael Heron, was published by Putnam in 1970.

Chapter 2. The Depth of the Atom

29 Coronado's adventures are described in numerous histories, including Herbert Eugene Bolton's *Coronado: Knight of Pueblos and Plains.*

30 For the theory that the Grand Canyon was carved by the Great Flood, see John C. Whitcomb and Henry M. Morris, *The Genesis Flood: The Biblical Record and Its Scientific Implications* (Phillipsburg, N.J.: Presbyterian and Reformed Publishing, 1961). The controversy between the geological catastrophists and the uniformitarians is nicely drawn in Chapter 33 of Stephen Mason's indispensable *A History of the Sciences.*

For an engaging account of a Grand Canyon raft trip and the philosophizing it evoked among a group of scientists, see William Calvin's *The River That Flows Uphill.*

32 Lake Peak and the stunning views it affords can be reached in half a day by driving up Hyde Park Road to the ski basin above Santa Fe. From there one can take a chairlift, which runs during the summer, to the top of the slopes, then hike north toward the microwave towers atop Tesuque Peak. From here the Skyline Trail be-

gins, entering the Pecos Wilderness and heading northeast toward Lake Peak. This is a fairly grueling climb, and the granite escarpment that forms Lake Peak is steep and dangerous to traverse. For those who have a full day of time and stamina, the Skyline Trail continues onward to Penitente Peak, then winds down into the peaceful meadow called Puerto Nambe, from which one can return by the Windsor Trail to the ski basin parking lot, a total distance of about eight miles. (For details, see the map *Pecos Wilderness* published by the United States Forest Service.) Avoid summer lightning storms, which can be deadly on these exposed heights, and bring a topographical map and compass. The first time I tried this hike I mistook a farther peak for Penitente, and my mental map of the area was so poorly drawn that I confused north and south, getting so lost that I eventually had to follow the Santa Fe River some twenty-five miles back to town, emerging onto Upper Canyon Road at one in the morning.

36–7 The history of particle physics is authoritatively told in Abraham Pais's *Inward Bound*. It is also nicely portrayed in Robert Crease and Charles Mann's *Second Creation*. The early years are described in Stephen Mason's *A History of the Sciences* and the early experiments with the electron and other particles in Steven Weinberg's *The Discovery of Subatomic Particles*. Hans Christian von Bayer's elegantly written *Taming the Atom* tells about very recent work, in which atoms are isolated and studied individually.

42 For the flavor of the early days of quantum theory, I recommend Werner Heisenberg's books *Physics and Philosophy* and *Physics and Beyond*, Abraham Pais's *Niels Bohr's Times*, and two fine histories: Barbara Lovett Cline's *Men Who Made a New Physics* and George Gamow's *Thirty Years That Shook Physics*. Many of the ideas of twentieth-century physics are nicely elaborated in books by physicists: John Barrow's *The World Within the World*, Richard Morris's *The Edges of Science*, Heinz Pagels's *Perfect Symmetry*, Frank Wilczek and Betsy Devine's *Longing for the Harmonies*, and Anthony Zee's *Fearful Symmetry*.

45 Sometimes when photons are spoken of as waves, spin can be considered analogous to polarization.

52 In addition to the general histories mentioned above, the story of the quark can be found in Harald Fritzsch's *Quarks* and Michael Riordan's *The Hunting of the Quark*.

53 Actually the colors the Tewa assigned to the cardinal directions might not be as arbitrary as the colors assigned to quarks. Blue (north) is the color of ice; red (south) is the color of fire; east (white) is where the sun rises; west (yellow) is where the sun sets.

56 The problem of retrospective realism is explored in *Constructing Quarks* by Andrew Pickering. The science of experimental particle physics is described in Christine Sutton's *The Particle Connection* and in the beautifully illustrated book *The*

Particle Explosion, which she wrote with Frank Close and Michael Marten. For a sociological study of the field, see Sharon Traweek's *Beamtimes and Lifetimes.*

The quest for a grand unified theory is elegantly described in Steven Weinberg's *Dreams of a Final Theory.* For skeptical and equally elegant views see John Barrow's *Theories of Everything* and David Lindley's *The End of Physics.*

Chapter 3. The Height of the Sky

60 The distances to stars vary from one guidebook to another. I mostly relied on Robert Burnham, Jr.'s, *Burnham's Celestial Handbook* and Jay Pasachoff and Donald Menzel's *A Field Guide to the Stars and Planets.* The story of how we developed our picture of the sky is engagingly told in Timothy Ferris's *Coming of Age in the Milky Way,* David Layzer's *Constructing the Universe,* Marcia Bartusiak's *Thursday's Universe,* and Dennis Overbye's *Lonely Hearts of the Cosmos.* A very good beginning astronomy text is Michael Zeilik and John Gaustad's *Astronomy: The Cosmic Perspective.*

69 The story of Penzias and Wilson is described in many books, including Jeremy Bernstein's *Three Degrees Above Zero.*

71 The story of Einstein and the theories of relativity is nicely told in Jeremy Bernstein's *Einstein,* in Lincoln Barnett's *The Universe and Dr. Einstein,* and, with great rigor, in Abraham Pais's biography *Subtle Is the Lord.* For a very clear mathematical treatment of the special theory of relativity, see Edwin Taylor and John Archibald Wheeler's *Spacetime Physics.*

"full of this wonderful medium": quoted in Pais, *Niels Bohr's Times,* p. 67.

74 Miller's attempt to revive the aether hypothesis and Einstein's quick rebuttal are described in Pais, *Subtle Is the Lord,* pp. 113–14.

75 In addition to the general histories of astronomy and particle physics mentioned above, good accounts of the development of the big bang theory include Alan Lightman's *Ancient Light* and Steven Weinberg's classic *The First Three Minutes.* The story of dark matter is authoritatively told in Lawrence Krauss's *The Fifth Essence.* An especially lucid and compact account is included in Alan Dressler's *Voyage to the Great Attractor.*

79 The Cobe story is told in *Wrinkles in Time* by George Smoot and Keay Davidson and *Through a Universe Darkly* by Marcia Bartusiak.

Tesuque Interlude: The Riddle of the Camel

86 Those tempted to try their hand at bingo should be warned that it is not as easy as it looks. Veteran players have developed impressive information-processing abilities, allowing them to simultaneously mark off half a dozen cards the instant each number is called, leaving us novices helplessly in the dust.

88 Farmer and Packard's attempts to beat the game of roulette are compellingly described in Thomas Bass's *The Eudaemonic Pie,* published in Britain as *The Newtonian Casino.* The book also describes the efforts of the Santa Cruz collective's illustrious predecessors, from the Bernoulli brothers to Claude Shannon; see pp. 119 and 126–31.

92 The contributions of Crutchfield, Farmer, Packard, and Shaw to the science of nonlinear dynamics are described on pages 243–72 of James Gleick's *Chaos.*

94 Crutchfield's calculation involving the electron at the edge of the galaxy is in "Chaos," by Crutchfield, Farmer, Packard, and Shaw, *Scientific American,* December 1986.

98 Chaitin's work on algorithmic information is described for a lay audience in Chapter 6 of John Casti's *Searching for Certainty* and in Chaitin, "Randomness in Arithmetic," *Scientific American,* July 1988. See also Chaitin's books *Information, Randomness, and Incompleteness,* 2nd ed. (Singapore: World Scientific, 1990); and *Algorithmic Information Theory* (Cambridge: Cambridge University Press, 1987).

"any more than a one-hundred-pound pregnant woman": Casti, *Searching for Certainty,* p. 355.

99 "God not only plays dice in quantum mechanics": ibid., p. 390.

100 The studies on the effects of random reinforcement on pigeons were done by Richard Herrnstein in the 1950s.

100–1 The view that the stock market is a perfectly efficient processor of information is argued in Burton Malkiel's *A Random Walk down Wall Street.* For a taste of some of the alternative theories being developed at the Santa Fe Institute and elsewhere, see the papers collected by Philip Anderson, Kenneth Arrow, and David Pines in *The Economy as an Evolving Complex System.* Also see Chapter 4 of Casti's *Searching for Certainty.*

104 if "we peel back the next layer of the scientific onion": Farmer and J. J. Sidorowich, "Can New Approaches to Nonlinear Modeling Improve Economic Forecasts?," in Anderson, Arrow, and Pines, *The Economy as an Evolving Complex System,* p. 101.

PART 2. "THE COLD, GRAY CAVE OF ABSTRACTION"

The quote from Poincaré can be found on page 68 of the English translation of his book *Science and Method.*

Chapter 4. The Demonology of Information

110 Zurek's manifesto and the other papers delivered at the conference are collected in the volume he edited: *Complexity, Entropy and the Physics of Information.*

112 Cosmologists have recently decided that the inevitable heat death of the universe may be a myth. While entropy will indeed continue to increase in an expanding universe, the argument goes, the maximum amount of entropy the universe can encompass will increase even faster. See Chapter 1 of John Barrow's *The Origin of the Universe.*

114 The Dallas conference "The Symbiosis of Physics and Information" was held in October 1992.

The trajectory of ideas leading from Maxwell's original demon to the work of Bennett, Landauer, Zurek, and others is laid out in the lucid introduction to Harvey Leff and Andrew Rex's *Maxwell's Demon,* which includes the important papers. The relationship between entropy and information is explored in three excellent books, Jeremy Campbell's *Grammatical Man,* Robert Wright's *Three Scientists and Their Gods,* and Peter Coveney and Roger Highfield's *The Arrow of Time.*

117 this "very observant and neat-fingered" being: Leff and Rex, *Maxwell's Demon,* p. 39.

thermodynamics "has the same degree of truth": ibid., p. 39.

118 "A memorandum-book does not": ibid., p. 41.

119 "I do not see why even intelligence might not be dispensed with": ibid., p. 43.

"such a device might, perhaps": quoted in Szilard's paper, ibid., p. 125.

120 It is not clear from Szilard's paper whether he anticipated, as Bennett and Landauer later did, that it was the erasure, not the gathering of information, in which the thermodynamic price must be paid.

121 Brillouin and Gabor's writings, "Maxwell's Demon Cannot Operate" and "Light and Information," also appear in Leff and Rex's anthology.

For an example of the attempts to circumvent Maxwell's demon without introducing information, see "Entropy, Information and Szilard's Paradox" by J. M. Jauch and J. G. Báron, collected in Leff and Rex.

124 Landauer's work is described in his paper "Irreversibility and Heat Generation in the Computing Process," in Leff and Rex. His ideas and those of Bennett are explored in Wright's *Three Scientists and Their Gods.*

125 Bennett's crucial realization about reversibility and erasure is described in his papers "Logical Reversibility of Computation," "The Thermodynamics of Computation—A Review," and "Notes on the History of Reversible Computation," all conveniently collected in Leff and Rex.

127 Some cosmologists have argued that an oscillating universe—one which cycles from big bang to big crunch and back to big bang again—would indeed be reset with each incarnation. But our own cycle would still be fated to increase in entropy.

127–8 Zurek's notion of physical entropy is introduced in his paper "Algorithmic Randomness and Physical Entropy," *Physical Review A* 40 (1989): 4731–51.

129 Zurek's superintelligent demon, who still can't beat the second law, is described in his paper "Thermodynamic Cost of Computation, Algorithmic Complexity, and the Information Metric," *Nature* 341 (1989): 119–24.

Chapter 5. The Undetermined World

132 QED is described nowhere more elegantly than in Richard Feynman's 1985 book by that name, which is an edited transcription of the first Alix G. Mautner lectures at UCLA.

137 "Physicists used to seek picturable mechanisms": Russell McCormmach, *Night Thoughts of a Classical Physicist*, pp. 133–34.

139 Having mathematics without understanding what it means: Joseph Schwartz considers this dilemma of modern physics in *The Creative Moment*.

"Over his lifetime, physics had taken a turn": ibid., p. 64.

140 Among the best popular accounts of quantum theory are Chapter 7 of John Casti's *Paradigms Lost*, Chapter 6 of Richard Feynman's *The Character of Physical Law*, Nick Herbert's *Quantum Reality*, George Greenstein's *The Symbiotic Universe*, and J. M. Jauch's Galilean dialogue, *Are Quanta Real?* Papers on the measurement problem are collected in John Archibald Wheeler and Wojciech Zurek's *Quantum Theory and Measurement*.

143 "It is as if the mere possibility": "Six Possible Worlds of Quantum Mechanics," in J. S. Bell, *Speakable and Unspeakable in Quantum Mechanics*, p. 185.

145 The analogy between quantum waves and musical waves is clearly drawn in Nick Herbert's *Quantum Reality*.

146 "We must be clear that, when it comes to atoms": Heisenberg, *Physics and Beyond*, p. 41.

147 "What we learn about is not nature itself": Herbert, *Quantum Reality*, p. 172.

149 "No question, no answer.": Wheeler, "Information, Physics, Quantum: The Search for Links," in Zurek, *Complexity, Entropy and the Physics of Information*, p. 11.

Hugh Everett's interpretation of quantum theory is explored in a collection of papers edited by Bryce DeWitt and Neill Graham: *The Many-Worlds Interpretation of Quantum Mechanics*.

150–1 The Zurek and Lloyd anecdote is recounted somewhat differently in Gell-Mann, *The Quark and the Jaguar*, p. 138.

152 Hidden variable theories and Bell's theorem are explained in the books by Casti and Herbert, mentioned above, and in Bell's *Speakable and Unspeakable in Quantum Mechanics*.

153 This version of the EPR paradox was devised by David Bohm.

Quantum teleportation is described in Charles H. Bennett, Gilles Brassard, Claude Crèpeau, Richard Jozsa, Asher Peres, and William Wootters, "Teleporting an Unknown Quantum State via Dual Classical and Einstein-Podolsky-Rosen Channels," *Phys. Rev. Lett.* 70 (1993):1895–99. The difference between classical bits and quantum bits is also explored in Bennett's work on quantum cryptography. For a semipopular review of the subject, see Bennett, "Quantum Cryptography: Uncertainty in the Service of Privacy," *Science* 257 (1992):752–53.

Crutchfield's remarks were made at a Santa Fe Institute conference called "Integrative Themes," at Sol y Sombra, an estate on Old Santa Fe Trail, July 1992. For a collection of papers by the participants, see George Cowan, David Pines, and David Meltzer, *Integrative Themes*.

Chapter 6. The Democracy of Measurement

160 "time and space are really only statistical concepts": Heisenberg's letter to Pauli, dated October 28, 1926, is quoted in Crease and Mann, *Second Creation*, p. 63.

162 Zurek's quantum version of the demon problem is described in his paper "Maxwell's Demon, Szilard's Engine, and Quantum Measurements," in Leff and Rex, *Maxwell's Demon*.

162–3 Seminal work on decoherence was done by Robert Griffiths, Roland Omnès, Erich Joos, and Dieter Zeh. (For references, see the citations in the following papers by Zurek, Hartle, and Gell-Mann.) Zurek's ideas on decoherence are introduced in his papers "Pointer Basis of Quantum Apparatus: Into What Mixture Does the Wavepacket Collapse?" *Physical Review D*, 24 (1981):1516–25, and "Environment-Induced Superselection," *Physical Review D* 26 (1982): 1862–80. For a fairly accessible review of the work, see "Decoherence and the Transition from Quantum to Classical," *Physics Today* 44 (October 1991): 36–44.

165 "The essence is that the environment 'knows' ": Unless otherwise indicated, all quotations are from interviews with Zurek conducted in Los Alamos and Santa Fe in September, October, and November of 1992.

"It is as if the 'watchful eye' of the environment": Zurek, *Complexity, Entropy and the Physics of Information*, p. viii.

166 "Our senses did not evolve for the purpose": Zurek, "Decoherence and the Transition from Quantum to Classical, p. 44."

167 Gell-Mann's ideas on quantum theory and decoherence are described for a lay audience in Chapter 11 of *The Quark and the Jaguar*.

168 For a description of Gell-Mann and Hartle's work on decoherence, see their paper "Quantum Mechanics in Light of Quantum Cosmology," in Zurek, *Complexity, Entropy and the Physics of Information*. Also see James Hartle, "The Quantum Mechanics of Closed Systems," in *Directions in General Relativity*, Volume 1: *A Symposium and Collection of Essays in Honor of Professor Charles W. Misner's 60th Birth-*

day, edited by B. L. Hu, M. P. Ryan, and C. V. Vishveshwara (Cambridge: Cambridge University Press, 1993).

168 "What was being measured when the universe": This and all subsequent quotes are from a conversation with Hartle in December 1992 in Santa Fe.

171 The anthropic principle is explored to great depths in John Barrow and Frank Tipler's *The Anthropic Cosmological Principle.*

173 Hawking and Hartle's origin theory is described in Chapter 2 of Paul Davies's fine book *The Mind of God* and toward the end of Chapter 8 in Hawking's *Brief History of Time.*

175 "unnecessary intellectual baggage": quoted in Hartle's paper "Excess Baggage," delivered at a sixtieth-birthday celebration for Gell-Mann at Caltech.

San Ildefonso Interlude: The Mystery of Other Minds

179 The description of San Ildefonso's feast day is based on a visit in January 1993. My description of the buffalo dance is supplemented by that of Edith Warner in Peggy Pond Church's evocative memoir *The House at Otowi Bridge* and Elsie Clews Parsons in *The Social Organization of the Tewa of New Mexico.* I also relied on Gertrude Kurath and Antonio Garcia's *Music and Dance of the Tewa Pueblos,* Jill Sweet's *Dances of the Tewa Pueblo Indians,* and Erna Fergusson's *Dancing Gods.*

182 My rendition of the Tewa creation myth is compiled from several sources, principally Alfonso Ortiz's *The Tewa World* and Parsons's *Social Organization of the Tewa.*

185 The intricate Tewa cosmology is beautifully delineated in Ortiz's *Tewa World,* which also lays out the clan system and the hierarchical social structure. Also see Parsons's book, William Whitman's *The Pueblo Indians of San Ildefonso,* and Edward Dozier's *The Pueblo Indians of North America.*

186 Bandelier's novel is called *The Delight Makers.*

187 Tewa constellations are described in John Peabody Harrington's Smithsonian report, "The Ethnogeography of the Tewa Indians," as are the finely textured geographical features. Harrington, who was also coauthor of Smithsonian reports on Tewa botany and zoology, was an obsessive compiler and hoarder of information, so afraid that others would steal his data that his secrecy was said to border on paranoia. In her haunting memoir *Encounter with an Angry God,* Carobeth Laird describes the ordeals of living and working with Harrington, whom she met and married when she was a young student.

189 "The difficulties encountered have been many": the quote is from Harrington's report, p. 37.

The anthropologists' theories of where the Tewa came from can be found in "Prehistory: Eastern Anasazi," by Linda S. Cordell, in *Handbook of North American*

Indians: Southwest, Vol. 9, edited by Alfonso Ortiz, pp. 131–51. This voluminous work is an indispensable reference on the history and the beliefs of all the Southwestern pueblos. For details about Chaco Canyon, see Kendrick Frazier's *People of Chaco.*

For a description of the geographical distribution of Pueblo languages and beliefs, see Fred Eggan, "Pueblos: Introduction," in Ortiz, *Handbook of North American Indians,* pp. 206–23.

The lost pueblo of Pecos is described in John Kessell's *Kiva, Cross, and Crown,* a well-written and lavishly illustrated account first published by the National Park Service.

191 The history of San Ildefonso is described in a chapter by Sandra A. Edelman in Ortiz's *Handbook,* pp. 308–16, and in Whitman's *Pueblo Indians of San Ildefonso.* Whitman, his wife, and their three children lived at the pueblo in 1936 and 1937, before distrust of anthropologists made such close contact impossible.

The Tewa names for the months are recorded in Harrington's "Ethnogeography of the Tewa Indians."

The encounters between the pueblo Indians and the Spanish are described in many accounts, including Ramón Gutiérrez's *When Jesus Came, the Corn Mothers Went Away* and Joe Sando's *Pueblo Nations.*

Parsons's report on the altered Tewa names is in *Social Organization of the Tewa.*

193 The story of Frank Hamilton Cushing is frequently recounted; see, for example, Barbara Tedlock's *The Beautiful and the Dangerous* and Ortiz's *Handbook of North American Indians,* which includes a chapter by Marc Simmons, "History of the Pueblos Since 1821," pp. 206–23, describing encounters between Indians and anthropologists.

194 "The women were particularly timid": Parsons, *Social Organization of the Tewa,* pp. 7–8.

María Martínez's story is told in Alice Marriott's *María: the Potter of San Ildefonso* and, to a lesser extent, in Whitman's *Pueblo Indians of San Ildefonso.*

195 Spanish and Indian beliefs in witches are the subject of Marc Simmons's *Witchcraft in the Southwest.*

The story of the factional wars of San Ildefonso is told in Whitman's *Pueblo Indians of San Ildefonso,* in the chapter on San Ildefonso in Ortiz's *Handbook of North American Indians,* and in other sources.

196–7 The story of Edith Warner and her friendships with the physicists of Los Alamos and the people of San Ildefonso became the basis for her friend Peggy Pond Church's book *The House at Otowi Bridge.* Frank Waters fictionalized Warner's story in his novel *The Woman at Otowi Crossing.*

197 Stories about San Ildefonso's claims against Los Alamos were regularly reported in Santa Fe's daily newspaper *The New Mexican* in 1992 and 1993.

200 "impinges on experience only along the edges": Quine is quoted in A. J. Ayer, *Philosophy in the Twentieth Century*, pp. 247–48.

202 My description of the Santo Domingo dances is from a visit to the pueblo in December 1992.

203 The account of the dance at Picuris is from a visit in August 1992.

The story of the former governor of Tesuque was told to me in an interview in November 1992.

204 For a brief account of Whorf's ideas, see his essay "An American Indian Model of the Universe" in Dennis and Barbara Tedlock's *Teachings from the American Earth*.

"Just as it is possible to have any number of geometries": ibid., p. 122.

205 "subjective from our viewpoint": ibid., p. 124.

PART 3. "A FEVER OF MATTER"

Thomas Mann's quote is in the chapter titled "Research" of *The Magic Mountain* and can be found on p. 275 of the Vintage paperback edition.

Chapter 7. The Dawn of Recognition

209 The biologists' various creation stories are outlined in Chapter 2 of John Casti's *Paradigms Lost* and analyzed in greater detail in Robert Shapiro's *Origins*.

213 The discovery of the genetic code is narrated with drama and finesse in the first part of Horace Freeland Judson's classic *The Eighth Day of Creation*. One of the discoverers of DNA's double helical structure, James Watson, gave his side of the story in *The Double Helix;* especially recommended is the 1980 Norton edition of the 1968 book, edited by Gunther Stent. Watson's book *Molecular Biology of the Gene* has become a standard reference on the subject. His partner Francis Crick gives his own viewpoint on the discovery in *What Mad Pursuit*.

214 The case for and case against the idea that life started with nucleic acids are described in the Casti and Shapiro books.

215 The biologist Richard Lewontin has written on the myth of DNA as the master controller of the cell. See his book *Biology as Ideology*.

217 The protein-first school is also examined in the Casti and Shapiro books.

218 For Hoyle's theory that life came from outer space, see his book *Evolution from Space : A Theory of Cosmic Creationism* (New York: Simon & Schuster, 1982).

219 Francis Crick's extraterrestrial-origins theory is described in his book *Life Itself*.

220–1 Graham Cairns-Smith's genetic takeover theory is described in his book

Seven Clues to the Origin of Life: A Scientific Detective Story (New York: Cambridge University Press, 1985) and distilled in Chapter 2 of Casti's *Paradigms Lost.*

221 Harold Morowitz's iconoclastic theory is described in his book *Beginnings of Cellular Life.*

227 The three seminal papers on autocatalytic sets: Manfred Eigen, "Self-Organization of Matter and the Evolution of Biological Macromolecules," *Naturwissenschaften* 58 (1971): 465; Stuart Kauffman, "Cellular Homeostasis, Epigenesis and Replication in Randomly Aggregated Macromolecular Systems," *J. Cybernetics* 1 (1971): 71; Otto Rössler, "A System-Theoretic Model of Biogenesis," *Z. Naturforsch.* B266 (1971): 741. For popular treatments of the subject, see Steven Levy's *Artificial Life* and the *Complexity* books by Waldrop and Lewin.

Hypercycles are described in a two-part paper by Manfred Eigen and Peter Schuster: "The Hypercycle: A Principle of Natural Self-Organization," *Naturwissenschaften* 64 (1977): 541, 65 (1978): 7.

For a recent popular account, see Manfred Eigen's book with Ruthild Winkler-Oswatitsch, *Steps Towards Life.*

228 Kauffman's theories of autocatalysis are described in his book *The Origins of Order* and in the forthcoming *At Home in the Universe.*

229 The research on autocatalysis at the Los Alamos Complex Systems Group is described in several papers, including J. Doyne Farmer, Stuart Kauffman, and Norman Packard, "Autocatalytic Replication of Polymers," *Physica D* 22 (1986): 50; Richard J. Bagley and Farmer, "Spontaneous Emergence of a Metabolism," and Bagley, Farmer, and Walter Fontana, "Evolution of a Metabolism," both in Christopher Langton, Charles Taylor, J. Doyne Farmer, and Steen Rasmussen, *Artificial Life II.*

Chapter 8. The Arrival of the Fittest

234 Lynn Margulis's ideas on the early evolution of cellular life are described in Margulis and Dorion Sagan, *Microcosmos.*

237 Stephen Jay Gould explores the implications of rewinding evolution's tape in his magisterial book *Wonderful Life.*

239 "Now they swarm": The quote is from pp. 19–20 of *The Selfish Gene.*

242 "Life, once established": Leo Buss, *Evolution of Individuality,* p. 186.

244 Fontana's original work on Alchemy is the 1992 paper "Algorithmic Chemistry," in Langton, Taylor, Farmer, and Rasmussen, *Artificial Life II.*

245 The 1992 conference at which Fontana and Buss gave their presentation is documented in the proceedings, *Artificial Life III,* edited by Christopher Langton. It was repeated at the Santa Fe Institute conference on Integrative Themes at Sol y Sombra, July 1992. My quotations are from this version, which was videotaped.

Fontana and Buss's ideas are described in two papers: "What Would Be Conserved if the Tape Were Played Twice?," *Proceedings of the National Academy of Science* 91 (1994): 757–61, and " 'The Arrival of the Fittest': Toward a Theory of Biological Organization," *Bulletin of Mathematical Biology* 56 (1994): 1–64.

245 The Hugo De Vries book Buss refers to is *Species and Varieties: Their Origin by Mutation* (Chicago: Open Court, 1904).

247 For a heady sense of the powerfully recursive nature of Lisp, see Douglas Hofstadter's *Gödel, Escher, Bach*. Lambda calculus is described in Chapter 2 of Roger Penrose's *The Emperor's New Mind*, pp. 66–70.

252 For an entry into the sprawling archives on modern literary theory, try Terry Eagleton, *Literary Theory*. For a taste of Umberto Eco, try (in addition to his novels) his essay "On Truth: A Fiction," in *The Limits of Interpretation*. Some early ideas on semiotics, pioneered by Charles Sanders Peirce, can be found in *Peirce on Signs*, edited by James Hoopes. Also see the acclaimed biography *Charles Sanders Peirce* by Joseph Brent.

254 Artificial life, including the work of Tom Ray and many others, is documented in Steven Levy's *Artificial Life*, in the journal *Artificial Life*, and in the proceedings of the Artificial Life conferences edited by Christopher Langton et al.: *Artificial Life, Artificial Life II*, and *Artificial Life III*. Langton's introduction to the first volume provides an especially lucid summary. Ray's ideas are described in detail in his paper "An Evolutionary Approach to Synthetic Biology: Zen and the Art of Creating Life," *Artificial Life* $1(\frac{1}{2})$ (1994): 195–226.

255 Von Neumann's proof is contained in his 1966 book *Theory of Self-Replicating Automata*, published by the University of Illinois Press.

256 The ideas of John Searle are described most recently in his book *The Rediscovery of the Mind*.

Chapter 9. In Search of Complexity

259 It is indeed possible to get to the top of Talaya in half a morning's hike. After pulling into the parking lot at the Randall Davies Center (be careful—the gate is closed and locked at a posted hour), ask whoever is on duty for advice. Failing that, make your way by trial and error through the nature center's maze of short trails until you find yourself on a path that heads away from the Santa Fe River and into a dark canyon. (For a God's-eye view, refer to the United States Geological Survey's quadrangle map for Santa Fe.) Start watching immediately for a side canyon heading off to the right. Follow it to the top of the ridge and you should be able to see Talaya and plot a course. If not, return to the main trail and try another side canyon.

264 Cuvier's classification scheme is in Chapter 31 of Mason's *A History of the Sciences*.

265 Two of the most influential books on biological structuralism are D'Arcy Thompson's *On Growth and Form*, published in 1917 by Cambridge University Press, and C. H. Waddington's *The Strategy of the Genes*, published in 1957 by Allen & Unwin.

Goodwin made his remarks at the Santa Fe Institute conference on Integrative Themes at Sol y Sombra, July 1992. For a taste of the ideas presented by the structuralists, see the papers collected in Goodwin and Peter Saunders, eds., *Theoretical Biology*. For some of the strongest counterarguments celebrating the power of natural selection, see Dawkins's *The Selfish Gene* and *The Blind Watchmaker*, Ernest Mayr's *The Growth of Biological Thought*, and John Maynard Smith's *The Theory of Evolution*.

266 Robert Wesson's Darwinian stories are on p. 17 of *Beyond Natural Selection*.

269 "The eye developed independently": the quote is from Goodwin's presentation at the 1992 conference on Integrative Themes in Santa Fe.

"as waves and spirals arise naturally in water": Goodwin, "A Structuralist Research Programme in Developmental Biology," in Goodwin and Saunders, *Theoretical Biology*, p. 50.

270–1 Kauffman's work is described in his book *The Origins of Order* and in *Artificial Life* by Levy, *Complexity* by Waldrop, and *Complexity* by Lewin. Also see Kauffman's forthcoming book for nonspecialists, *At Home in the Universe*.

271 The story of Monod and Jacob is beautifully told in Part 2 of Judson's *Eighth Day of Creation*.

Chapter 10. In the Eye of the Beholder

279 The edge of chaos is described in the popular treatments written by Waldrop, Levy, and Lewin, all listed directly above. For more details, see Kauffman's two books and Langton's "Computation at the Edge of Chaos: Phase Transitions and Emergent Computation," in Stephanie Forrest, *Emergent Computation*.

281 Cris Moore made his remarks during a discussion on the edge of chaos at the Santa Fe Institute, June 29, 1993.

For an excellent introduction to cellular automata, see William Poundstone's *The Recursive Universe*.

281–2 Wolfram's paper dividing cellular automata into four categories: Wolfram, "Universality and Complexity in Cellular Automata," *Physica D* 10 (1984): 1–35.

283 Langton's scale from order to chaos: Langton, "Computation at the Edge of Chaos: Phase Transitions and Emergent Computation," in Forrest, *Emergent Com-*

putation. He introduced his ideas in a 1988 thesis with the same title at the University of Michigan Department of Computer Science; it was accepted in 1991.

283 Packard's attempt to show that evolution takes cellular automata to the edge of chaos: "Adaptation Toward the Edge of Chaos," in J. A. S. Kelso, A. J. Mandell, and M. F. Shlesinger, eds., *Dynamic Patterns in Complex Systems* (Singapore: World Scientific, 1988), pp. 293–301.

The Gacs rule is described in P. Gacs, G. L. Kurdyumov, and L. A. Levin, "One-Dimensional Uniform Arrays That Wash Out Finite Islands," *Probl. Peredachi. Inform.* 14 (1978): 92–98.

284 The classic work on genetic algorithms is John Holland's 1975 book, reissued in a second edition in 1992, *Adaptation in Natural and Artificial Systems.* For a more recent treatment, see Melanie M. Mitchell, "An Introduction to Genetic Algorithms" (Cambridge: MIT Press, forthcoming).

Mitchell, Hraber, and Crutchfield's reexamination of Packard's work: "Revisiting the Edge of Chaos: Evolving Cellular Automata to Perform Computations," *Complex Systems* 7 (1993): 89–130; "Evolving Cellular Automata to Perform Computations: Mechanisms and Impediments," *Physica D* (1994), in press; "Dynamics, Computation, and the 'Edge of Chaos': A Reexamination," Santa Fe Institute Working Paper 93-06-040.

285 For the "onset of chaos," see the 1990 paper by Crutchfield and Karl Young, "Computation at the Onset of Chaos," in Zurek, *Complexity, Entropy, and the Physics of Information,* pp. 223–69.

286 In addition to Langton's papers, research lending support to the notion that selection pressures can drive systems to a transition regime between orderly and disorderly behavior includes Kunihiko Kaneko, "Chaos as a Source of Complexity and Diversity in Evolution," *Artificial Life* I (1994): 163–77; Kaneko and Junji Suzuki, "Evolution to the Edge of Chaos in an Imitation Game," in Langton, *Artificial Life III;* and Stuart Kauffman and Sonke Johnson, "Coevolution to the Edge of Chaos: Coupled Fitness Landscapes, Poised States, and Coevolutionary Avalanches," *J. Theoret. Biol.* 149 (1991): 467.

a hierarchy of computational devices: The power of a computer depends, for example, on whether its memory is finite or unlimited and how it is arranged— in what computer scientists call stacks, queues, or parallel arrays. Crutchfield's ideas on complexity and computation are described in "Observing Complexity and the Complexity of Observation," Santa Fe Institute Working Paper 93-06-035, and in "Is Anything Ever New?: Considering Emergence," in Cowan, Pines, and Meltzer, *Integrative Themes.*

For a discussion of the various measures of complexity, see Charles Bennett, "How to Define Complexity in Physics, and Why," in Zurek, *Complexity, Entropy, and the Physics of Information.*

289 Gell-Mann's "effective complexity" is described in Chapter 8 of *The Quark and the Jaguar.*

290 "Subjects would form a hypothesis": Arthur, "On Learning and Adaptation in the Economy," Santa Fe Institute Working Paper 92-07-038, p. 12. The work he describes is reported in Julian Feldman, "Computer Simulation of Cognitive Processes," in Harold Borko, ed., *Computer Application in the Behavioral Sciences* (Englewood Cliffs, N.J.: Prentice-Hall, 1962).

For an attempt to explain increasing biological complexity in the Darwinian framework, see John Tyler Bonner, *The Evolution of Complexity.*

291 "the fundamental ground plans of anatomy" etc.: Gould, *Wonderful Life,* p. 99.

293 "The modern order": ibid., pp. 288–89.

"virtually inevitable" etc.: ibid., p. 289.

"Whether the evolutionary origin": ibid., p. 291.

294 Alvarez's lecture at the Santa Fe Institute was on August 26, 1993.

295 "This situation prevailed" etc.: Gould, *Wonderful Life,* p. 318.

Truchas Interlude: The Leap into the Unknown

297 For the history of Chimayo and its Santuario, see Stephen de Borhegyi, *El Santuario de Chimayo,* and Elizabeth Kay, *Chimayo Valle Traditions.*

299 The story of Los Alamos's role in the study of the Shroud of Turin is told in Cullen Murphy, "Shreds of Evidence," *Harper's,* November 1981, pp. 42–47.

301 Escápula: A similar legend told at Picuris pueblo, northeast of Chimayo, tells of a St. Istípula. See de Borhegyi, *El Santuario de Chimayo,* pp. 17–18.

302 For information on the Penitentes, see Marta Weigle's *Brothers of Light, Brothers of Blood.* For personal reminiscences, see Angelico Chavez, *My Penitente Land,* and Lorenzo de Cordova, *Echoes of the Flute.* For more recent scholarship, see Thomas J. Steele and Rowena A. Rivera, *Penitente Self-Government,* and William Wroth, *Images of Penance, Images of Mercy,* which includes pictures of northern New Mexican religious art.

305 My description of Holy Week in Truchas is based on visits there in 1993.

307 Holy Week Tinieblas services are also commonly held on Thursday nights. The Friday service is sometimes called Los Maitines.

308 For a study of Protestant missionaries in northern New Mexico, see Randi Jones Walker, *Protestantism in the Sangre de Cristos.* There are many disagreements and schisms among fundamentalists, pentacostalists, and evangelicals. Much enlightenment is provided by two standard references on the history of fundamentalism and the intricacies of fundamentalist thought: George M. Marsden's *Fundamentalism and American Culture* and *Understanding Fundamentalism and Evangeli-*

calism. Also see the authoritative book edited by Martin E. Marty and R. Scott Appleby, *Fundamentalisms Observed.*

309 A typical and influential example of fundamentalist prophecy is Hal Lindsey's *The Late Great Planet Earth* (New York: Bantam, 1970).

311 The Great Chain of Being is explored in Arthur O. Lovejoy's book of that name.

312 The standard account of creationism is John C. Whitcomb and Henry M. Morris, *The Genesis Flood* (see full reference in notes for Chapter 2). I first encountered creationist science when I was researching my book *Architects of Fear: Conspiracy Theories and Paranoia in American Politics.* For an article on radioactive decay rates, see the August 1981 issue of *Bible-Science Newsletter,* published by the Bible-Science Association of Minneapolis. For a creationist atomic model, see the issue of December 1981. For an attempt to disprove Hubble and Einstein, see the issues of August 1981 and December 1981. In *A New Interest in Geocentricity,* also published by the association, an engineer named James Hanson contends that the Michelson-Morley experiment can be reinterpreted to establish that the earth sits still with everything else revolving around it.

Conclusion

317 The ruins of Otowi can be reached by driving from Santa Fe toward Los Alamos. At the interchange beyond San Ildefonso pueblo, take the option marked "Los Alamos" and after three-tenths of a mile watch for a service road splitting off to the right. Park outside the gate (lest you find it locked behind you) and walk down a gravel road about a mile and a half until you cross a bridge. Turn right onto an old dirt road and follow it north half a mile to the Otowi ruins. For details, see the Sierra Club's *Day Hikes in the Santa Fe Area,* 3rd ed. Alex Patterson's *Field Guide to Rock Art of the Southwest* is a fine introduction to the subject. For further exploration of the Pajarito Plateau, see cartographer Andrea Kron's *Hiking Trails & Jeep Roads of Los Alamos County, Bandelier National Monument and Vicinity,* published by Otowi Station Science Museum Shop and Book Store in Los Alamos, and the United States Forest Service map of Santa Fe National Forest. Also recommended is Roland A Pettitt's *Exploring the Jemez Country.*

320 I first encountered the idea of representing all human knowledge with a notch on a stick in Richard Powers's novel *The Gold Bug Variations.*

321 Details of the attempts to contact extraterrestrials are given in Frank Drake and Dava Sobel's *Is Anyone Out There?*

323 For thorough and lucid explorations of the question of whether mathematics is discovered or invented, see John Barrow's *Pi in the Sky* and Morris Kline's *Mathematics and the Search for Knowledge.*

325 Seminal papers on quantum computation include Richard Feynman, "Simulating Physics with Computers," *International Journal of Theoretical Physics* 21 (1982), No. 6/7: 467–88, and "Quantum Mechanical Computers," *Foundations of Physics* 16 (1986): 507–31; P. Benioff, "Quantum Mechanical Hamiltonian Models of Turing Machines," *Journal of Statistical Physics* 29 (1982): 515–46; David Deutsch, "Quantum Theory, the Church-Turing Principle and the Universal Quantum Computer," *Proc. R. Soc. Lond.* A400 (1985): 96–117, and "Quantum Computational Networks," *Proc. R. Soc. Lond.* A425 (1989): 73–90; and David Deutsch and R. Jozsa, "Rapid Solution of Problems by Quantum Computation," *Proc. R. Soc. Lond.* A439 (1992): 553–58. Seth Lloyd speculates on how to make a quantum computer in "A Potentially Realizable Quantum Computer," *Science* 261 (1993): 1569–71. Rolf Landauer has argued, however, that in the real world it would be impossible to maintain the perfect coherence needed to get a quantum computer to compute. See his forthcoming paper "Is Quantum Mechanics Useful?," *Proc. R. Soc. Lond.* Even if a quantum computer could be built, it remains controversial whether one would be generally more powerful than a Turing machine. One sign for optimism is work by Peter Shor of AT&T Bell Laboratories showing that a quantum computer could be used to factor integers and find discrete logarithms—two problems that easily swamp the resources of a conventional Turing machine. See his paper "Algorithms for Quantum Computation: Factoring and Discrete Logarithms," *35th IEEE Symposium on Foundations of Computer Science,* November 1994.

John Searle makes a strong argument that information is an artifact in *Rediscovering the Mind.*

BIBLIOGRAPHY

BOOKS ABOUT SCIENCE AND PHILOSOPHY

Anderson, Philip W., Kenneth J. Arrow, and David Pines, eds. *The Economy as an Evolving Complex System*. Santa Fe Institute Studies in the Science of Complexity. Vol. 5. Redwood City, Calif.: Addison-Wesley, 1988.

Appleyard, Bryan. *Understanding the Present: Science and the Soul of Man*. New York: Doubleday, 1993.

Ayer, A.J. *Philosophy in the Twentieth Century*. New York: Random House, 1982.

Barnett, Lincoln Kinnear. *The Universe and Dr. Einstein*. New York: W. Sloane Associates, 1957.

Barrow, John D. *The Origin of the Universe*. New York: Basic Books, 1994.

————. *Pi in the Sky: Counting, Thinking, and Being*. New York: Oxford University Press, 1992.

————. *Theories of Everything: The Quest for Ultimate Explanation*. New York: Oxford University Press, 1991.

————. *The World Within the World*. New York: Oxford University Press, 1991.

————, and Frank J. Tipler. *The Anthropic Cosmological Principle.* New York: Oxford University Press, 1986.

Bartusiak, Marcia. *Through a Universe Darkly: A Cosmic Tale of Ancient Ethers, Dark Matter, and the Fate of the Universe.* New York : HarperCollins, 1993.

————. *Thursday's Universe.* New York: Times Books, 1986.

Bass, Thomas A. *The Eudaemonic Pie.* Boston: Houghton Mifflin, 1985.

Bell, J. S. *Speakable and Unspeakable in Quantum Mechanics: Collected Papers on Quantum Philosophy.* Cambridge, England: Cambridge University Press, 1987.

Bernstein, Jeremy. *Einstein.* New York: Viking, 1973.

————. *Three Degrees Above Zero: Bell Labs in the Information Age.* New York: Scribner's, 1984.

Bonner, John Tyler. *The Evolution of Complexity by Means of Natural Selection.* Princeton, N.J.: Princeton University Press, 1988.

Brent, Joseph. *Charles Sanders Peirce: A Life.* Bloomington: Indiana University Press, 1993.

Burnham, Robert, Jr. *Burnham's Celestial Handbook: An Observer's Guide to the Universe Beyond the Solar System.* 3 vols. New York: Dover, 1978.

Buss, Leo W. *The Evolution of Individuality.* Princeton, N.J.: Princeton University Press, 1987.

Calvin, William H. *The River That Flows Uphill: A Journey from the Big Bang to the Big Brain.* San Francisco: Sierra Club, 1986.

Campbell, Jeremy. *Grammatical Man: Information, Entropy, Language, and Life.* New York: Simon & Schuster, 1982.

Casti, John L. *Complexification: Explaining a Paradoxical World Through the Science of Surprise.* New York: HarperCollins, 1994.

————. *Paradigms Lost: Images of Man in the Mirror of Science.* New York: William Morrow, 1989.

————. *Searching for Certainty: What Scientists Can Know About the Future.* New York: William Morrow, 1990.

————, and Anders Karlqvist, eds. *Beyond Belief: Randomness, Prediction and Explanation in Science.* Boca Raton, Fla.: CRC Press, 1991.

Changeux, Jean-Pierre, and Alain Connes. *Conversations on Mind, Matter, and Mathematics.* Edited and translated by M. B. DeBevoise. Princeton, N.J.: Princeton University Press, 1995.

Cline, Barbara Lovett. *Men Who Made a New Physics: Physicists and the Quantum Theory.* New York: Thomas Y. Crowell, 1965.

Close, Frank, Michael Marten, and Christine Sutton. *The Particle Explosion.* New York: Oxford University Press, 1987.

Cohen, I. Bernard. *Revolution in Science.* Cambridge, Mass.: Belknap/Harvard University Press, 1985.

Cooper, Leon N. *An Introduction to the Meaning and Structure of Physics*. Short ed. New York: Harper & Row, 1970.

Coveney, Peter, and Roger Highfield. *The Arrow of Time: A Voyage Through Science to Solve Time's Greatest Mystery*. New York: Fawcett Columbine, 1990.

Cowan, George, David Pines, and David Meltzer, eds. *Integrative Themes*. Santa Fe Institute Studies in the Science of Complexity. Vol. 19. Reading, Mass.: Addison-Wesley, 1994.

Crease, Robert P., and Charles C. Mann. *The Second Creation: Makers of the Revolution in 20th-Century Physics*. New York: Macmillan, 1986.

Crick, Francis. *Life Itself: Its Origin and Nature*. New York: Simon & Schuster, 1981.

————. *What Mad Pursuit: A Personal View of Scientific Discovery*. New York: Basic Books, 1988.

Davies, Paul. *The Cosmic Blueprint: New Discoveries in Nature's Creative Ability to Order the Universe*. New York: Simon & Schuster, 1988.

————. *The Last Three Minutes: Conjecture About the Ultimate Fate of the Universe*. New York: Basic Books, 1994.

————. *The Mind of God: The Scientific Basis for a Rational World*. New York: Simon & Schuster, 1992.

————, and John Gribbin. *The Matter Myth: Dramatic Discoveries That Challenge Our Understanding of Physical Reality*. New York: Touchstone/Simon & Schuster, 1992.

Dawkins, Richard. *The Blind Watchmaker*. New York: W. W. Norton, 1986.

————. *The Selfish Gene*. New ed. Oxford: Oxford University Press, 1989.

DeDuve, Christian. *Blueprint for a Cell: The Nature and Origin of Life*. Burlington, N.C.: Neil Patterson Publishers, 1991.

Dennett, Daniel C. *Consciousness Explained*. Boston: Little, Brown, 1991.

DeWitt, Bryce S., and Neill Graham, eds. *The Many-Worlds Interpretation of Quantum Mechanics*. Princeton, N.J.: Princeton University Press, 1973.

Drake, Frank, and Dava Sobel. *Is Anyone Out There? The Scientific Search for Extraterrestrial Intelligence*. New York: Delacorte, 1992.

Dressler, Alan. *Voyage to the Great Attractor: Exploring Intergalactic Space*. New York: Alfred A. Knopf, 1994.

Eagleton, Terry. *Literary Theory: An Introduction*. Minneapolis: University of Minnesota Press, 1983.

Eco, Umberto. *The Limits of Interpretation*. Bloomington: Indiana University Press, 1990.

Edelman, Gerald. *Bright Air, Brilliant Fire: On the Matter of the Mind*. New York: Basic Books, 1992.

Eigen, Manfred, with Ruthild Winkler-Oswatitsch. *Steps Towards Life: A Perspective on Evolution*. New York: Oxford University Press, 1992.

Ferris, Timothy. *Coming of Age in the Milky Way*. New York: William Morrow, 1988.

————. *The Mind's Sky: Human Intelligence in a Cosmic Context*. New York: Bantam, 1992.

Feynman, Richard P. *QED: The Strange Theory of Light and Matter*. Princeton, N.J.: Princeton University Press, 1985.

————. *The Character of Physical Law*. Cambridge, Mass.: MIT Press, 1967.

Forrest, Stephanie, ed. *Emergent Computation*. Cambridge, Mass.: MIT Press, 1991.

Fritzsch, Harald. *Quarks: The Stuff of Matter*. New York: Basic Books, 1983.

Gamow, George. *Thirty Years That Shook Physics: The Story of Quantum Theory*. Garden City, N.Y.: Doubleday, 1966.

Gell-Mann, Murray. *The Quark and the Jaguar: Adventures in the Simple and the Complex*. New York: W. H. Freeman, 1994.

Gilovich, Thomas. *How We Know What Isn't So: The Fallibility of Human Reason in Everyday Life*. New York: Free Press, 1991.

Gleick, James. *Chaos: Making a New Science*. New York: Viking, 1987.

————. *Genius: The Life and Science of Richard Feynman*. New York: Pantheon, 1992.

Goodwin, Brian C., and Peter Saunders, eds. *Theoretical Biology: Epigenetic and Evolutionary Order from Complex Systems*. Edinburgh: Edinburgh University Press, 1989.

Greenstein, George. *The Symbiotic Universe: Life and Mind in the Cosmos*. New York: William Morrow, 1988.

Gould, Stephen Jay. *Wonderful Life: The Burgess Shale and the Nature of History*. New York: W. W. Norton, 1989.

Hall, Stephen. *Mapping the Next Millennium: The Discovery of New Geographies*. New York: Random House, 1992.

Hartle, James. "Excess Baggage." In *Elementary Particle Physics and the Universe: Essays in Honour of Gell-Mann*. Cambridge, England: Cambridge University Press, 1991.

Hawking, Stephen W. *A Brief History of Time: From the Big Bang to Black Holes*. New York: Bantam, 1988.

Heisenberg, Werner. *Physics and Beyond: Encounters and Conversations*. New York: Harper & Row, 1971.

————. *Physics and Philosophy: The Revolution in Modern Science*. New York: Harper & Row, 1958.

Herbert, Nick. *Quantum Reality: Beyond the New Physics*. New York: Anchor/ Doubleday, 1985.

Hofstadter, Douglas R. *Gödel, Escher, Bach: An Eternal Golden Braid*. New York: Basic Books, 1979.

Holland, John. *Adaptation in Natural and Artificial Systems*. 2nd ed. Cambridge, Mass.: MIT Press, 1992.

Hoopes, James, ed. *Peirce on Signs: Writings on Semiotic by Charles Sanders Peirce*. Chapel Hill: University of North Carolina Press, 1991.

Jauch, J. M. *Are Quanta Real? A Galilean Dialogue*. Bloomington: Indiana University Press, 1973.

Judson, Horace Freeland. *The Eighth Day of Creation: Makers of the Revolution in Biology*. New York: Simon & Schuster, 1979.

Kauffman, Stuart A. *At Home in the Universe: The Search for Laws of Self-Organization and Complexity*. New York: Oxford University Press, 1995.

————. *The Origins of Order: Self-Organization and Selection in Evolution*. New York: Oxford University Press, 1993.

Kline, Morris. *Mathematics and the Search for Knowledge*. New York: Oxford University Press, 1985.

————. *Mathematics: The Loss of Certainty*. New York: Oxford University Press, 1980.

————. *Mathematics in Western Culture*. New York: Oxford University Press, 1953.

Krauss, Lawrence. *The Fifth Essence: The Search for Dark Matter in the Universe*. New York: Basic Books, 1989.

Lakoff, George. *Women, Fire, and Dangerous Things: What Categories Reveal About the Mind*. Chicago: University of Chicago Press, 1987.

Langton, Christopher G., ed. *Artificial Life*. Santa Fe Institute Studies in the Science of Complexity. Vol. 6. Redwood City, Calif.: Addison-Wesley, 1989.

————, ed. *Artificial Life III*. Santa Fe Institute Studies in the Science of Complexity. Vol. 6. Redwood City, Calif.: Addison-Wesley, 1994.

————, Charles Taylor, J. Doyne Farmer, and Steen Rasmussen. *Artificial Life II*. Santa Fe Institute Studies in the Science of Complexity. Vol. 10. Redwood City, Calif.: Addison Wesley, 1992.

Laudan, Larry. *Science and Relativism: Some Key Controversies in the Philosophy of Science*. Chicago: University of Chicago Press, 1990.

Layzer, David. *Constructing the Universe*. New York: Scientific American Library, 1984.

Leff, Harvey S., and Andrew F. Rex, eds. *Maxwell's Demon: Entropy, Information, Computing*. Princeton, N.J.: Princeton University Press, 1990.

Levy, Steven. *Artificial Life: The Quest for a New Creation*. New York: Pantheon, 1992.

Lewin, Roger. *Complexity: Life at the Edge of Chaos*. New York: Macmillan, 1992.

Lewontin, Richard C. *Biology as Ideology: The Doctrine of DNA*. New York: HarperPerennial, 1992.

Lightman, Alan. *Ancient Light: Our Changing View of the Universe.* Cambridge, Mass.: Harvard University Press, 1991.

————, and Roberta Brauwer. *Origins: The Lives and Worlds of Modern Cosmologists.* Cambridge, Mass.: Harvard University Press, 1990.

Lindley, David. *The End of Physics: The Myth of a Unified Theory.* New York: Basic, 1993.

Lockwood, Michael. *Mind, Brain and the Quantum: The Compound "I."* Oxford: Basil Blackwell, 1989.

McCormmach, Russell. *Night Thoughts of a Classical Physicist.* Cambridge, Mass.: Harvard University Press, 1982.

Malkiel, Burton Gordon. *A Random Walk down Wall Street.* New York: W. W. Norton, 1973.

Margulis, Lynn, and Dorion Sagan. *Microcosmos: Four Billion Years of Evolution from Our Microbial Ancestors.* New York: Simon & Schuster, 1986.

Markson, David. *Wittgenstein's Mistress.* Elmwood Park, Ill.: Dalkey Archive Press, 1988.

Marsden, George M. *Fundamentalism and American Culture: The Shaping of Twentieth Century Evangelism, 1870–1925.* New York: Oxford University Press, 1980.

Marty, Martin E., and R. Scott Appleby, eds. *Fundamentalisms Observed: A Study Conducted by the American Academy of Arts and Sciences.* Chicago: University of Chicago Press, 1991.

————. *Understanding Fundamentalism and Evangelicalism.* Grand Rapids, Mich.: William B. Eerdmans, 1991.

Mason, Stephen F. *A History of the Sciences.* New York: Collier, 1962.

Mayr, Ernst. *The Growth of Biological Thought: Diversity, Evolution, and Inheritance.* Cambridge, Mass.: Belknap Press, 1982.

Monod, Jacques. *Chance and Necessity: An Essay on the Natural Philosophy of Modern Biology.* New York: Alfred A. Knopf, 1971.

Morowitz, Harold J. *Beginnings of Cellular Life: Metabolism Recapitulates Biogenesis.* New Haven: Yale University Press, 1992.

Morris, Richard. *The Edges of Science: Crossing the Boundary from Physics to Metaphysics.* New York: Prentice-Hall, 1990.

Overbye, Dennis, *Lonely Hearts of the Cosmos: The Scientific Quest for the Secret of the Universe.* New York: HarperCollins, 1991.

Pagels, Heinz. *The Dreams of Reason: The Computer and the Rise of the Sciences of Complexity.* New York: Simon & Schuster, 1988.

————. *Perfect Symmetry: The Search for the Beginning of Time.* New York: Simon & Schuster, 1985.

Pais, Abraham. *Inward Bound: Of Matter and Forces in the Physical World.* New York: Oxford University Press, 1986.

————. *Niels Bohr's Times: In Physics, Philosophy, and Polity.* New York: Oxford University Press, 1991.

————. *"Subtle Is the Lord . . .": The Science and the Life of Albert Einstein.* New York: Oxford University Press, 1982.

Pasachoff, Jay M., and Donald H. Menzel. *A Field Guide to the Stars and Planets.* 3rd ed. The Peterson Field Guide Series. Boston: Houghton Mifflin, 1992.

Penrose, Roger. *The Emperor's New Mind: Concerning Computers, Minds, and the Laws of Physics.* New York: Oxford University Press, 1989.

Pickering, Andrew. *Constructing Quarks: A Sociological History of Particle Physics.* Edinburgh: Edinburgh University Press, 1984.

Pines, David, ed. *Emerging Syntheses in Science.* Santa Fe Institute Studies in the Science of Complexity. Vol. 1. Redwood City, Calif.: Addison-Wesley, 1988.

Pirsig, Robert. *Lila: An Inquiry into Morals.* New York: Bantam, 1991.

————. *Zen and the Art of Motorcycle Maintenance: An Inquiry into Values.* New York: William Morrow, 1974.

Poincaré, Henri. *Science and Method.* Trans. Francis Maitland. New York: Dover, 1952.

Poundstone, William. *Labyrinths of Reason: Paradox, Puzzles, and the Frailty of Knowledge.* New York: Doubleday/Anchor, 1988.

————. *The Recursive Universe: Cosmic Complexity and the Limits of Scientific Knowledge.* New York: William Morrow, 1985.

Powers, Richard. *The Gold Bug Variations.* New York: Morrow, 1991.

Preston, Richard. *First Light: The Search for the Edge of the Universe.* New York: Atlantic Monthly, 1987.

Putnam, Hilary. *Representation and Reality.* Cambridge, Mass.: MIT Press, 1988.

Rhodes, Richard. *The Making of the Atomic Bomb.* New York: Simon & Schuster, 1987.

Riordan, Michael. *The Hunting of the Quark: A True Story of Modern Physics.* New York: Simon & Schuster, 1987.

Rorty, Richard. *Philosophy and the Mirror of Nature.* Princeton, N.J.: Princeton University Press, 1979.

Rosen, Joe. *The Capricious Cosmos: Universe Beyond Law.* New York: Macmillan, 1991.

Ruelle, David. *Chance and Chaos.* Princeton, N.J.: Princeton University Press, 1991.

Schrödinger, Erwin. *What Is Life?* Cambridge, England: Cambridge University Press, 1944.

Schwartz, Joseph. *The Creative Moment: How Science Made Itself Alien to Modern Culture.* New York: HarperCollins, 1992.

Searle, John R. *The Rediscovery of the Mind.* Cambridge, Mass.: MIT Press, 1992.

Shapiro, Robert. *Origins: A Skeptic's Guide to the Creation of Life on Earth.* New York: Summit Books, 1986.

Smith, John Maynard. *The Theory of Evolution.* 3rd ed. Harmondsworth, England: Penguin, 1975.

Smoot, George, and Keay Davidson. *Wrinkles in Time.* New York : William Morrow, 1993.

Stewart, Ian. *Does God Play Dice? The Mathematics of Chaos.* Oxford, England: Basil Blackwell, 1989.

————, and Martin Golubitsky. *Fearful Symmetry: Is God a Geometer?* Oxford, England: Basil Blackwell, 1992.

Sutton, Christine. *The Particle Connection: The Most Exciting Scientific Chase Since DNA and the Double Helix.* New York: Simon & Schuster, 1984.

Taylor, Edwin F., and John Archibald Wheeler. *Spacetime Physics.* San Francisco: W. H. Freeman, 1966.

Traweek, Sharon. *Beamtimes and Lifetimes: The World of High Energy Physicists.* Cambridge, Mass.: Harvard University Press, 1988.

Trefil, James S. *The Dark Side of the Universe : A Scientist Explores the Mysteries of the Cosmos.* New York: Scribner's, 1988.

————. *Reading the Mind of God: In Search of the Principle of Universality.* New York: Scribner's, 1989.

Updike, John. *Roger's Version.* New York: Alfred A. Knopf, 1986.

von Bayer, Hans Christian. *Taming the Atom: The Emergence of the Visible Microworld.* New York: Random House, 1992.

Waldrop, M. Mitchell. *Complexity: The Emerging Science at the Edge of Order and Chaos.* New York: Simon & Schuster, 1992.

Watson, James D. *The Double Helix : A Personal Account of the Discovery of the Structure of DNA.* Ed. Gunther S. Stent. New York: W. W. Norton, 1980.

————. *Molecular Biology of the Gene.* 2nd ed. New York: Benjamin, 1970.

Weinberg, Steven. *The Discovery of Subatomic Particles.* New York: W. H. Freeman, 1983.

————. *Dreams of a Final Theory.* New York: Pantheon, 1992.

————. *The First Three Minutes: A Modern View of the Origin of the Universe.* Updated ed. New York: Basic Books, 1988.

Wesson, Robert. *Beyond Natural Selection.* Cambridge, Mass.: MIT Press, 1991.

Weyl, Hermann. *Symmetry.* Princeton, N.J.: Princeton University Press, 1969.

Wheeler, John Archibald, and Wojciech Hubert Zurek, eds. *Quantum Theory and Measurement.* Princeton, N.J.: Princeton University Press, 1983.

Wilczek, Frank, and Betsy Devine. *Longing for the Harmonies: Themes and Variations from Modern Physics.* New York: W. W. Norton, 1987.

Wright, Robert. *Three Scientists and Their Gods: Looking for Meaning in an Age of Information.* New York: Times Books, 1986.

Zee, A. *Fearful Symmetry: The Search for Beauty in Modern Physics.* New York: Macmillan, 1986.

Zeilik, Michael, and John Gaustad. *Astronomy: The Cosmic Perspective.* 2nd ed. New York: John Wiley & Sons, 1990.

Zurek, Wojciech H., ed. *Complexity, Entropy and the Physics of Information.* Santa Fe Institute Studies in the Science of Complexity. Vol. 8. Redwood City, Calif.: Addison-Wesley, 1990.

BOOKS ABOUT NEW MEXICO AND THE SOUTHWEST

Bandelier, Adolph. *The Delight Makers.* New York: Dodd, Mead, 1918.

Benedek, Emily. *The Wind Won't Know Me: A History of the Navajo-Hopi Land Dispute.* New York: Alfred A. Knopf, 1992.

Bolton, Herbert Eugene. *Coronado: Knight of Pueblos and Plains.* Albuquerque: University of New Mexico Press, 1949.

Calvin, Ross. *Sky Determines: An Interpretation of the Southwest.* Illus. by Peter Hurd. Albuquerque: University of New Mexico Press, 1965.

Cather, Willa. *Death Comes for the Archbishop.* New York: Alfred A. Knopf, 1927.

Chavez, Angelico. *My Penitente Land: Reflections on Spanish New Mexico.* Albuquerque: University of New Mexico Press, 1974.

Chew, Joe. *Storms Above the Desert: Atmospheric Research in New Mexico, 1935–1985.* Albuquerque: University of New Mexico Press, 1987.

Chronic, Halka. *Roadside Geology of New Mexico.* Missoula, Mont.: Mountain Press Publishing Company, 1987.

Church, Peggy Pond. *The House at Otowi Bridge: The Story of Edith Warner and Los Alamos.* Albuquerque: University of New Mexico Press, 1959.

Coles, Robert. *The Old Ones of New Mexico.* San Diego: Harcourt Brace Jovanovich, 1984.

Day Hikes in the Santa Fe Area. 3rd ed. Santa Fe: Santa Fe Group of the Sierra Club, 1990.

de Borhegyi, Stephen F. *El Santuario de Chimayo.* Spanish Colonial Arts Society. Santa Fe: Ancient City Press, 1956.

deBuys, William. *Enchantment and Exploitation: The Life and Hard Times of a New Mexico Mountain Range.* Albuquerque: University of New Mexico Press, 1985.

————, and Alex Harris. *River of Traps: A Village Life.* Albuquerque: University of New Mexico Press, 1990.

de Cordova, Lorenzo. *Echoes of the Flute.* Santa Fe: Ancient City Press, 1972.

Dockstader, Frederick J. *The Kachina and the White Man: The Influences of White Culture on the Hopi Kachina Religion.* Rev. and expanded ed. Albuquerque: University of New Mexico Press, 1985.

Dozier, Edward P. *Hano: A Tewa Indian Community in Arizona.* New York: Holt, Rinehart & Winston, 1966.

——. *The Pueblo Indians of North America.* New York: Holt, Rinehart & Winston, 1970.

Fergusson, Erna. *Dancing Gods: Indian Ceremonials of New Mexico and Arizona.* Albuquerque: University of New Mexico Press, 1966.

Frazier, Kendrick. *People of Chaco: A Canyon and Its Culture.* New York: W. W. Norton, 1986.

Griffin-Pierce, Trudy. *Earth Is My Mother, Sky Is My Father: Space, Time, and Astronomy in Navajo Sandpainting.* Albuquerque: University of New Mexico Press, 1992.

Grimes, Ronald L. *Symbol and Conquest: Public Ritual and Drama in Santa Fe, New Mexico.* Ithaca, N.Y.: Cornell University Press, 1976.

Gutiérrez, Ramón A. *When Jesus Came, the Corn Mothers Went Away: Marriage, Sexuality, and Power in New Mexico, 1500–1846.* Stanford, Calif.: Stanford University Press, 1991.

Hall, Edward T. *West of the Thirties: Discoveries Among the Navajo and Hopi.* New York: Doubleday, 1994.

Harrington, John P. "The Ethnogeography of the Tewa Indians." *Bureau of American Ethnology, Annual Report.* Vol. 29. Washington, D.C.: U.S. Government Printing Office, 1916.

Hill, W. W. *An Ethnography of Santa Clara Pueblo, New Mexico.* Ed. and annotated by Charles H. Lange. Albuquerque: University of New Mexico Press, 1982.

Kay, Elizabeth. *Chimayo Valley Traditions.* Santa Fe: Ancient City Press, 1987.

Kessell, John. *Kiva, Cross, and Crown.* Washington, D.C.: National Park Service, 1979.

Kunetka, James W. *City of Fire: Los Alamos and the Atomic Age, 1943–1945.* Rev. ed. Albuquerque: University of New Mexico Press, 1979.

Kurath, Gertrude P., and Antonio Garcia. *Music and Dance of the Tewa Pueblos.* Santa Fe: Museum of New Mexico Press, 1970.

Laird, Carobeth. *Encounter with an Angry God.* Albuquerque: University of New Mexico Press, 1975.

Lovejoy, Arthur O. *The Great Chain of Being: A Study of the History of an Idea.* Cambridge, Mass.: Harvard University Press, 1936.

Marriott, Alice. *María: the Potter of San Ildefonso.* Norman: University of Oklahoma Press, 1948.

Matthews, Kay. *Hiking in the Wilderness: A Backpacking Guide to the Wheeler Peak, Pecos, and San Pedro Parks Wilderness Areas.* El Valle, N.M.: Acequia Madre, 1992.

Ortiz, Alfonso, ed. *Handbook of North American Indians: Southwest*. Vol. 9. Washington, D.C.: Smithsonian Institution, 1979.

———, ed. *New Perspectives on the Pueblos*. Albuquerque: University of New Mexico Press, 1972.

———. *The Tewa World: Space, Time, Being and Becoming in a Pueblo Society*. Chicago: University of Chicago Press, 1969.

Parsons, Elsie Clews. *The Social Organization of the Tewa of New Mexico*. Menasha, Wis.: American Anthropological Association, 1929.

Patterson, Alex. *Field Guide to Rock Art of the Southwest*. Boulder, Colo.: Johnson Books, 1992.

Pearce, T. M., ed. *New Mexico Place Names: A Geographical Dictionary*. Albuquerque: University of New Mexico Press, 1965.

Pettitt, Roland A. *Exploring the Jemez Country*. Los Alamos, N.M.: Pajarito Publications, 1975.

Rosenthal, Debra. *At the Heart of the Bomb: The Dangerous Allure of Weapons Work*. New York: Addison-Wesley, 1990.

Sando, Joe S. *Pueblo Nations: Eight Centuries of Pueblo Indian History*. Santa Fe: Clear Light, 1992.

Simmons, Marc. *Witchcraft in the Southwest: Spanish and Indian Supernaturalism on the Rio Grande*. Flagstaff, Ariz.: Northland Press, 1974.

Steele, Thomas J., and Rowena A. Rivera. *Penitente Self-Government: Brotherhoods and Councils, 1797–1947*. Santa Fe: Ancient City Press, 1985.

Sweet, Jill D. *Dances of the Tewa Pueblo Indians*. Santa Fe: School of American Research Press, 1985.

Tedlock, Barbara. *The Beautiful and the Dangerous: Encounters with the Zuni Indians*. New York: Viking, 1992.

Tedlock, Dennis, and Barbara Tedlock, eds. *Teachings from the American Earth: Indian Religion and Philosophy*. New York: Liveright, 1975.

Walker, Randi Jones. *Protestantism in the Sangre de Cristos, 1850–1920*. Albuquerque: University of New Mexico Press, 1991.

Waters, Frank. *Masked Gods: Navaho and Pueblo Ceremonialism*. Albuquerque: University of New Mexico Press, 1950.

———. *The Woman at Otowi Crossing*. Athens: Ohio University Press, 1966.

Weigle, Marta. *Brothers of Light, Brothers of Blood: The Penitentes of the Southwest*. Santa Fe: Ancient City Press, 1976.

Whitman, William. *The Pueblo Indians of San Ildefonso: A Changing Culture*. New York: Columbia University Press, 1947.

Wroth, William. *Images of Penance, Images of Mercy: Southwestern Santos in the Late Nineteenth Century*. With an Introduction to Part 2 by Marta Weigle. Norman: University of Oklahoma Press, 1991.

INDEX

ALSO BY

GEORGE JOHNSON

IN THE PALACES OF MEMORY

How We Build the Worlds Inside Our Heads

"An eloquent foray into how our brains convert experience into knowledge. Written with all the alacrity of a
detective gathering clues . . . with a lucidity that at
times approaches artistry." —*Boston Phoenix*

Even as you read these words, a tiny portion of your
brain is physically changing. New connections are being
sprouted—a circuit that will create a jab of recognition
if you encounter the words again. That is one of the theories of memory presented in this intriguing and splen-
didly readable book, which distills three researchers'
inquiries into the processes that enable us to recognize a
face that has aged ten years or remember a melody for
decades.

Science/0-679-73759-6